STUDENT SOLUTIONS MANUAL

Bob Martin

INTERMEDIATE ALGEBRA FOURTH EDITION

John Tobey   Jeffrey Slater

Prentice
Hall

Upper Saddle River, NJ 07458

Executive Editor: Karin E. Wagner
Supplements Editor: Elizabeth Covello
Editorial Assistant: Rudy Leon
Assistant Managing Editor, Math Media Production: John Matthews
Production Editor: Wendy A. Perez
Supplement Cover Manager: Paul Gourhan
Supplement Cover Designer: PM Workshop Inc.
Manufacturing Buyer: Ilene Kahn

© 2002 by Prentice-Hall, Inc.
Upper Saddle River, NJ 07458

Printed in the United States of America

10 9 8 7 6 5 4 3 2 1

ISBN   0-13-034201-7

Prentice-Hall International (UK) Limited, London
Prentice-Hall of Australia Pty. Limited, Sydney
Prentice-Hall Canada, Inc., Toronto
Prentice-Hall Hispanoamericana, S.A., Mexico City
Prentice-Hall of India Private Limited, New Delhi
Pearson Education Asia Pte. Ltd., Singapore
Prentice-Hall of Japan, Inc., Tokyo
Editora Prentice-Hall do Brazil, Ltda., Rio de Janeiro

Table of Contents

Diagnostic Pretest

Chapter 1

1. $3-(-4)^2+16-(-2)=3-16+16+2$
$$=3+2=5$$

2. $(3xy^{-2})^3(2x^2y)=3^3x^3y^{-6}(2x^2y)$
$$=27\cdot2x^5y^{-5}=\frac{54x^5}{y^5}$$

3. $3x-4x[x-2(3-x)]$
$$=3x-4x[x-6+2x]$$
$$=3x-4x[3x-6]=3x-12x^2+24x$$
$$=-12x^2+27x$$

4. $F=\dfrac{9}{5}C+32=\dfrac{9}{5}(-35)+32=-31°$

Chapter 2

5. $-8+2(3x+1)=-3(x-4)$
$$-8+6x+2=-3x+12$$
$$9x=18$$
$$x=2$$

6. $\dfrac{1}{9}h=23\Rightarrow h=207$

$\dfrac{8}{9}(207)=184$ ft below water line

7. $4+5(x+3)\geq x-1$
$$4+5x+15\geq x-1$$
$$4x\geq-20$$
$$x\geq-5$$

8. $\left|3\left(\dfrac{2}{3}x-4\right)\right|\leq12$

$$-12\leq3\left(\dfrac{2}{3}x-4\right)\leq12$$
$$-12\leq2x-12\leq12$$
$$0\leq2x\leq24$$
$$0\leq x\leq12$$

Chapter 3

9. $3x-5y=-7$
$$5y=3x+7$$
$$y=\dfrac{3}{5}x+\dfrac{7}{5}$$
$$m=\dfrac{3}{5},\ b=\dfrac{7}{5}$$

10. $(-5,6),\ (2,-3)$

$$m=\dfrac{y_2-y_1}{x_2-x_1}=\dfrac{-3-6}{2-(-5)}=-\dfrac{9}{7}$$
$$y-y_1=m(x-x_1)$$
$$y-6=-\dfrac{9}{7}(x-(-5))$$
$$7y-42=-9x-45$$
$$9x+7y=-3$$

11. Yes, $\{(5,6),(-6,5),(6,5),(-5,6)\}$ is a function since no two ordered pairs have the same first coordinate.

12. $f(x)=-2x^2-6x+1$
$$f(-3)=-2(-3)^2-6(-3)+1=1$$

Chapter 4

13. Solve by addition method.

$$3x + 5y = 30 \xrightarrow{\times 3} \quad 9x + 15y = 90$$

$$5x + 3y = 34 \xrightarrow{\times -5} \underline{-25x - 15y = -170}$$

$$-16x = -80$$

$$x = 5$$

$$3(5) + 5y = 30$$

$$5y = 15$$

$$y = 3$$

$$x = 5, \ y = 3 \text{ is the solution.}$$

14. Switch the first and second equation

$$x + y + z = 7$$

$$2x - y + 2z = 8$$

$$4x + y - 3x = -6$$

Add -2 times first equation to second equation -4 times first equation to second equation

$$x + y + z = 7$$

$$-3y \qquad = -6 \Rightarrow y = 2$$

$$-3y - 7z = -34 \Rightarrow -3(2) - 7z = -34$$

$$-7z = -28 \Rightarrow z = 4$$

$$x + 2 + 4 = 7 \Rightarrow x = 1$$

$$x = 1, \ y = 2, \ z = 4 \text{ is the solution.}$$

15. Using $d = rt$,

$$90 = (v + w) \cdot 2 \Rightarrow v + w = 45$$

$$105 = (v - w) \cdot 3 \Rightarrow \underline{v - w = 35}$$

$$2v \quad = 80$$

$v = 40$ mph is the boat's speed

$w = 45 - v = 5$ mph is the current

16. $x - y \leq -4$

$(0,0)$ test point:

$0 - 0 \leq -4$ false.

$2x + y \leq 0$

$(0,2)$ test point:

$2(0) + 2 \leq 0, \ 2 \leq 0$ false

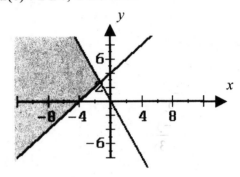

Chapter 5

17. $(3x - 4)(2x^2 - x + 3)$

$$= 6x^3 - 3x^2 + 9x - 8x^2 + 4x - 12$$

$$= 6x^3 - 11x^2 + 13x - 12$$

18. $(2x^3 + 7x^2 - 4x - 21) \div (x + 3)$

$$\begin{array}{r} 2x^2 + x - 7 \\ x + 3 \overline{)\,2x^3 + 7x^2 - 4x - 21} \\ \underline{2x^3 + 6x^2} \\ x^2 - 4x \\ \underline{x^2 + 3x} \\ -7x - 21 \\ \underline{-7x - 21} \end{array}$$

$$(2x^3 + 7x^2 - 4x - 21) \div (x + 3)$$

$$= 2x^2 + x - 7$$

19. $125x^3 - 8y^3$

$$= (5x - 2y)(25x^2 + 10xy + 4y^2)$$

20. Solve by factoring,

$$2x^2 - 7x - 4 = 0$$

$$(2x+1)(x-4) = 0$$

$$2x+1=0 \text{ or } x-4=0$$

$$2x = -1 \qquad\qquad x = 4$$

$$x = -\frac{1}{2}$$

Chapter 6

21. $\dfrac{10x-5y}{12x+36y} \cdot \dfrac{8x+24y}{20x-10y}$

$$= \dfrac{5(2x-y)}{12(x+3y)} \cdot \dfrac{8(x+3y)}{10(2x-y)}$$

$$= \dfrac{40}{120}$$

$$= \dfrac{1}{3}$$

22. $2x - 1 + \dfrac{2}{x+2} = \dfrac{(2x-1)(x+2)+2}{x+2}$

$$= \dfrac{2x^2 + 3x - 2 + 2}{x+2}$$

$$= \dfrac{2x^2 + 3x}{x+2}$$

23. $\dfrac{\dfrac{1}{x+h} - \dfrac{1}{x}}{h} \cdot \dfrac{x(x+h)}{x(x+h)}$

$$= \dfrac{x - (x+h)}{xh(x+h)}$$

$$= \dfrac{-h}{xh(x+h)}$$

$$= \dfrac{-1}{x(x+h)}$$

24. $\dfrac{x}{x-2} + \dfrac{3x}{x+4} = \dfrac{6}{x^2+2x-8}$

$$\dfrac{x}{x-2} + \dfrac{3x}{x+4} = \dfrac{6}{(x-2)(x+4)}$$

$$x(x+4) + 3x(x-2) = 6$$

$$x^2 + 4x + 3x^2 - 6x = 6$$

$$4x^2 - 2x - 6 = 0$$

$$2x^2 - x - 3 = 0$$

$$(2x-3)(x+1) = 0$$

$$2x-3 = 0 \text{ or } x+1 = 0$$

$$2x = 3 \qquad\qquad x = -1$$

$$x = \dfrac{3}{2}$$

Chapter 7

25. $(\sqrt{3} + \sqrt{2x})(\sqrt{7} - \sqrt{2x^3})$

$$= \sqrt{21} - \sqrt{6x^3} + \sqrt{14x} - \sqrt{4x^4}$$

$$= \sqrt{21} - x\sqrt{6x} + \sqrt{14x} - 2x^2$$

26. $\dfrac{3\sqrt{x} + \sqrt{y}}{\sqrt{x} - \sqrt{y}} \cdot \dfrac{\sqrt{x} + \sqrt{y}}{\sqrt{x} + \sqrt{y}} = \dfrac{3x + 4\sqrt{xy} + y}{x-y}$

27. $2\sqrt{x-1} = x - 4$

$$4(x-1) = x^2 - 8x + 16$$

$$x^2 - 12x + 20 = 0$$

$$(x-10)(x-2) = 0$$

$$x-10 = 0 \text{ or } x-2 = 0$$

$$x = 10 \qquad\qquad x = 2$$

check:

$$2\sqrt{10-1} \overset{?}{=} 10 - 4, \quad 2\sqrt{2-1} \overset{?}{=} 2 - 4$$

$$6 = 6 \qquad\qquad 2 \neq -2$$

$x = 10$ is the solution.

Chapter 8

28. Solve with quadratic equation.

$x^2 - 2x - 4 = 0$

$$x = \frac{-(-2) \pm \sqrt{(-2)^2 - 4(1)(-4)}}{2(1)}$$

$$x = \frac{2 \pm \sqrt{20}}{2} = \frac{2 \pm 2\sqrt{5}}{2} = 1 \pm \sqrt{5}$$

29. Let $x^2 = w$.

$x^4 - 12x^2 + 20 = 0$

$w^2 - 12w + 20 = 0$

$(w - 10)(w - 2) = 0$

$w - 10 = 0$ or $w - 2 = 0$

$\quad w = 10 \qquad\qquad w = 2$

$\quad x^2 = 10 \qquad\qquad x^2 = 2$

$\quad x = \pm\sqrt{10} \qquad\quad x = \pm\sqrt{2}$

30. The graph of $f(x) = (x-2)^2 + 3$ is the graph of $f(x) = x^2$ shifted right 2 units and up 3 units.

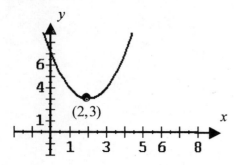

Chapter 9

31. $(x - x_1)^2 + (y - y_1)^2 = r^2$

$(x - 5)^2 + (y - (-2))^2 = 6^2$

$(x - 5)^2 + (y + 2)^2 = 36$

32. $a = 3, \quad b = 4$

$$\frac{x^2}{a^2} + \frac{y^2}{b^2} = 1$$

$$\frac{x^2}{3^2} + \frac{y^2}{4^2} = 1$$

$$\frac{x^2}{9} + \frac{y^2}{16} = 1$$

33. $x + 2y = 3 \Rightarrow x = 3 - 2y$

$x^2 + 4y^2 = 9 \Rightarrow (3 - 2y)^2 + 4y^2 = 9$

$9 - 12y + 4y^2 + 4y^2 = 9$

$8y^2 - 12y = 0$

$y(2y - 3) = 0$

$y = 0 \qquad$ or $\quad 2y - 3 = 0, \ y = \dfrac{3}{2}$

$x = 3 - 2(0) \qquad\quad x = 3 - 2\left(\dfrac{3}{2}\right)$

$x = 3 \qquad\qquad\qquad x = 0$

$(3, 0), \ \left(0, \dfrac{3}{2}\right)$ is the solution.

Chapter 10

34. $\quad f(x) = 2x^2 - 3x + 4$

$f(a + 2) = 2(a + 2)^2 - 3(a + 2) + 4$

$f(a + 2) = 2(a^2 + 4a + 4) - 3a - 6 + 4$

$f(a + 2) = 2a^2 + 8a + 8 - 3a - 2$

$f(a + 2) = 2a^2 + 5a + 6$

35. The graph of $f(x) = |x + 3|$ is the graph of $y = |x|$ shifted left 3 units. The graph of $g(x) = |x + 3| - 3$ is the graph of $f(x)$ shifted down 3 units.

(continued)

35. (continued)

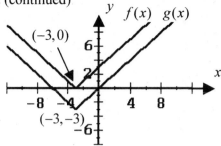

36. $f(x) = \dfrac{3}{x+2}, \quad g(x) = 3x^2 - 1$

$g[f(x)] = g\left[\dfrac{3}{x+2}\right]$

$g[f(x)] = 3\left(\dfrac{3}{x+2}\right)^2 - 1 = \dfrac{27}{(x+2)^2} - 1$

37. $f(x) = -\dfrac{1}{2}x - 5, \quad f(x) \rightarrow y$

$y = -\dfrac{1}{2}x - 5, \quad x \leftrightarrow y$

$x = -\dfrac{1}{2}y - 5$

$2x = -y - 10$

$y = -2x - 10, \quad y \rightarrow f^{-1}(x)$

$f^{-1}(x) = -2x - 10$

Chapter 11

38. $\log_5 125 = y \Leftrightarrow 5^y = 125 = 5^3$

$y = 3$

39. $\log_b 4 = \dfrac{2}{3} \Leftrightarrow b^{2/3} = 4 = 2^2$

$(b^{1/3})^2 = 2^2$

$b^{1/3} = 2$

$b = 8$

40. $\log 10,000 = N \Leftrightarrow 10^N = 10,000 = 10^4$

$N = \log 10,000 = 4$

41. $\log_6 (5 + x) + \log_6 x = 2$

$\log_6 [x(x+5)] = 2 \Leftrightarrow 6^2 = x(x+5)$

$x^2 + 5x = 36$

$x^2 + 5x - 36 = 0$

$(x-4)(x+9) = 0$

$x - 4 = 0 \ \text{ or } \ x + 9 = 0$

$x = 4 \qquad\qquad x = -9, \text{ reject, } x > 0$

$x = 4$ is the solution.

Chapter 1

Pretest Chapter 1

1. $\pi, \sqrt{7}$

2. $\sqrt{9}, 3, \dfrac{6}{2}, 0$

3. Associative property of addition.

4. $30 \div (-6) + 3 - 2(-5)$

$= -5 + 3 - (-10)$

$= -2 + 10$

$= 8$

5. $\dfrac{\dfrac{2}{3} - \dfrac{1}{4}}{\dfrac{1}{2} + \dfrac{1}{4}} = \dfrac{\dfrac{5}{12}}{\dfrac{3}{4}} = \dfrac{5}{12} \cdot \dfrac{4}{3} = \dfrac{5}{9}$

6. $3^3 - \sqrt{4(-5) + 29}$

$= 27 - \sqrt{-20 + 29}$

$= 27 - \sqrt{9}$

$= 27 - 3$

$= 24$

7. $(-4)^3 + 2(3^2 - 2^2)$

$= -64 + 2(3^2 - 2^2)$

$= -64 + 2(9 - 4)$

$= -64 + 2(5)$

$= -64 + 10$

$= -54$

8. $\left(-7a^2 b\right)\left(-2a^0 b^3 c^2\right)$

$= 14a^2 b^4 c^2$

9. $\dfrac{\left(3x^{-1} y^2\right)^3}{\left(4x^2 y^{-2}\right)^2} = \dfrac{27x^{-3} y^6}{16x^4 y^{-4}} = \dfrac{27 y^{10}}{16x^7}$

10. $\dfrac{4x^3 y^2}{-16x^2 y^{-3}} = -\dfrac{xy^5}{4}$

11. $\left(\dfrac{3a^{-5} b^0}{2a^{-2} b^3}\right)^2 = \left(\dfrac{3}{2a^3 b^3}\right)^2 = \dfrac{9}{4a^6 b^6}$

12. $\left(-2x^3 y^{-2}\right)^{-2}$

$= \dfrac{1}{(-2)^2 \left(x^3\right)^2 \left(y^{-2}\right)^2}$

$= \dfrac{1}{4x^6 y^{-4}}$

$= \dfrac{y^4}{4x^6}$

13. $0.000058 = 5.8 \times 10^{-5}$

14. $8.95 \times 10^7 = 89{,}500{,}000$

15. $3x + 5x^2 - 2x^3 - 6x - 8x^3$

$= -10x^3 + 5x^2 - 3x$

16. $3ab^2 \left(-2a^2 + 3ab^2 - 1\right)$

$= -6a^3 b^2 + 9a^2 b^4 - 3ab^2$

17. $-2\left\{x+3\left[y-5(x+y)\right]\right\}$

$\quad = -2\left\{x+3\left[y-5x-5y\right]\right\}$

$\quad = -2\left\{x+3\left[-5x-4y\right]\right\}$

$\quad = -2\left\{x-15x-12y\right\}$

$\quad = -2\left\{-14x-12y\right\}$

$\quad = 28x+24y$

18. $5a^2 - 3ab + 4b$

$\quad = 5(-3)^2 - 3(-3)(-2) + 4(-2)$

$\quad = 5(9) - 3(6) + (-8)$

$\quad = 45 - 18 + (-8)$

$\quad = 27 + (-8)$

$\quad = 19$

19. $A = \pi r^2 = 3.14(4)^2$

$\quad = 50.24$ square meters

20. $T = 2\pi\sqrt{L/g}$

$\quad T = 2(3.14)\sqrt{512/32}$

$\quad = 25.12$ seconds

1.1 Exercises

1. Integers include negative numbers. Whole numbers are positive integers.

3. A terminating decimal is one that comes to an end.

5. 13,001: Natural, Whole, Integer, Rational, Real

7. -42 : Integer, Rational, Real

9. $0.\overline{34}$: Rational, Real

11. $-\dfrac{8}{7}$: Rational, Real

13. $\dfrac{\pi}{5}$: Irrational, Real

15. 0.79: Rational, Real

17. 7.040040004... : Irrational, Real

19. $-25, -\dfrac{28}{7}$

21. $-\dfrac{28}{7}, -25, -\dfrac{18}{5}, -\pi - 0.763, -0.333...$

23. $\dfrac{1}{10}, \dfrac{2}{7}, \dfrac{283}{5}, 52.8$

25. $-\dfrac{18}{5}, -0.763, -0.333...$

27. $1, 3, 5, 7, ...$

29. Associative property of addition.

31. Distributive property of multiplication over addition.

33. Inverse property of addition.

35. Distributive property of multiplication over addition.

37. Associative property of multiplication.

39. Identity property of addition.

41. Associative property of addition.

43. Identity property of multiplication.

45. Distributive property of multiplication over addition.

47. Mount Kea:
$13,784 - (-18,000) = 31,784$
Mount Everest:
$29,022 - 12,000 = 17,022$. The set of such mountains is the empty set.

49. $1,000,000 + 20(1,000,000)$
$= \$21,000,000$

1.2 Exercises

1. To add two real numbers with the same sign, add their absolute values. The sum takes the common sign. To add two real numbers with different signs, find the difference of their absolute values. The answer takes the sign of the number with the larger absolute value.

3. $|-9| = -(-9) = 9$

5. $|8.3| = 8.3$

7. $|9 - 14| = |-5| = 5$

9. $|-b| = -(-b) = b$

11. $-6 + (-12) = -18$

13. $5 - 6 + 2 - 4 = 7 - 10 = -3$

15. $5\left(-\dfrac{1}{3}\right) = -\dfrac{5}{3} = -1\dfrac{2}{3}$

17. $\dfrac{-18}{-2} = 9$

19. $(-0.3)(0.1) = -0.03$

21. $-4.9 + 10.5 = 5.6$

23. $-\dfrac{5}{12} + \dfrac{7}{18} = -\dfrac{15}{36} + \dfrac{14}{36} = -\dfrac{1}{36}$

25. $(-2.4) \div (6) = \dfrac{-2.4}{6} = -0.4$

27. $-7(-2)(-1)(3) = 14(-3) = -42$

29. $-8 + (6)\left(-\dfrac{1}{3}\right) = -8 + (-2) = -10$

31. $-3 + 3 \div (1.5) = -3 + 2 = -1$

33. $12 + (-12) = 0$

35. $\dfrac{-5}{0}$ Undefined

37. $\dfrac{0}{-3} = 0$

39. $\dfrac{3-3}{-8} = \dfrac{0}{-8} = 0$

41. $\dfrac{-7 + (-7)}{-14} = \dfrac{-14}{-14} = 1$

43. $5+6-(-3)-8+4-3$

$\quad =11+3-8+1$

$\quad =14+(-8)+1$

$\quad =6+1=7$

45. $\dfrac{12-4(3)}{2-6}=\dfrac{12-12}{-4}=\dfrac{0}{-4}=0$

47. $6(-2)+3-5(-3)-4$

$\quad =-12+3-(-15)-4$

$\quad =-9+15+(-4)$

$\quad =6+(-4)$

$\quad =2$

49. $15+20\div2-4(3)$

$\quad =15+10-12$

$\quad =24-12$

$\quad =13$

51. $\dfrac{1+49\div(-7)-(-3)}{-1-2}$

$\quad =\dfrac{1+(-7)+3}{-3}$

$\quad =\dfrac{-6+3}{-3}$

$\quad =\dfrac{-3}{-3}$

$\quad =1$

53. $4\left(-\dfrac{1}{2}\right)+\left(\dfrac{2}{3}\right)9$

$\quad =-2+6$

$\quad =4$

55. $\dfrac{1.63482-2.48561}{(16.05436)(0.07814)}$

$\quad =\dfrac{-0.85079}{1.25448769}$

$\quad =-0.678197169$

57. One or three of the quantities a,b,c must be negative. In other words, when multiplying an odd number of negative numbers, the answer will be negative.

59. Commutative property of addition.

Cumulative Review Problems

61. $\left\{-\dfrac{1}{2}\pi,\sqrt{3}\right\}$

63. $0.125+2(0.06)+5(0.1875)$

$\quad +7(0.10)+3(0.25)=\$2.63\dfrac{1}{4}$

65. $1,300,000(0.75)=975,000$

1.3 Exercises

1. The base is a and the exponent is 3.

3. Positive

5. -11 and 11, $(-11)(-11)=121$

$\quad (11)(11)=121$

7. $9\cdot9\cdot9\cdot9=9^4$

9. $x\cdot x\cdot x\cdot x\cdot x\cdot x=x^6$

11. $(-6)(-6)(-6)(-6)(-6)=(-6)^5$

13. $3^5 = 3 \cdot 3 \cdot 3 \cdot 3 \cdot 3 = 243$

15. $(-5)^2 = (-5)(-5) = 25$

17. $-6^2 = -(6)(6) = -36$

19. $-1^4 = -(1)(1)(1)(1) = -1$

21. $\left(\dfrac{2}{3}\right)^2 = \left(\dfrac{2}{3}\right)\left(\dfrac{2}{3}\right) = \dfrac{4}{9}$

23. $\left(-\dfrac{1}{6}\right)^3 = \left(-\dfrac{1}{6}\right)\left(-\dfrac{1}{6}\right)\left(-\dfrac{1}{6}\right) = -\dfrac{1}{216}$

25. $(0.8)^2 = (0.8)(0.8) = 0.64$

27. $(0.04)^3 = (0.04)(0.04)(0.04)$
         $= 0.000064$

29. $\sqrt{121} = 11$

31. $-\sqrt{64} = -8$

33. $\sqrt{\dfrac{1}{36}} = \dfrac{1}{6}$

35. $\sqrt{0.25} = 0.5$

37. $\sqrt{4+12} = \sqrt{16} = 4$

39. $\sqrt{\dfrac{5}{36} + \dfrac{31}{36}} = \sqrt{\dfrac{36}{36}} = \sqrt{1} = 1$

41. $-\sqrt{-0.36}$ not a real number

43. $8 \div (4 \cdot 2) - 6$
     $= 8 \div 8 - 6$
     $= 1 - 6$
     $= -5$

45. $\sqrt{16} + 8(-3) = 4 + (-24) = -20$

47. $(5 + 2 - 8)^3 - (-7)$
     $= (7 - 8)^3 + 7$
     $= (-1)^3 + 7$
     $= -1 + 7$
     $= 6$

49. $\sqrt{8^2 + 9 \cdot 4} = \sqrt{64 + 36} = \sqrt{100} = 10$

51. $2\sqrt{16} + (-4)^2 - 3$
     $= 2 \cdot 4 + 16 - 3$
     $= 8 + 16 - 3$
     $= 24 - 3$
     $= 21$

53. $3^2 - 2[6 + (7 - 8)]$
     $= 9 - 2[6 + (-1)]$
     $= 9 - 2(5)$
     $= 9 - 10$
     $= -1$

55. $5[(1.2 - 0.4) - 0.8]$
     $= 5(0.8 - 0.8)$
     $= 5(0)$
     $= 0$

57.
$$\frac{7+2(-4)+5}{8-6}$$
$$=\frac{7+(-8)+5}{2}$$
$$=\frac{-1+5}{2}$$
$$=\frac{4}{2}$$
$$=2$$

59.
$$\frac{-3(2^3-1)}{-7}$$
$$=\frac{-3(8-1)}{-7}$$
$$=\frac{-3(7)}{-7}$$
$$=\frac{-21}{-7}$$
$$=3$$

61.
$$\frac{|2^2-5|-3^2}{-5+3}$$
$$=\frac{|4-5|-9}{-2}$$
$$=\frac{|-1|-9}{-2}$$
$$=\frac{1+(-9)}{-2}$$
$$=\frac{-8}{-2}$$
$$=4$$

63.
$$\frac{|3-2^3|-5}{2+3}$$
$$=\frac{|3-8|-5}{5}$$
$$=\frac{|-5|-5}{5}$$
$$=\frac{5-5}{5}$$
$$=\frac{0}{5}$$
$$=0$$

65. $(5.986)^5 = 7685.702373$

67. $2^{11} - 2^9 = 2048 - 512 = 1536$

Cumulative Review Problems

69. Associative property of addition.

71.
$$\frac{40,000-5000}{5000} = \frac{35,000}{5000} = 7$$
An increase of 700%

73. $1 + 0.1 + 1.1(0.1) = 1.21$
An increase of 21%.

1.4 Exercises

1. $3^{-2} = \dfrac{1}{3^2} = \dfrac{1}{9}$

3. $x^{-5} = \dfrac{1}{x^5}$

5. $\left(\dfrac{3}{4}\right)^{-3} = \dfrac{1}{\left(\dfrac{3}{4}\right)^3} = \dfrac{1}{\dfrac{27}{64}} = \dfrac{64}{27}$

7. $\left(-\dfrac{1}{9}\right)^{-1} = \dfrac{1}{-\dfrac{1}{9}} = -9$

9. $x^4 \cdot x^8 = x^{4+8} = x^{12}$

11. $12^5 \cdot 12^9 = 12^{5+9} = 12^{14}$

13. $(3x)(-2x^5) = -6x^{1+5} = -6x^6$

15. $(-12x^3 y)(-3x^5 y^2) = 36x^8 y^3$

17. $(2x^0 y^5 z)(-5xy^0 z^8) = -10xy^5 z^9$

19. $(-3m^{-2}n^4)(5m^2 n^{-5}) = -15m^0 n^{-1}$
$$= -\dfrac{15}{n}$$

21. $\dfrac{x^{16}}{x^5} = x^{16-5} = x^{11}$

23. $\dfrac{a^{20}}{a^{25}} = a^{20-25} = a^{-5} = \dfrac{1}{a^5}$

25. $\dfrac{2^8}{2^5} = 2^{8-5} = 2^3 = 8$

27. $\dfrac{2x^3}{x^8} = 2x^{3-8} = 2x^{-5} = \dfrac{2}{x^5}$

29. $\dfrac{10ab^5 c}{-2ab^2} = -5b^3 c$

31. $\dfrac{-20a^{-3}b^{-8}}{14a^{-5}b^{-12}} = -\dfrac{10}{7} a^{-3-(-5)} b^{-8-(-12)}$
$$= -\dfrac{10}{7} a^2 b^4$$

33. $\left(x^2\right)^8 = x^{2 \cdot 8} = x^{16}$

35. $\left(3a^5 b\right)^4 = 3^4 a^{5 \cdot 4} b^4 = 81a^{20} b^4$

37. $\left(\dfrac{x^2 y^3}{z}\right)^6 = \dfrac{x^{2 \cdot 6} y^{3 \cdot 6}}{z^6} = \dfrac{x^{12} y^{18}}{z^6}$

39. $\left(\dfrac{3ab^{-2}c^3}{4a^0 b^4}\right)^2 = \left(\dfrac{3ac^3}{4b^6}\right)^2 = \dfrac{3^2 a^2 c^{3 \cdot 2}}{4^2 b^{6 \cdot 2}}$
$$= \dfrac{9a^2 c^6}{16b^{12}}$$

41. $\left(\dfrac{2xy^2}{x^{-3} y^{-4}}\right)^{-3} = \left(2x^4 y^6\right)^{-3}$
$$= \dfrac{1}{\left(2x^4 y^6\right)^3}$$
$$= \dfrac{1}{2^3 x^{4 \cdot 3} y^{6 \cdot 3}}$$
$$= \dfrac{1}{8x^{12} y^{18}}$$

43. $\left(x^{-1} y^3\right)^{-2} \left(2xy^4\right)^2$
$$= x^{(-1)(-2)} y^{3(-2)} 2^2 x^2 y^{4 \cdot 2}$$
$$= 4x^2 y^{-6} x^2 y^8$$
$$= 4x^4 y^2$$

45. $\dfrac{\left(-3m^5 n^{-1}\right)^3}{\left(-mn\right)^2} = \dfrac{(-3)^3 m^{5 \cdot 3} n^{-1(3)}}{(-m)^2 n^2}$
$$= \dfrac{-27m^{15} n^{-3}}{m^2 n^2}$$
$$= \dfrac{-27m^{13}}{n^5}$$

47. $\dfrac{2^{-3}a^2b^{-4}}{2^{-4}a^{-2}b^3} = 2^{-3-(-4)}a^{2-(-2)}b^{-4-3}$

$$= 2a^4b^{-7} = \dfrac{2a^4}{b^7}$$

49. $\left(\dfrac{2}{3}x^7y^{-1}\right)^{-2}$

$$= \left(\dfrac{2}{3}\right)^{-2}\left(x^7\right)^{-2}\left(y^{-1}\right)^{-2}$$

$$= \dfrac{1}{\left(\dfrac{2}{3}\right)^2}x^{7(-2)}y^{(-1)(-2)}$$

$$= \dfrac{1}{\dfrac{4}{9}}x^{-14}y^2$$

$$= \dfrac{9y^2}{4x^{14}}$$

51. $\left(\dfrac{x^{-2}z}{y^{-5}}\right)^{-2} = \dfrac{x^{(-2)(-2)}z^{-2}}{y^{(-5)(-2)}}$

$$= \dfrac{x^4z^{-2}}{y^{10}}$$

$$= \dfrac{x^4}{y^{10}z^2}$$

53. $\dfrac{a^{-2}b^0c}{ab^{-5}c} = \dfrac{b^5}{a^3}$

55. $\left(\dfrac{14x^{-3}y^{-3}}{7x^{-4}y^{-3}}\right)^{-2}$

$$= (2x)^{-2}$$

$$= \dfrac{1}{(2x)^2}$$

$$= \dfrac{1}{4x^2}$$

57. $\dfrac{7^{-8}\cdot 5^{-6}}{7^{-9}\cdot 5^{-5}\cdot 6^0}$

$$= 7^{-8-(-9)}5^{-6-(-5)}$$

$$= 7^1 5^{-1} = \dfrac{7}{5}$$

59. $\left(9x^{-2}y\right)\left(-\dfrac{2}{3}x^3y^{-2}\right)$

$$= -6xy^{-1} = \dfrac{-6x}{y}$$

61. $\left(-3.6982x^3y^4\right)^7$

$$= \left(-3.6982\right)^7 x^{3\cdot 7}y^{4\cdot 7}$$

$$= 9460.906704x^{21}y^{28}$$

63. $7,200,000\dfrac{\text{ft}^3}{\text{s}}\cdot\dfrac{0.305^3\,\text{m}^3}{\text{ft}^3}$

$$= 204,283\dfrac{\text{m}^3}{\text{s}}$$

65. $470 = 4.7\times 10^2$

67. $1,730,000 = 1.73\times 10^6$

69. $0.017 = 1.7\times 10^{-2}$

71. $0.0000529 = 5.29\times 10^{-5}$

73. $7.13 \times 10^5 = 713,000$

75. $1.863 \times 10^{-2} = 0.01863$

77. $9.01 \times 10^{-7} = 0.000000901$

79. $\left(7.2 \times 10^{-3}\right)\left(5.0 \times 10^{-5}\right)$
$= (7.2)(5.0)\left(10^{-3}\right)\left(10^{-5}\right)$
$= 36 \times 10^{-8}$
$= 3.6 \times 10^{-7}$

81. $\dfrac{3.6 \times 10^{-5}}{1.2 \times 10^{-6}} = 3 \times 10^1$

83. $\left(5.1 \times 10^4\right)\left(2 \times 10^2\right) = 10.2 \times 10^6$
$= 1.02 \times 10^7$

85. $\left(5.3 \times 10^{-23}\right)\left(2 \times 10^4\right) = 10.6 \times 10^{-19}$
$= 1.06 \times 10^{-18}$

Cumulative Review Problems

87. $5 + 2(-3) + 12 \div (-6)$
$= 5 + (-6) + (-2)$
$= -1 + (-2)$
$= -3$

89. $250\left(1 - \left(0.62 + 0.26 + 0.04\right)\right)$
$= 250(1 - 0.92)$
$= 250(0.08)$
$= 20$ receive no newspapers

1.5 Exercises

1. It is $-5x^2$ since y is multiplied by $-5x^2$.

3. $5x^3, -6x^2, 4x, 8$

5. $1, 2, 3$

7. $5, -3, 1$

9. $6.5, -0.02, 3.05$

11. $3ab + 8ab = 11ab$

13. $4y - 7x + 2x - 6y$
$= -2y - 5x$

15. $x^2 - 3x - 4x^2 + 2x$
$= -3x^2 - x$

17. $4ab - 3b^2 - 5ab + 3b^2 = -ab$

19. $0.1x^2 + 3x - 0.5x^2 = -0.4x^2 + 3x$

21. $\dfrac{2}{3}m + \dfrac{5}{6}n - \dfrac{1}{3}m + \dfrac{1}{3}n$
$= \dfrac{1}{3}m + \dfrac{7}{6}n$

23. $\dfrac{1}{5}a^2 - 3b - \dfrac{1}{2}a^2 + 5b$
$= a^2\left(\dfrac{2}{10} - \dfrac{5}{10}\right) + 2b$
$= -\dfrac{3}{10}a^2 + 2b$

25. $1.2x^2 - 5.6x - 8.9x^2 + 2x$
$= -7.7x^2 - 3.6x$

27. $6x(3x+y)$

$\quad = (6x)(3x)+(6x)(y)$

$\quad = 18x^2 + 6xy$

29. $-x(-x^3+3x^2+5x)$

$\quad = (-x)(-x^3)+(-x)(3x^2)+(-x)(5x)$

$\quad = x^4 - 3x^3 - 5x^2$

31. $2xy(x^2-3xy+4y^2)$

$\quad = 2x^3y - 6x^2y^2 + 8xy^3$

33. $\dfrac{2}{3}(4x-3y+6)$

$\quad = \dfrac{8}{3}x - 2y + 4$

35. $\dfrac{x}{6}(x^2+5x-9)$

$\quad = \dfrac{x^3}{6} + \dfrac{5x^2}{6} - \dfrac{3x}{2}$

37. $3ab(a^4b-3a^2+a-b)$

$\quad = 3a^5b^2 - 9a^3b + 3a^2b - 3ab^2$

39. $1.5x^2(x-2y+3y^2)$

$\quad = 1.5x^3 - 3x^2y + 4.5x^2y^2$

41. $2(x-1)-x(x+1)+3(x^2+2)$

$\quad = 2x - 2 - x^2 - x + 3x^2 + 6$

$\quad = 2x^2 + x + 4$

43. $4\left[x+\left(\dfrac{1}{2}x-y\right)\right]-3\left[\dfrac{2}{3}x-(x-y)\right]$

$\quad = 4\left[x+\dfrac{1}{2}x-y\right]-3\left[\dfrac{2}{3}x-x+y\right]$

$\quad = 4\left[\dfrac{3}{2}x-y\right]-3\left[-\dfrac{1}{3}x+y\right]$

$\quad = 6x - 4y + x - 3y$

$\quad = 7x - 7y$

45. $2\{3x-2[x-4(x+1)]\}$

$\quad = 2\{3x-2[x-4x-4]\}$

$\quad = 2\{3x-2[-3x-4]\}$

$\quad = 2\{3x+6x+8\}$

$\quad = 2\{9x+8\}$

$\quad = 18x + 16$

Cumulative Review Problems

47. $2(-3)^2 + 4(-2)$

$\quad = 2(9) - 8$

$\quad = 18 - 8$

$\quad = 10$

49. $\dfrac{5(-2)-8}{3+4-(-3)}$

$\quad = \dfrac{-10-8}{7+3}$

$\quad = \dfrac{-18}{10} = \dfrac{-9}{5}$

$\quad = -1.8$

51. $\dfrac{1,893,500 \text{ organisms}}{\text{inch}} \cdot \dfrac{1 \text{ inch}}{0.0254 \text{ meters}}$

$= \dfrac{1,893,500 \text{ organisms}}{0.0254 \text{ meters}} \cdot \dfrac{1000 \text{ meter}}{\text{kilometer}}$

$= \dfrac{7.4547 \times 10^{10} \text{ organisms}}{\text{kilometer}}$

53. $35,861 \cdot \dfrac{1}{7} = 5123$

1.6 Exercises

1. $11x - 7 = 11(3) - 7 = 33 - 7 = 26$

3. $x^2 + 5x - 6$
$= (-3)^2 + 5(-3) - 6$
$= 9 + (-15) - 6$
$= -6 - 6 = -12$

5. $-x^2 + 5x + 3$
$= -(-1)^2 - 2(-1) + 3$
$= -1 + 2 + 3$
$= 1 + 3 = 4$

7. $-2x^2 + 5x - 3$
$= -2(-4)^2 + 5(-4) - 3$
$= -2(16) + (-20) - 3$
$= -32 + (-20) - 3$
$= -52 - 3 = -55$

9. $2ax - by - a$
$= 2(1)\left(\dfrac{1}{2}\right) - (-2)(3) - 1$
$= 1 - (-6) - 1$
$= 1 + 6 - 1 = 7 - 1 = 6$

11. $\sqrt{b^2 - 4ac}$
$= \sqrt{5^2 - 4(1)(-14)}$
$= \sqrt{24 + 56} = \sqrt{81} = 9$

13. $2x^2 - 5x + 6$
$= 2(-3.52176)^2 - 5(-3.52176) + 6$
$= 48.41439$

15. $F = \dfrac{9}{5}(-60) + 32 = -76°\text{F}$

17. $C = \dfrac{5(122) - 160}{9} = 50°\text{C}$

19. $T = 2\pi\sqrt{L/g}$
$T = 2(3.14)\sqrt{32/32}$
$T = 6.28 \text{ seconds}$

21. $A = \$4800\left[1 + (0.12)(1.5)\right]$
$A = \$5664$

23. $A = \$1200\left[1 + (0.11)(3)\right]$
$A = \$1596$

25. $S = \dfrac{1}{2}(32)(6)^2 = 576 \text{ feet}$

27. $S = \dfrac{1}{2}(32)(4)^2 = 256 \text{ feet}$

29. $z = \dfrac{(36)(4)}{36 + 4} = \dfrac{144}{40} = \dfrac{18}{5}$

31. $m = \dfrac{9 \cdot 150}{9 + 12} = \dfrac{1350}{21} = 64 \text{ milligrams}$

33. $A = \pi r^2 = 3.14 \cdot 0.5^2 = 0.785$ sq. in.

35. $A = \dfrac{1}{2}ab = \dfrac{1}{2} \cdot 12 \cdot 14 = 84$ sq. meters

37. $A = ab = 4\left(\dfrac{7}{8}\right) = \dfrac{7}{2}$ sq. yards

39. $A = lw = (0.3)(0.08) = 0.024$ sq. mi.

41. $P = 4b = 4(0.38) = 1.52$ meters

43. (a) $V = \pi r^2 h = 3.14(4)^2(11)$
  $V = 552.64$ cubic centimeters
  (b) $S = 2\pi rh + 2\pi r^2$
  $S = 2(3.14)(4)(11) + 2(3.14)(4)^2$
  $S = 376.8$ square centimeters

45. $A = \pi r^2 = 3.14(6)^2 = 113.04$ sq cm
  $C = \pi(12) = 3.14(12) = 37.68$ cm

47. (a) $\dfrac{45{,}032 \text{ cal}}{\text{hr}} \dfrac{120 \text{ watt}}{104 \dfrac{\text{cal}}{\text{hr}}} = 51{,}960$ watt

  (b)
  $\dfrac{104 \text{ cal}}{\text{hr} \cdot \text{person}} 433 \text{ persons} = \dfrac{45{,}032 \text{ cal}}{\text{hr}}$

Cumulative Review Problems

49. $-2a - b - 5a + 6b = -7a + 5b$

51. $\left(\dfrac{-5x^2}{2y^3}\right)^2 = \dfrac{(-5)^2(x^2)^2}{(2)^2(y^3)^2} = \dfrac{25x^4}{4y^6}$

53. $12{,}000{,}000 - 0.30(19{,}000{,}000)$
  $= 6{,}300{,}000$ people

Putting Your Skills To Work

1.

Fraction	Decimal
1/16	0.06
1/8	0.13
3/16	0.19
1/4	.0.25
5/16	0.31
3/8	0.38
7/16	0.44
1/2	0.50
9/16	0.56
5/8	0.63
11/16	0.69
3/4	0.75
13/16	0.81
7/8	0.88
15/16	0.94

2. $61\dfrac{1}{4} = \$61.25,\ 53\dfrac{11}{16} = \$53.69$

3. $54\dfrac{5}{8}(50) = 54.63(50) = \$2731.50$

4. $90\dfrac{3}{16}(38) = 90.81(50) = \$3450.78$

5. $\dfrac{1}{4}, \dfrac{1}{2}, \dfrac{3}{4}$ are exact, others rounded.

6. $\dfrac{1}{4}, \dfrac{1}{2}, \dfrac{3}{4}$ have 0 error
  $\dfrac{1}{16}, \dfrac{3}{16}, \dfrac{5}{16}, \dfrac{7}{16}, \dfrac{9}{16}, \dfrac{11}{16}, \dfrac{13}{16}, \dfrac{15}{16}$ have 0.0025 error.
  $\dfrac{1}{8}, \dfrac{3}{8}, \dfrac{5}{8}, \dfrac{7}{8}$ have the most error with 0.005.

7. $10,000\left(87.38 - 87\frac{3}{8}\right) = 50$. The

quoted price was $50 too high.

8. $200,000\left(55.81 - 55\frac{13}{16}\right) = -500$.

The sales price was $500 too low.

Chapter 1 Review Problems

1. -5: Integer, Rational, Real

2. $\frac{7}{8}$: Rational, Real

3. 3: Natural, Whole, Integer, Rational, Real

4. $0.\overline{3}$: Rational, Real

5. $2.1652384\ldots$: Irrational, Real

6. Commutative property of addition

7. Associative property of multiplication

8. Yes, all rational numbers are real numbers.

9. $-15 - (-20) = -15 + 20 = 5$

10. $-7.3 + (-16.2) = -23.5$

11. $-8(-6) = 48$

12. $-12 \div 3 = -4$

13. $-\frac{5}{7} \div \left(-\frac{5}{13}\right) = \frac{5}{7} \cdot \frac{13}{5} = \frac{13}{7}$

14. $-\frac{3}{5}\left(\frac{2}{3}\right) = -\frac{2}{5}$

15. $4(-3)(-10) = -12(-10) = 120$

16. $5 + 6 - 2 - 5$
$= 11 - 2 - 5$
$= 9 - 5$
$= 4$

17. $-3.6(-1.5) = 5.4$

18. $0 \div (-14) = 0$

19. $7 \div 0$ undefined

20. $-17 + (+17) = 0$

21. $2 - 3\left[(-4) + 6\right] \div (-2)$
$= 2 - 3[2] \div (-2)$
$= 2 - 6 \div (-2)$
$= 2 - (-3) = 2 + 3 = 5$

22. $\frac{5 - 8}{2 - 7 - (-2)} = \frac{-3}{-5 + 2} = \frac{-3}{-3} = 1$

23. $4\sqrt{16} + 2^3 - 6 = 4 \cdot 4 + 8 - 6$
$= 16 + 2 = 18$

24. $3 - |-4| + (-2)^3$
$= 3 - 4 + (-8)$
$= -1 + (-8) = -9$

25. $4 - 2 + 6\left(-\dfrac{1}{3}\right) = 2 + (-2) = 0$

26. $\sqrt{(-3)^2} + (-2)^3 = \sqrt{9} + (-8)$

$\qquad = 3 + (-8)$

$\qquad = -5$

27. $\sqrt{\dfrac{25}{36}} - 2\left(\dfrac{1}{12}\right) = \dfrac{5}{6} - \left(\dfrac{1}{6}\right) = \dfrac{4}{6} = \dfrac{2}{3}$

28. $2\sqrt{16} + 3(-4)(0)(2) - 2^2$

$\qquad = 2 \cdot 4 + 0 - 4 = 8 - 4 = 4$

29. $(-0.4)^3 = -0.064$

30. $\left(3xy^2\right)\left(-2x^{0y}\right)\left(4x^3y^3\right)$

$\qquad = -24x^{1+0+3}y^{2+1+3} = -24x^4y^6$

31. $\left(2^4 ab\right)\left(2^{-3} a^{-5} b^6\right)$

$\qquad = 2^{4-3} a^{1-5} b^{1+6}$

$\qquad = 2a^{-4}b^7 = \dfrac{2b^7}{a^4}$

32. $\dfrac{5^{-3} x^{-3} y^6}{5^{-5} x^{-5} y^8} = 5^{-3+5} x^{-3+5} y^{6-8}$

$\qquad = 5^2 x^2 y^{-2} = \dfrac{25x^2}{y^2}$

33. $\dfrac{27ab^3c}{81a^5bc^0} = \dfrac{b^2c}{3a^4}$

34. $\left(\dfrac{-3x^3 y}{2x^4 z^2}\right)^4 = \dfrac{(-3)^4 \left(x^3\right)^4 y^4}{2^4 \left(x^4\right)^4 \left(z^2\right)^4}$

$\qquad = \dfrac{81x^{12} y^4}{16x^{16} z^8} = \dfrac{81y^4}{16x^4 z^8}$

35. $\dfrac{\left(-2a^{-3}b^{-4}\right)^3}{\left(-3a^{-4}b^2 c\right)^{-2}}$

$\qquad = \dfrac{(-2)^3 \left(a^{-3}\right)^3 \left(b^{-4}\right)^3}{(-3)^{-2} \left(a^{-4}\right)^{-2} \left(b^2\right)^{-2} (c)^{-2}}$

$\qquad = \dfrac{-8a^{-9}b^{-12}}{\dfrac{1}{(-3)^2} a^8 b^{-4} c^{-2}}$

$\qquad = \dfrac{-8a^{-9-8}b^{-12+4}}{\dfrac{1}{9} c^{-2}} = \dfrac{-72a^{-17}b^{-8}}{c^{-2}}$

$\qquad = \dfrac{-72c^2}{a^{17}b^8}$

36. $\left(2^{-1} a^2 b^{-4}\right)^3 = \left(2^{-1}\right)^3 \left(a^2\right)^3 \left(b^{-4}\right)^3$

$\qquad = 2^{-3} a^6 b^{-12} = \dfrac{a^6}{2^3 b^{12}} = \dfrac{a^6}{8b^{12}}$

37. $\dfrac{3^{-2} x^5 y^{-6}}{3x^{-4} y^{-5}} = 3^{-2-1} x^{5-(-4)} y^{-6-(-5)}$

$\qquad = 3^{-3} x^9 y^{-1} = \dfrac{x^9}{3^3 y} = \dfrac{x^9}{27y}$

38. $\dfrac{\left(3^{-1} x^{-2} y\right)^{-2}}{\left(4^{-1} xy^{-2}\right)^{-1}} = \dfrac{\left(3^{-1}\right)^{-2} \left(x^{-2}\right)^{-2} (y)^{-2}}{\left(4^{-1}\right)^{-1} (x)^{-1} \left(y^{-2}\right)^{-1}}$

$\qquad = \dfrac{3^2 x^4 y^{-2}}{4^1 x^{-1} y^2} = \dfrac{9x^{4+1}}{4y^{2+2}} = \dfrac{9x^5}{4y^4}$

39. $\dfrac{\left(2a^{-2}b\right)^{-3}}{\left(3a^{-3}b\right)^{-1}} = \dfrac{2^{-3}a^6b^{-3}}{3^{-1}a^3b^{-1}} = \dfrac{3a^3}{8b^2}$

40. $\left(\dfrac{a^5b^2}{3^{-1}a^{-5}b^{-4}}\right)^3 = \dfrac{a^{15}b^6}{3^{-3}a^{-15}b^{-12}}$

$$= 27a^{30}b^{18}$$

41. $\left(\dfrac{x^3y^4}{5x^6y^8}\right)^3 = \dfrac{x^9y^{12}}{5^3x^{18}y^{24}} = \dfrac{1}{125x^9y^{12}}$

42. $0.00721 = 7.21\times10^{-3}$

43. $\left(5,300,000\right)\left(2,000,000,000\right)$

$$= 5.3\times10^6\left(2.0\times10^9\right)$$

$$= 10.6\times10^{15} = 1.06\times10^{16}$$

44. $3.4\times10^{-7} = 0.000000348$

45. $5.82\times10^{13} = 58,200,000,000,000$

46. $2ab - 4a^2b - 6b^2 - 3ab + 2a^2b + 5b^3$

$$= -ab - 2a^2b - 6b^2 + 5b^3$$

47. $-5ab^2\left(a^3 + 2a^2b - 3b - 4\right)$

$$= -5a^4b^2 - 10a^3b^3 + 15ab^3 + 20ab^2$$

48. $3a\left[2a - 3(a+4)\right]$

$$= 3a\left[2a - 3a - 12\right]$$

$$= 3a\left[-a - 12\right] = -3a^2 - 36a$$

49. $2x^2 - \left\{2 + x\left[3 - 2(x-1)\right]\right\}$

$$= 2x^2 - \left\{2 + x\left[3 - 2x + 2\right]\right\}$$

$$= 2x^2 - \left\{2 + x\left[5 - 2x\right]\right\}$$

$$= 2x^2 - \left\{2 + 5x - 2x^2\right\}$$

$$= 2x^2 - 2 - 5x + 2x^2$$

$$= 4x^2 - 5x - 2$$

50. $5(2)^2 - 3(2)(-1) - 2(-1)^3$

$$= 5(4) - 6(-1) - 2(-1)$$

$$= 20 + 6 + 2 = 28$$

51. $V = \pi r^2 h$

$$V = 3.14(3)^2(2)$$

$$V = 56.52 \text{ cubic meters}$$

52. $A = \dfrac{1}{2}bh = \dfrac{1}{2}(46)(58)$

$$A = 1334 \text{ square yards}$$

53. $A = \dfrac{1}{2}\left(b_1 + b_2\right)h$

$$A = \dfrac{1}{2}(26 + 34)14$$

$$A = 420 \text{ square inches}$$

Chapter 1 Test

1. $\pi, 2\sqrt{5}$

2. $-2, 12, \dfrac{9}{3}, \dfrac{25}{25}, 0, \sqrt{4}$

3. Commutative property of multiplication.

4. $(7-5)^3 - 18 \div (-3) + 3\sqrt{10+6}$

$= 2^3 - (-6) + 3\sqrt{16}$

$= 8 + 6 + 3(4)$

$= 14 + 12 = 26$

5. $\dfrac{-4 + 2\sqrt{9} - (-2)^3}{|8-13|}$

$= \dfrac{-4 + 2(3) - (-8)}{|-5|}$

$= \dfrac{-4+6+8}{5} = \dfrac{2+8}{5}$

$= \dfrac{10}{5} = 2$

6. $\dfrac{12a^{-2}b^3c^{-1}}{15a^{-4}b^{-1}c^2} = \dfrac{4a^2b^4}{5c^3}$

7. $\left(5x^{-3}y^{-5}\right)\left(-3xy\right)\left(-2x^3y^0\right)$

$= 30xy^{-4} = \dfrac{30x}{y^4}$

8. $\left(\dfrac{2x^{-3}y^{-1}}{-8x^2y^{-4}}\right)^{-2} = \dfrac{2^{-2}x^6y^2}{(-8)^{-2}x^{-4}y^8}$

$= \dfrac{(-8)^2 x^{10}}{2^2 y^6} = \dfrac{64x^{10}}{4y^6} = \dfrac{16x^{10}}{y^6}$

9. $7x - 9x^2 - 12x - 8x^2 + 5x$

$= -17x^2$

10. $5a + 4b - 6a^2 + b - 7a - 2a^2$

$= -8a^2 - 2a + 5b$

11. $3xy^2\left(4x - 3y + 2x^2\right)$

$= 12x^2y^2 - 9xy^3 + 6x^3y^2$

12. $0.000002186 = 2.186 \times 10^{-6}$

13. $2.158 \times 10^9 = 2,158,000,000$

14. $\left(3.8 \times 10^{-5}\right)\left(4 \times 10^{-2}\right) = 15.2 \times 10^{-7}$

$= 1.52 \times 10^{-6}$

15. $\dfrac{3.6 \times 10^8}{1.2 \times 10^2} = 3 \times 10^6$

16. $2\{3x - 2[x - 3(x+5)]\}$

$= 2\{3x - 2[x - 3x - 15]\}$

$= 2\{3x - 2[-2x - 15]\}$

$= 2\{3x + 4x + 30\}$

$= 2\{7x + 30\} = 14x + 60$

17. $2(-4)^2 - 3(-4) - 6$

$= 2(16) - (-12) - 6$

$= 32 + 12 - 6 = 44 - 6 = 38$

18. $5(3)^2 + 3(3)(-3) - (-3)^2$

$= 5(9) + (-27) - 9$

$= 45 - 27 - 9 = 45 - 36 = 9$

19. $A = \dfrac{1}{2}\left(b_1 + b_2\right)$

$A = \dfrac{1}{2}(6+7)(12) = 78$ sq. m

20. $A = \pi r^2 = 3.14(6)^2 = 113.04$ sq. m

21. $A = p(1+rt)$

$A = \$8000\left(1 + 0.05(3)\right) = \$9200$

Chapter 2

Pretest Chapter 2

1. $\dfrac{1}{3}(2x-1)=4(x+3)$

$2x-1=12(x+3)$

$2x-1=12x+36$

$10x=-37$

$x=-3.7$

2. $\dfrac{x-2}{4}=\dfrac{1}{2}x+4$

$x-2=2x+16$

$x=-18$

3. $4(x-3)=x+2(5x-1)$

$4x-12=x+10x-2$

$7x=-10$

$x=-\dfrac{10}{7}$

4. $0.6x+3=0.5x-7$

$0.1x=-10$

$x=-100$

5. $5x-8y=15$

$8y=5x-15$

$y=\dfrac{5x-15}{8}$

6. $5ab-2b=16ab-24-3b$

$11ab=b+24$

$a=\dfrac{b+24}{11b}$

7. (a) $A=P+\text{Pr}t$

$\text{Pr}t=A-P$

$r=\dfrac{A-P}{Pt}$

(b) $r=\dfrac{118-100}{100(3)}=\dfrac{3}{50}=0.06$

8. $|3x-2|=7$

$3x-2=7$ or $3x-2=-7$

$3x=9 \qquad 3x=-5$

$x=3 \qquad x=-\dfrac{5}{3}$

9. $|8-x|-3=1$

$|8-x|=4$

$8-x=4$ or $8-x=-4$

$x=4 \qquad x=12$

10. $\left|\dfrac{2x+3}{4}\right|=2$

$\dfrac{2x+3}{4}=2$ or $\dfrac{2x+3}{4}=-2$

$2x+3=8 \qquad 2x+3=-8$

$2x=5 \qquad 2x=-11$

$x=2.5 \qquad x=-5.5$

11. $|5x-8|=|3x+2|$

$5x-8=3x+2$ or $5x-8=-(3x+2)$

$2x=10 \qquad 5x-8=-3x-2$

$x=5 \qquad 8x=6$

$x=0.75$

12. $P = 2l + 2w$

$64 = 2(3w - 4) + 2w$

$64 = 6w - 8 + 2w$

$8w = 72$

$w = 9$

$l = 3(9) - 4 = 23$

The rectangle is 9 cm. by 23 cm.

13. $9.12 = 6 + 0.12x$

$0.12x = 3.12$

$x = 26$

Jose used 26 checks.

14. $100(0.8) = 0.77x + 0.92(100 - x)$

$80 = 0.77x + 92 - 0.92x$

$-12 = -0.15x$

$x = 80$

80 gm of 77% pure copper and 20 gm of 92% pure copper are needed.

15. $4725 = 0.10x + 0.15(40,000 - x)$

$4725 = 0.10x + 6000 - 0.15x$

$-1275 = -0.05x$

$x = 25,500$

The couple invested $25,500 at 10% and $14,500 at 15%.

16. $7x + 12 < 9x$

$-2x < -12$

$x > 6$

17. $3(x - 5) - 5x \geq 2x + 9$

$3x - 15 - 5x \geq 2x + 9$

$-2x - 15 \geq 2x + 9$

$-4x \geq 24$

$x \leq -6$

18. $\dfrac{2}{3}x - \dfrac{5}{6}x - 3 \leq \dfrac{1}{2}x - 5$

$-\dfrac{2}{3}x \leq -2$

$x \geq 3$

19. $-2 \leq x < 5$

20. $x < -3$ or $x > 0$

21. $-2 \leq x + 1 \leq 4$

$-3 \leq x \leq 3$

22. $2x + 3 < -5$ or $x - 2 > 1$

$2x < -8 \qquad x > 3$

$x < -4$

23. $|3x+2|<8$

$$-8<3x+2<8$$

$$-10<3x<6$$

$$-\frac{10}{3}<x<2$$

24. $\left|\dfrac{2}{3}x-\dfrac{1}{2}\right|\le 3$

$$-3\le \frac{2}{3}x-\frac{1}{2}\le 3$$

$$-18\le 4x-3\le 18$$

$$-15\le 4x\le 21$$

$$-\frac{15}{4}\le x\le \frac{21}{4}$$

25. $|2-5x-4>13|$

$$2-5x-4<-13 \text{ or } 2-5x-4>13$$

$$-5x-2<-13 \qquad -5x-2>13$$

$$-5x<-11 \qquad\qquad -5x>15$$

$$x>\frac{11}{5} \qquad\qquad x<-3$$

26. $|2x-7|\le 11$

$$-11\le 2x-7\le 11$$

$$-4\le 2x\le 18$$

$$-2\le x\le 9$$

2.1 Exercises

1. $3x-15=3(-20)-15=-75\ne 45$

No: when you replace x by -20 in the equation, you do not get a true statement.

3. Multiply each term of the equation by the LCD, 12, to clear the fractions.

5. No; it would be easier to subtract 3.6 from both sides of the equation since the coefficient of x is 1.

7. $-15+x=37$

$$x=37+15$$

$$x=52$$

check: $-15+52\overset{?}{=}37$

$$37=37$$

9. $-9x=45$

$$x=-5$$

check: $-9(-5)\overset{?}{=}45$

$$45=45$$

11. $9x-4=32$

$$9x=36$$

$$x=4$$

check: $9(4)-4\overset{?}{=}32$

$$36-4\overset{?}{=}32$$

$$32=32$$

13. $9x+3=5x-9$

$$4x=-12$$

$$x=-3$$

check: $9(-3)+3\overset{?}{=}5(-3)-9$

$$-27+3\overset{?}{=}-15-9$$

$$-24=-24$$

15. $16 - 2x + 5 = 5x$

$$21 = 7x$$

$$x = 3$$

check: $16 - 2(3) + 5 \overset{?}{=} 5(3)$

$$16 - 6 + 5 \overset{?}{=} 15$$

$$10 + 5 \overset{?}{=} 15$$

$$15 = 15$$

17. $3a - 5 - 2a = 2a - 3$

$$a - 5 = 2a - 3$$

$$-a = 2$$

$$a = -2$$

check: $3(-2) - 5 - 2(-2) \overset{?}{=} 2(-2) - 3$

$$-6 - 5 + 4 \overset{?}{=} -4 - 3$$

$$-11 - 4 \overset{?}{=} -7$$

$$-7 = -7$$

19. $4(y - 1) = -2(3 + y)$

$$4y - 4 = -6 - 2y$$

$$6y = -2$$

$$y = -\frac{1}{3}$$

check: $4\left(-\frac{1}{3} - 1\right) \overset{?}{=} -2\left(3 + \left(-\frac{1}{3}\right)\right)$

$$4\left(\frac{-4}{3}\right) \overset{?}{=} -2\left(\frac{8}{3}\right)$$

$$-\frac{16}{3} = -\frac{16}{3}$$

21. $6 - (y + 4) = 5 - 2(y - 1)$

$$6 - y - 4 = 5 - 2y + 2$$

$$2 - y = 7 - 2y$$

$$y = 5$$

check: $6 - (5 + 4) \overset{?}{=} 5 - 2(5 - 1)$

$$6 - 9 \overset{?}{=} 5 - 2(4)$$

$$-3 \overset{?}{=} 5 - 8$$

$$-3 = -3$$

23. $\dfrac{y}{4} + \dfrac{1}{2} = \dfrac{2}{3}$

$$3y + 6 = 8$$

$$3y = 2$$

$$y = \frac{2}{3}$$

check: $\dfrac{\frac{2}{3}}{4} + \dfrac{1}{2} \overset{?}{=} \dfrac{2}{3}$

$$\frac{2}{3} \cdot \frac{1}{4} + \frac{1}{2} \overset{?}{=} \frac{2}{3}$$

$$\frac{1}{6} + \frac{1}{2} \overset{?}{=} \frac{2}{3}$$

$$\frac{2}{3} = \frac{2}{3}$$

25. $\dfrac{y}{2}+4=\dfrac{1}{6}$

$3y+24=1$

$3y=-23$

$y=-\dfrac{23}{3} \text{ or } -7\dfrac{2}{3}$

check: $\dfrac{-\dfrac{23}{3}}{2}+4\overset{?}{=}\dfrac{1}{6}$

$-\dfrac{23}{3}\cdot\dfrac{1}{2}+\dfrac{24}{6}\overset{?}{=}\dfrac{1}{6}$

$\dfrac{1}{6}=\dfrac{1}{6}$

27. $\dfrac{2}{3}-\dfrac{x}{6}=1$

$4-x=6$

$x=4-6$

$x=-2$

check: $\dfrac{2}{3}-\dfrac{-2}{6}\overset{?}{=}1$

$\dfrac{4}{6}+\dfrac{2}{6}\overset{?}{=}1$

$\dfrac{6}{6}=1$

29. $\dfrac{1}{2}(x+3)-2=1$

$x+3-4=2$

$x-1=2$

$x=3$

check: $\dfrac{1}{2}(3+3)-2\overset{?}{=}1$

$\dfrac{1}{2}\cdot6-2\overset{?}{=}1$

$3-2\overset{?}{=}1$

$1=1$

31. $5-\dfrac{2x}{7}=1-(x-4)$

$35-2x=7-7x+28$

$35-2x=35-7x$

$5x=0$

$x=0$

check: $5-\dfrac{2\cdot0}{7}\overset{?}{=}1-(0-4)$

$5-0\overset{?}{=}1+4$

$5=5$

33. $0.3x+0.4=0.5x-0.8$

$3x+4=5x-8$

$-2x=-12$

$x=6$

check: $0.3(6)+0.4\overset{?}{=}0.5(6)-0.8$

$1.8+0.4\overset{?}{=}3-0.8$

$2.2=2.2$

35. $0.2(x-4)=3$

$2(x-4)=30$

$2x-8=30$

$2x=38$

$x=19$

check: $0.2(19-4)\overset{?}{=}3$

$0.2(15)\overset{?}{=}3$

$3=3$

37. $0.6 - 0.02x = 0.4x - 0.03$
$$60 - 2x = 40x - 3$$
$$42x = 63$$
$$x = 1.5$$
check: $0.6 - 0.02(1.5) \overset{?}{=} 0.4(1.5) - 0.03$
$$0.6 - 0.03 \overset{?}{=} 0.6 - 0.03$$
$$0.57 = 0.57$$

39. $0.08 = 0.4x + 2$
$$8 = 40x + 200$$
$$40x = -192$$
$$x = -4.8$$
check: $0.08 \overset{?}{=} 0.4(-4.8) + 2$
$$0.08 \overset{?}{=} -1.92 + 2$$
$$0.08 = 0.08$$

41. $\dfrac{2}{3}(x+6) = 1 + \dfrac{4x-7}{3}$
$$2(x+6) = 3 + 4x - 7$$
$$2x + 12 = 4x - 4$$
$$2x = 16$$
$$x = 8$$

43. $\dfrac{x-2}{7} + \dfrac{5}{2} = \dfrac{9}{2}$
$$2(x-2) + 35 = 63$$
$$2x - 4 = 28$$
$$2x = 32$$
$$x = 16$$

45. $\dfrac{1}{3} - \dfrac{x+1}{5} = \dfrac{x}{3}$
$$5 - 3x - 3 = 5x$$
$$8x = 2$$
$$x = \dfrac{1}{4}$$

47. $2 + 0.1(5 - x) = 1.3x - (0.4x - 2.5)$
$$20 + 5 - x = 13x - 4x + 25$$
$$10x = 0$$
$$x = 0$$

49. $\dfrac{1}{2}(x+2) = \dfrac{2}{3}(x-1) - \dfrac{3}{4}$
$$6x + 12 = 8x - 8 - 9$$
$$2x = 29$$
$$x = \dfrac{29}{2}$$

51. $9x - 12 = 3x + 12 + 5x$
$$9x - 12 = 8x + 12$$
$$x = 24$$

53. $30x + 2 - 6x = 15x + 3 + 9x$
$$24x + 2 = 24x + 3$$
$$2 = 3$$
No solution.

55. $7(x+1) - 4 = 10x + 3(1 - x)$
$$7x + 7 - 4 = 10x + 3 - 3x$$
$$7x + 3 = 7x + 3$$
Any real number is a solution.

57. $6 + 8(x-2) = 10x - 2(x+4)$
$$6 + 8x - 16 = 10x - 2x - 8$$
$$8x - 10 = 8x - 8$$
$$-10 = -8, \quad \text{No solution.}$$

59. $x - 2 + \dfrac{2x}{5} = -2 + \dfrac{7x}{5}$

$5x - 10 + 2x = -10 + 7x$

$7x - 10 = 7x - 10$

Any real number is a solution.

Cumulative Review Problems

61. $5 - (4 - 2)^2 + 3(-2)$

$= 5 - (2)^2 + (-6)$

$= 5 - 4 + (-6)$

$= 1 + (-6)$

$= -5$

63. $(-2)^4 - 12 - 6(-2)$

$= 16 - 12 - (-12)$

$= 4 + 12$

$= 16$

65. (a) $\dfrac{27,000,000}{4000} = 6750$

(b) $\dfrac{6200 + 8420 + 12,065}{3} = 8895$

2.2 Exercises

1. $5x - 2(x - y) = 3y$

$5x - 2x + 2y = 3y$

$3x = y$

$x = \dfrac{y}{3}$

3. $\dfrac{x}{3} + \dfrac{y}{2} = 4 - x$

$2x + 3y = 24 - 6x$

$8x = 24 - 3y$

$x = \dfrac{24 - 3y}{8}$

5. $y = \dfrac{2}{3}x - 4$

$3y = 2x - 12$

$2x = 3y + 12$

$x = \dfrac{3y + 12}{2}$

7. $A = lw$

$l = \dfrac{A}{w}$

9. $A = \dfrac{h}{2}(B + b)$

$2A = hB + hb$

$hB = 2A - hb$

$B = \dfrac{2A - hb}{h}$

11. $0.2(a - 3x) = 0.5a - 1.2x$

$2(a - 3x) = 5a - 12x$

$2a - 6x = 5a - 12x$

$6x = 3a$

$x = 0.5a$

13. $\dfrac{2A - B}{5} = \dfrac{B}{2} - 1$

$4A - 2B = 5B - 10$

$7B = 4A + 10$

$B = \dfrac{4A + 10}{7}$

15. (a) $A = \dfrac{1}{2}ab$

$2A = ab$

$a = \dfrac{2A}{b}$

(b) $a = \dfrac{2(20)}{2.5} = 16$

17. (a) $A = a + d(n-1)$

$A = a + dn - d$

$dn = A - a + d$

$n = \dfrac{A - a + d}{d}$

(b) $n = \dfrac{A - a + d}{d} = \dfrac{28 - 3 + 15}{15} = \dfrac{8}{3}$

19. $t = -0.7x + 160$

$10t = -7x + 1600$

$7x = 1600 - 10t$

$x = \dfrac{1600 - 10t}{7} = \dfrac{1600 - 10(139)}{7} = 30$

$1975 + 30 = 2005$

21. (a) $\dfrac{m}{1.15} = k$

$m = 1.15k$

(b) $m = 1.15(29) = 33.35$ knots

23. (a) $C = 0.6547D + 5.8263$

$D = \dfrac{C - 5.8623}{0.6547}$

(b) $D = \dfrac{9.56 - 5.8263}{0.6547} = \$5.7$ billion

25. $\left(2x^{-3}y\right)^{-2} = 2^{-2}x^6 y^{-2} = \dfrac{x^6}{4y^2}$

27. $5000(0.05) + 4000(0.09) = 610$

$5000 + 4000 + 610 = \$9610$ at the end of one year.

29. $562 \text{ USD} \cdot \left(\dfrac{1.93 \text{ marks}}{\text{USD}} - \dfrac{1.73 \text{ marks}}{\text{USD}}\right)$

$= 112.40$ German marks

2.3 Exercises

1. $|x| = 30$

$x = 30$ or $x = -30$

check: $|30| \overset{?}{=} 30$ $|-30| \overset{?}{=} 30$

$30 = 30$ $30 = 30$

3. $|2x - 5| = 13$

$2x - 5 = 13$ or $2x - 5 = -13$

$2x = 18$ $2x = -8$

$x = 9$ $x = -4$

check: $|2 \cdot 9 - 5| \overset{?}{=} 13$ $|2(-4) - 5| \overset{?}{=} 13$

$|18 - 5| \overset{?}{=} 13$ $|-8 - 5| \overset{?}{=} 13$

$|13| \overset{?}{=} 13$ $|-13| \overset{?}{=} 13$

$13 = 13$ $13 = 13$

5. $|5 - 4x| = 11$

$5 - 4x = 11$ or $5 - 4x = -11$

$-4x = 6$ $-4x = -16$

$x = -\dfrac{3}{2}$ $x = 4$

(continued)

5. (continued)

check: $\left|5-4\left(-\dfrac{3}{2}\right)\right|\overset{?}{=}11$ $\left|5-4(4)\right|\overset{?}{=}11$

$\left|5+6\right|\overset{?}{=}11$ $\left|5-16\right|\overset{?}{=}11$

$\left|11\right|\overset{?}{=}11$ $\left|-11\right|\overset{?}{=}11$

$11=11$ $11=11$

7. $\left|\dfrac{1}{2}x-3\right|=2$

$\dfrac{1}{2}x-3=2$ or $\dfrac{1}{2}x-3=-2$

$x-6=4$ $x-6=-4$

$x=10$ $x=2$

check: $\left|\dfrac{1}{2}\cdot10-3\right|\overset{?}{=}2$ $\left|\dfrac{1}{2}\cdot2-3\right|\overset{?}{=}2$

$\left|5-3\right|\overset{?}{=}2$ $\left|1-3\right|\overset{?}{=}2$

$\left|-2\right|\overset{?}{=}2$ $\left|1-3\right|\overset{?}{=}2$

$2=2$ $2=2$

9. $\left|2.3-0.3x\right|=1$

$2.3-0.3x=1$ or $2.3-0.3x=-1$

$23-3x=10$ $23-3x=-10$

$-3x=-13$ $-3x=-33$

$x=\dfrac{13}{3}$ $x=11$

check: $\left|2.3-0.3\left(\dfrac{13}{3}\right)\right|\overset{?}{=}1$ $\left|2.3-0.3(11)\right|\overset{?}{=}1$

$\left|2.3-1.3\right|\overset{?}{=}1$ $\left|2.3-3.3\right|\overset{?}{=}1$

$\left|1\right|\overset{?}{=}1$ $\left|-1\right|\overset{?}{=}1$

$1=1$ $1=1$

11. $\left|x+2\right|-1=7$

$\left|x+2\right|=8$

$x+2=8$ or $x+2=-8$

$x=6$ $x=-10$

check: $\left|6+2\right|-1\overset{?}{=}7$ $\left|-10+2\right|-1\overset{?}{=}7$

$\left|8\right|-1\overset{?}{=}7$ $\left|-8\right|-1\overset{?}{=}7$

$8-1\overset{?}{=}7$ $8-1\overset{?}{=}7$

$7=7$ $7=7$

13. $\left|\dfrac{x-1}{3}\right|-2=1,$ $\left|\dfrac{x-1}{3}\right|=3$

$\dfrac{x-1}{3}=3$ or $\dfrac{x-1}{3}=-3$

$x-1=9$ $x-1=-9$

$x=10$ $x=-8$

check: $\left|\dfrac{10-1}{3}\right|-2\overset{?}{=}1$ $\left|\dfrac{-8-1}{3}\right|-2\overset{?}{=}1$

$\left|\dfrac{9}{3}\right|-2\overset{?}{=}1$ $\left|\dfrac{-9}{3}\right|-2\overset{?}{=}1$

$\left|3\right|-2\overset{?}{=}1$ $\left|-3\right|-2\overset{?}{=}1$

$3-2\overset{?}{=}1$ $3-2\overset{?}{=}1$

$1=1$ $1=1$

15. $\left|1-\dfrac{3}{4}x\right|+4=7$

$\left|1-\dfrac{3}{4}x\right|=3$

(continued)

15. (continued)

$$1 - \frac{3}{4}x = 3 \text{ or } 1 - \frac{3}{4}x = -3$$

$$4 - 3x = 12 \qquad 4 - 3x = -12$$

$$-3x = 8 \qquad -3x = -16$$

$$x = -\frac{8}{3} \qquad x = \frac{16}{3}$$

check: $\left|1 - \frac{3}{4} \cdot \frac{-8}{3}\right| + 4 \overset{?}{=} 7 \qquad \left|1 - \frac{3}{4} \cdot \frac{16}{3}\right| + 4 \overset{?}{=} 7$

$$|1 + 2| + 4 \overset{?}{=} 7 \qquad |1 - 4| + 4 \overset{?}{=} 7$$

$$|3| + 4 \overset{?}{=} 7 \qquad |-3| + 4 \overset{?}{=} 7$$

$$3 + 4 \overset{?}{=} 7 \qquad 3 + 4 \overset{?}{=} 7$$

$$7 = 7 \qquad 7 = 7$$

17. $|x + 6| = |2x - 3|$

$$x + 6 = 2x - 3 \text{ or } x + 6 = -2x + 3$$

$$-x = 9 \qquad 3x = -3$$

$$x = 9 \qquad x = -1$$

check: $|9 + 6| \overset{?}{=} |2 \cdot 9 - 3| \qquad |-1 + 6| \overset{?}{=} |2(-1) - 3|$

$$|15| \overset{?}{=} |18 - 3| \qquad |-5| \overset{?}{=} |-2 - 3|$$

$$15 \overset{?}{=} |15| \qquad 5 \overset{?}{=} |-5|$$

$$15 = 15 \qquad 5 = 5$$

19. $\left|\frac{x-1}{2}\right| = |2x + 3|$

$$\frac{x-1}{2} = 2x + 3 \text{ or } \frac{x-1}{2} = -2x - 3$$

$$x - 1 = 4x + 6 \qquad x - 1 = -4x - 6$$

$$-3x = 7 \qquad 5x = -5$$

$$x = -\frac{7}{3} \qquad x = -1$$

21. $|1.5x - 2| = |x - 0.5|$

$$1.5x - 2 = x - 0.5 \text{ or } 1.5x - 2 = -x + 0.5$$

$$15x - 20 = 10x - 5 \qquad 15x - 20 = -10x + 5$$

$$5x = 15 \qquad 25x = 25$$

$$x = 3 \qquad x = 1$$

23. $|3 - x| = \left|\frac{x}{2} + 3\right|$

$$3 - x = \frac{x}{2} + 3 \text{ or } 3 - x = -\frac{x}{2} - 3$$

$$6 - 2x = x + 6 \qquad 6 - 2x = -x - 6$$

$$-3x = 0 \qquad -x = -12$$

$$x = 0 \qquad x = 12$$

25. $|-0.74x - 8.26| = 5.36$

$$-0.74x - 8.26 = 5.36 \text{ or } -0.74x - 8.26 = -5.36$$

$$-0.74x = 13.62 \qquad -0.74x = 2.9$$

$$x = -18.41 \qquad x = -3.92$$

27. $|4(x - 2)| + 1 = 19$

$$|4x - 8| = 18$$

$$4x - 8 = 18 \text{ or } 4x - 8 = -18$$

$$4x = 26 \qquad 4x = 10$$

$$x = 6.5 \qquad x = 2.5$$

check: $|4(6.5 - 2)| + 1 \overset{?}{=} 19 \quad |4(2.5 - 2)| + 1 \overset{?}{=} 19$

$$|18| + 1 \overset{?}{=} 19 \qquad |-18| + 1 \overset{?}{=} 19$$

$$18 + 1 \overset{?}{=} 19 \qquad 18 + 1 \overset{?}{=} 19$$

$$19 = 19 \qquad 19 = 19$$

29. $\left|\dfrac{3}{4}x+9\right|=0$

$\dfrac{3}{4}x+9=0$ check: $\left|\dfrac{3}{4}(-12)+9\right|\overset{?}{=}0$

$3x+36=0$ $\left|-9+9\right|\overset{?}{=}0$

$\qquad 3x=-36$ $\left|0\right|\overset{?}{=}0$

$\qquad\quad x=-12$ $0=0$

31. $\left|\dfrac{3}{4}x-\dfrac{2}{3}\right|=-8.$ No solution.

33. $\left|\dfrac{3x+1}{2}\right|=\dfrac{3}{4}$

$\dfrac{3x+1}{2}=\dfrac{3}{4}$ or $\dfrac{3x+1}{2}=-\dfrac{3}{4}$

$6x+2=3 \qquad\quad 6x+2=-3$

$\quad 6x=1 \qquad\qquad\quad x=-5$

$\quad\ x=\dfrac{1}{6} \qquad\qquad\quad x=-\dfrac{6}{5}$

check: $\left|\dfrac{3\cdot\frac{1}{6}+1}{2}\right|\overset{?}{=}\dfrac{3}{4}$ $\left|\dfrac{3\left(-\frac{5}{6}\right)+1}{2}\right|\overset{?}{=}\dfrac{3}{4}$

$\left|\dfrac{\frac{3}{2}}{2}\right|\overset{?}{=}\dfrac{3}{4}$ $\left|\dfrac{-\frac{3}{2}}{2}\right|\overset{?}{=}\dfrac{3}{4}$

$\left|\dfrac{\frac{3}{2}}{2}\right|\overset{?}{=}\dfrac{3}{4}$ $\left|-\dfrac{\frac{3}{2}}{2}\right|\overset{?}{=}\dfrac{3}{4}$

$\left|\dfrac{3}{4}\right|\overset{?}{=}\dfrac{3}{4}$ $\left|-\dfrac{3}{4}\right|\overset{?}{=}\dfrac{3}{4}$

$\dfrac{3}{4}=\dfrac{3}{4}$ $\dfrac{3}{4}=\dfrac{3}{4}$

35. $\left|6x-0.3\right|=\left|5.6x+11.9\right|$

$6x-0.3=5.6x+11.9$ or $6x-0.3=-5.6x=11.9$

$\quad 0.4x=12.2 \qquad\qquad\qquad 11.6x=-11.6$

$\qquad\ x=30.5 \qquad\qquad\qquad\quad x=-1$

check: $\left|6(30.5)-0.3\right|\overset{?}{=}\left|5.6(30.5)+11.9\right|$

$\left|182.7\right|\overset{?}{=}\left|182.7\right|$

$182.7=182.7$

$\left|6(-1)-0.3\right|\overset{?}{=}\left|5.6(-1)+11.9\right|$

$\left|-6.3\right|\overset{?}{=}\left|6.3\right|$

$6.3=6.3$

37. $\dfrac{\left|x-3\right|}{-4}=-2,\ \left|x-3\right|=8$

$x-3=8$ or $x-3=-8$

$\quad x=11 \qquad\quad x=-5$

check: $\dfrac{\left|11-3\right|}{-4}\overset{?}{=}-2$ $\dfrac{\left|-5-3\right|}{-4}\overset{?}{=}-2$

$\dfrac{\left|8\right|}{-4}\overset{?}{=}-2$ $\dfrac{\left|-8\right|}{-4}\overset{?}{=}-2$

$\dfrac{8}{-4}\overset{?}{=}-2$ $\dfrac{8}{-4}\overset{?}{=}-2$

$-2=-2$ $-2=-2$

39. $\left(\dfrac{2x^{-2}y}{z^{-1}}\right)^{3}=\dfrac{2^{3}x^{-6}y^{3}}{z^{-3}}=\dfrac{8y^{3}z^{3}}{x^{6}}$

41. Pennsylvania:
$3(3.29-1.5)=\$5.37$
South Carolina:
$2(2.69)=\$5.38$. Pennsylvania has the better buy.

Putting Your Skills to Work

1. $506 - 396 = 110$
    $706 - 506 = 200$
    $894 - 706 = 188$
    $1172 - 894 = 278$
    The greatest increase in retail sales occurred between 1995 and 2000.

2. $\dfrac{1172 - 894}{894} = 0.31 = 31\%$
    between 1995 and 2000.
    $\dfrac{x - 1172}{1172} = \dfrac{1172 - 894}{894}$
    $x - 1172 = \dfrac{1172 - 894}{894} \cdot 1172$
    $x = \dfrac{1172 - 894}{894} \cdot 1172 + 1172$
    $x = \$1536$ billion, approximately.

3. $R = 38.8x + 346.8$
    $R = 38.8(2008 - 1980) + 346.8$
    $R = \$1433.2$ billion

4. $38.8x = R - 346.8$
    $x = \dfrac{R - 346.8}{38.8} = \dfrac{1472 - 346.8}{38.8} = 29$
    $1980 + 29 = 2009$. In 2009 sales will reach $1472 billion.
    $x = \dfrac{R - 346.8}{38.8} = \dfrac{463.2 - 346.8}{38.8} = 3$
    $1980 + 3 = 1983$. In 1983 the sales were $463.2 billion.

5. Answers will vary by state.

6. Answers will vary by state.

2.4 Exercises

1. Let $x =$ the number.
    $\dfrac{3}{5}x = -54$
    $3x = -270$
    $x = -90$
    The number is -90.

3. Let $x =$ the original population.
    $2x + 1.9 = 4.9$
    $2x = 3$
    $x = 1.5$
    The original population of the city was 1.5 million.

5. Let $x =$ the weight of the package.
    $3 + 0.8x = 17.40$
    $0.8x = 14.4$
    $x = 18$
    The package weighed 18 pounds.

7. Let $x =$ the number of checks.
    $8.00(4) + 0.10x = 39.7$
    $0.10x = 39.7 - 32 = 7.7$
    $x = 77$
    She wrote 77 checks.

9. Profit = Revenue − Cost. For one year the profit must be
    $120,000 \cdot 3 = 360,000$.
    The revenue for one week is
    $(5000 \cdot 4 \cdot 18) = 360,000$
    The cost for one week is
    $55,000 \cdot 4 + 110,000 = 330,000$.

(continued)

28

9. (continued)

The profit for one week is

$360,000 - 330,000 = 30,000$

Let $x =$ the number of weeks

on tour, then

$30,000x = 360,000$

$$x = 12$$

They need to be on tour

12 weeks each year.

11. Let $x =$ miles Melissa drives each day.

$0.5x =$ miles Marcia drives each day.

$x + 17 =$ miles John drives each day.

$x + 0.5x + x + 17 = 112$

$$2.5x = 95$$

$$x = 38$$

$$0.5x = 19$$

$$x + 17 = 55$$

Each day Melissa drives 38 miles, Marcia drives 19 miles, and John drives 55 miles.

13. Let $x =$ the width of the field.

$2x + 2(3x - 6) = 340$

$$2x + 6x - 12 = 340$$

$$8x = 352$$

$$x = 44$$

$$3x - 6 = 126$$

The width of the field is 44 yards and the length of the field is 126 yards.

15. Let $x =$ the length of third side.

$2x - 3 + x + 12 + x = 156 - 3$

$$4x = 144$$

$$x = 36$$

$$x + 12 = 48$$

$$2x - 3 = 69$$

The first side is 69 centimeters, the second side is 48 centimeters, and the third side is 36 centimeters.

17. Let $x =$ yearly phone bill for Saugus.

$x + 2x - 610 = 2504$

$$3x = 2664$$

$$x = 888$$

$$2x - 610 = 1166$$

The yearly telephone bill for the Saugus shop is $888. The yearly phone bill for the Salem shop is $1166.

19. (a) Let $t =$ the climbing time from Logan.

$t + 3t + t - 4 = 126$

$$5t = 130$$

$$t = 26$$

$$t - 4 = 22$$

Climbing time from Logan was 26 minutes. Descending time to O'Hare was 22 minutes.

(b) Let $R =$ climbing rate in feet per minute.

$R \cdot 26 = 33,000$

$$R = 1269.230769$$

$$R \approx 1270$$

The climbing rate was approximately 1270 feet per minute.

Cumulative Review Problems

21. Identity property of addition.

2.5 Exercises

1. Let $x =$ population in 1969.
$$x + 0.34x = 273.1$$
$$1.34x = 273.1$$
$$x = 203.8059701$$
$$x \approx 203.8$$
The population in 1969 was approximately 203.8 million people.

3. Let $x =$ the original price.
$$0.80x = 340$$
$$x = 425$$
The original price was $425.

5. Let $x =$ the number of fish in pond.
$$0.12x = 72$$
$$x = 600$$
There are approximately 600 fish in the pond.

7. Let $x =$ the number of hemlocks.
$$x + 2x + 3x + 20 = 1400$$
$$6x = 1380$$
$$x = 230$$
$$2x = 460$$
$$3x + 20 = 710$$
230 hemlocks, 460 spruces, and 710 balsams were planted.

9. Let $x =$ length of shorter piece.
$$4x + 6(16 - x) = 82$$
$$4x + 96 - 6x = 82$$
$$-2x = -14$$
$$x = 7$$
$$16 - x = 9$$
The shorter piece was 7 linear feet and the longer piece was 9 linear feet.

11. $I = \Pr t = 600(0.12)(3) = \$216$

13. $I = \Pr t = 3000(0.055)(0.5) = \$82.50$

15. Let $x =$ amount invested at 13%.
$$0.13x + 0.16(45,000 - x) = 6570$$
$$0.13x + 7200 - 0.16x = 66570$$
$$-0.3x = -630$$
$$x = 21,000$$
$$45,000 - x = 24,000$$
She invested $21,000 at 13% and $24,000 at 16%.

17. Let $x =$ amount invested at 9%.
$$0.09x + 0.07(25,000 - x) = 1930$$
$$0.09x + 1750 - 0.07x = 1930$$
$$0.02x = 180$$
$$x = 9000$$
$$25,000 - x = 16,000$$
They invested $9000 at 9% and $16,000 at 7%.

19. Let $x =$ number of grams with 45% fat.

$$0.45x + 0.20(30 - x) = 0.30(30)$$
$$0.45x + 6 - 0.2x = 9$$
$$0.25x = 3$$
$$x = 12$$
$$30 - x = 18$$

She should use 12 grams of the 45% fat cheese and 18 grams of the 20% fat cheese.

21. Let $x =$ number of pounds with 30% fat.

$$0.30x + 0.10(100 - x) = 0.25(100)$$
$$0.3x + 10 - 0.1x = 25$$
$$0.2x = 15$$
$$x = 75$$
$$100 - 75 = 25$$

She should use 75 pounds of the 30% fat hamburger and 25 pounds of the 10% fat hamburger.

23. Let $x =$ number of pounds of imported candy.

$$1.80(200) + 3x = 2.40(x + 200)$$
$$360 + 3x = 2.4x + 480$$
$$0.6x = 120$$
$$x = 200$$

200 pounds of imported candy should be used.

25. Let $x =$ speed on secondary roads.

$$4x + 2(x + 20) = 250$$
$$4x + 2x + 40 = 250$$
$$6x = 210$$
$$x = 35$$

Speed on the secondary roads was 35 miles per hour.

27. Let $x =$ time they walked on treadmill.

$$5x = 4.2x + 0.6$$
$$0.8x = 0.6$$
$$x = \frac{3}{4}$$

They each walked $\frac{3}{4}$ of an hour.

29. Let $x =$ profit two years ago.

$1.65x =$ profit last year.

$(0.6)(1.65x) =$ profit last year.

$$0.6(1.65x) = 17,820,000$$
$$x = 18,000,000$$

Profit was $18,000,000 two years ago.

31.
$$A_{\text{triangle}} = \frac{1}{2}bh = \frac{1}{2}(3)h = 6$$
$$3h = 12$$
$$h = 4$$
$$A_{\text{parallelogram}} = bh = 8 \cdot 4 = 32 \text{ cm}^2$$

Cumulative Review Problems

33.
$$2(-2)^2 - 3(-2) + 1$$
$$= 2 \cdot 4 + 6 + 1$$
$$= 8 + 6 + 1$$
$$= 14 + 1$$
$$= 15$$

35.
$$\frac{2 - 6(-1) + 5^2}{|4 - 7|} = \frac{2 + 6 + 25}{|-3|}$$
$$= \frac{8 + 25}{3}$$
$$= \frac{33}{3}$$
$$= 11$$

2.6 Exercises

1. True, $6 < 8$ and $8 > 6$ convey the same information.

3. True, dividing both sides of an inequality by -4 reverses the direction of the inequality.

5. False, the graph of $x \leq 6$ does include the point at 6 on the number line.

7. $6 > -3$

9. $-7 < -2$

11. $\dfrac{3}{4} > \dfrac{2}{3}$

13. $-3.4 > -3.41$

15. $x \geq -2$

-2

17. $x < 15$

15

19. $2x - 7 \leq -5$
$2x \leq 2$
$x \leq 1$

1

21. $5x + 6 > -7x - 6$
$12x > -12$
$x > -1$

-1

23. $0.5x + 0.1 < 1.1x + 0.7$
$5x + 1 < 11x + 7$
$-6x < x$
$x > -1$

-1

25. $\dfrac{2}{3}x - 1 \leq x - \dfrac{1}{2}$

$-\dfrac{1}{3}x \leq \dfrac{1}{2}$

$x \geq \dfrac{3}{2}$

27. $-2(x + 3) < -9$
$-2x - 6 < -9$
$-2x < -3$
$x > \dfrac{3}{2}$

29. $\dfrac{2}{3}x - (x - 2) \geq 4$
$2x - 3x + 6 \geq 12$
$-x \geq 6$
$x \leq -6$

31. $0.5x - 0.6 \leq 1.4$
$5x - 6 \leq 14$
$5x \leq 20$
$x \leq 4$

33. $0.3(x + 3) \geq 0.5x + 1.4$
$3(x + 3) \geq 5x + 14$
$3x + 9 \geq 5x + 14$
$-2x \geq 5$
$x \leq -2.5$

35. $1 + \dfrac{1}{8}x \le \dfrac{1}{4}x - \dfrac{3}{4} + \dfrac{1}{8}x$

$\qquad 8 + x \le 2x - 6 + x$

$\qquad\quad -2x \le -14$

$\qquad\qquad x \ge 7$

37. $\dfrac{2x - 3}{5} + 1 \ge \dfrac{1}{2}x + 3$

$2(2x - 3) + 10 \ge 5x + 30$

$\quad 4x - 6 + 10 \ge 5x + 30$

$\qquad\qquad -x \ge 26$

$\qquad\qquad\; x \ge -26$

39. Let $x =$ number of tables.

$\quad 4 \cdot 3 + 4x > 52$

$\quad\; 12 + 4x > 52$

$\qquad\quad 4x > 40$

$\qquad\quad\; x > 10$

She would have to serve more than ten tables.

41. Let $x =$ the number of minutes he talks after the first minute.

$\quad 3.95 + 0.55x \le 13.30$

$\qquad\; 0.55x \le 9.35$

$\qquad\qquad x \le 17$

He can talk for a maximum of $17 + 1 = 18$ minutes.

43. Let $x =$ number of computers

$130 + 155 + 59x \le 1100$

$\qquad\qquad 59x \le 815$

$\qquad\qquad\; x \le 13.81355932$

A maximum of 13 computers can be taken up.

45. Let $x =$ the number of time-shares sold each month.

$12,000 - (500,000 + 6000x) \ge 100,000$

$12,000x - 500,000 - 6000x \ge 100,000$

$\qquad\qquad\quad 6000x \ge 600,000$

$\qquad\qquad\qquad\quad x \ge 100$

At least 100 time-shares must be sold each month.

47. Let $x =$ the number of weeks for the washing machine to become the less expensive choice.

$5.75x \ge 295$

$\qquad x \ge \dfrac{295}{52} = 51.30434783$

It will take at least 52 weeks.

Cumulative Review Problems

49. $3xy(x + 2) - 4x^2(y - 1)$

$= 3x^2y + 6xy - 4x^2y + 4x^2$

$= 6xy - x^2y + 4x^2$

51. $\left(\dfrac{3x}{2y^2w^{-4}}\right)^3 = \dfrac{3^3 x^3}{2^3 y^6 w^{-12}} = \dfrac{27x^3 w^{12}}{8y^6}$

2.7 Exercises

1. $3 < x \text{ and } x < 8$

3. $-4 < x \text{ and } x < 2$

5. $7 < x < 9$

7. $-1 < x \le 3$

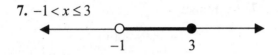

9. $x > 8 \ or \ x < 2$

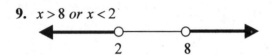

11. $x \le -\dfrac{5}{2} \ or \ x > 4$

13. $x \le -10 \ or \ x \ge 40$

15. $x \le -20 \ or \ x \ge 10$

17. $2x - 3 > 0 \ or \ x - 2 < -7$
$\qquad 2x > 3 \qquad\qquad x < -5$

$\qquad\qquad x > \dfrac{3}{2}$

19. $x < 8 \ and \ x > 10$
No solution.

21. $t < 10.9 \ or \ t > 11.2$

23. $5000 \le c \le 12,000$

25. $-20 \le C \le 11$

$\qquad -20 \le \dfrac{5}{9}(F - 32) \le 11$

$\qquad -180 \le 5F - 160 \le 99$

$\qquad -20 \le 5F \le 259$

$\qquad -4° \le F \le 51.8°$

27. $18,000 \le Y \le 33,000$

$\qquad 18,000 \le 129(d - 4) \le 33,000$

$\qquad \dfrac{18,000}{129} \le d - 4 \le \dfrac{33,000}{129}$

$\qquad \dfrac{18,000}{129} + 4 \le d \le \dfrac{33,000}{129} + 4$

$\qquad 143.53 \le d \le \$259.81$

29. $x - 3 > -5 \ and \ 2x + 4 < 8$
$\qquad x > -2 \qquad\qquad 2x < 4$
$\qquad -2 < x \qquad\qquad x < 2$
$\qquad -2 < x < 2$ is the solution.

31. $5x - 7 \ge 3 \ and \ 4x - 8 \le 0$
$\qquad 5x \ge 10 \qquad\qquad 4x \le 8$
$\qquad x \ge 2 \qquad\qquad x \le 2$
$\qquad 2 \le x \le 2$
$\qquad x = 2$ is the solution.

33. $2x - 5 < -11 \ or \ 5x + 1 \ge 6$
$\qquad 2x < -6 \qquad\qquad 5x \ge 5$
$\qquad x < -3 \qquad\qquad x \ge 1$
$\qquad x < -3 \ or \ x \ge 1$ is the solution.

35. $-0.3x + 1 \ge 0.2x \ or \ -0.2x + 0.5 > 0.7$
$\qquad -3x + 10 \ge 2x \qquad\qquad -2x + 5 > 7$
$\qquad -5x \ge -10 \qquad\qquad -2x > -2$
$\qquad x \le 2 \qquad\qquad x < 1$
$\qquad x \le 2$ is the solution.

37. $3x-4>-1$ *or* $2x+1<15$

$\quad\quad 3x>3 \quad\quad\quad 2x<14$

$\quad\quad\quad x>1 \quad\quad\quad\quad x<7$

All real numbers is the solution.

39. $\dfrac{5x}{2}+1\geq 3$ *and* $x-\dfrac{2}{3}\geq\dfrac{4}{3}$

$\quad\quad 5x+2\geq 6 \quad\quad 3x-2\geq 4$

$\quad\quad\quad 5x\geq 4 \quad\quad\quad 3x\geq 6$

$\quad\quad\quad x\geq\dfrac{4}{5} \quad\quad\quad\quad x\geq 2$

$x\geq 2$ is the solution.

41. $2x+5<3$ *and* $3x-1>-1$

$\quad\quad 2x<-2 \quad\quad\quad 3x>0$

$\quad\quad\quad x<-1 \quad\quad\quad\quad x>0$

No solution.

43. $2x-1>-9$ *and* $3x+1<13+x$

$\quad\quad 2x>-8 \quad\quad\quad\quad 2x<12$

$\quad\quad\quad x>-4 \quad\quad\quad\quad\quad x<6$

$-4<x<6$ is the solution.

45. $\dfrac{2+3x}{4}\leq -2$ *or* $\dfrac{2x-7}{3}<1$

$\quad\quad 2+3x\leq -8 \quad\quad 2x-7<3$

$\quad\quad\quad 3x\leq -10 \quad\quad\quad 2x<10$

$\quad\quad\quad x\leq -\dfrac{10}{3} \quad\quad\quad\quad x<5$

$x<5$ is the solution.

47.

$\dfrac{1}{4}(x+2)+\dfrac{1}{8}(x-3)\leq 1$ *and* $\dfrac{3}{4}(x-1)>-\dfrac{1}{4}$

$\quad\quad 2x+4+x-3\leq 8 \quad\quad 3x-3>-1$

$\quad\quad\quad\quad 3x+1\leq 8 \quad\quad\quad\quad 3x>2$

$\quad\quad\quad\quad\quad 3x\leq 7 \quad\quad\quad\quad\quad x>\dfrac{2}{3}$

(continued)

47. (continued)

$\dfrac{2}{3}<x\leq\dfrac{7}{3}$ is the solution.

49. $3y-5x=8$

$\quad\quad -5x=8-3y$

$\quad\quad\quad 5x=3y-8$

$\quad\quad\quad x=\dfrac{3y-8}{5}$

51. $3d-4a=5x+2a$

$\quad\quad -6a=5x-3d$

$\quad\quad\quad 6a=3d-5x$

$\quad\quad\quad a=\dfrac{3d-5x}{6}$

53. $40 to $50 is an increase of $10. $50 to $120 is an increase of $70. $120 to $190 is an increase of $70. $190 to $230 is an increase of $40. The next four bids will be:

1) $230 + $10 = $240

2) $240 + $70 = $310

3) $310 + $70 = $380

4) $380 + $40 = $420

2.8 Exercises

1. $|x|\leq 8$

$-8\leq x\leq 8$

$-8 \quad\quad\quad 8$

3. $|x|>5$

$x<-5$ *or* $x>5$

$-5 \quad\quad\quad 5$

5. $|x+4.5|<5$

$$-5<x+4.5<5$$
$$-9.5<x<0.5$$

-9.5 0.5

7. $|x-3|\le 5$

$$-5\le x-3\le 5$$
$$-2\le x\le 8$$

9. $|2x-5|\le 7$

$$-7\le 2x-5\le 7$$
$$-2\le 2x\le 12$$
$$-1\le x\le 6$$

11. $|0.2x-0.7|\le 0.3$

$$-0.3\le 0.2x-0.7\le 0.3$$
$$-3\le 2x-7\le 3$$
$$4\le 2x\le 10$$
$$2\le x\le 5$$

13. $\left|x-\dfrac{3}{2}\right|<\dfrac{1}{2}$

$$-\dfrac{1}{2}<x-\dfrac{3}{2}<\dfrac{1}{2}$$
$$-1<2x-3<1$$
$$2<2x<4$$
$$1<x<2$$

15. $\left|\dfrac{1}{4}x+2\right|<6$

$$-6<\dfrac{1}{4}x+2<6$$
$$-24<x+8<24$$
$$-32<x<16$$

17. $\left|\dfrac{3}{4}(x-1)\right|<6$

$$-6<\dfrac{3}{4}(x-1)<6$$
$$-24<3x-3<24$$
$$-21<3x<27$$
$$-7<x<9$$

19. $\left|\dfrac{3x-2}{4}\right|<3$

$$-3<\dfrac{3x-2}{4}<3$$
$$-12<3x-2<12$$
$$-10<3x<14$$
$$-\dfrac{10}{3}<x<\dfrac{14}{3}$$

21. $|x+2|>5$

$$x+2<-5 \ or \ x+2>5$$
$$x<-7 \qquad x>3$$

23. $|x-1|\ge 2$

$$x-1\le -2 \ or \ x-1\ge 2$$
$$x\le -1 \qquad x\ge 3$$

25. $|3x-8|\ge 7$

$$3x-8\le -7 \ or \ 3x-8\ge 7$$
$$3x\le 1 \qquad\qquad 3x\ge 15$$
$$x\le \dfrac{1}{3} \qquad\qquad x\ge 5$$

27. $\left|3-\dfrac{3}{4}x\right|>9$

$3-\dfrac{3}{4}x<-9 \ or \ 3-\dfrac{3}{4}x>9$

$12-3x<-36 \quad 12-3x>36$

$-3x<-48 \qquad -3x>24$

$x>16 \qquad\qquad x<-8$

29. $\left|\dfrac{1}{5}x-\dfrac{1}{10}\right|>2$

$\dfrac{1}{5}x-\dfrac{1}{10}<-2 \ or \ \dfrac{1}{5}x-\dfrac{1}{10}>2$

$2x-1<-20 \qquad 2x-1>20$

$2x<-19 \qquad\quad 2x>21$

$x<-\dfrac{19}{2} \qquad\quad x>\dfrac{21}{2}$

31. $\left|\dfrac{1}{3}(x-2)\right|<5$

$-5<\dfrac{1}{3}(x-2)<5$

$-15<x-2<5$

$-13<x<7$

33. $|m-s|\le 0.12$

$|m-18.65|\le 0.12$

$-0.12\le m-18.65\le 0.12$

$18.53\le m\le 18.77$

35. $|n-p|\le 0.05$

$|n-9.68|\le 0.05$

$-0.05\le n-9.68\le 0.05$

$9.63\le n\le 9.73$

37. $12<4x-8<-12$ implies $12<-12$
which is a false statement.

Cumulative Review Problems

39. $(6-4)^3\div(-4)+2^2$

$=2^3\div(-4)+2^2$

$=8\div(-4)+4$

$=-2+4$

$=2$

41.

$\text{distance} = \text{rate}\cdot\text{time}$

$2\left[\dfrac{1}{8}\text{circumference}\right] = \text{rate}\cdot\text{time}$

$2\left[\dfrac{1}{8}(2\pi\cdot\text{radius})\right] = \text{rate}\cdot\text{time}$

$2\left[\dfrac{1}{3}(2\cdot 3.14\cdot 19)\right]=3\cdot t$

$t=9.94$ seconds

43. First rack:
cost per CD

$=\dfrac{39.95+6.50}{160}=\$0.29=29$ cents

Second rack:
cost per CD

$=\dfrac{24.95+5.95}{120}=\$0.2575=26$ cents

Third rack:
cost per CD

$=\dfrac{18.95+4.75}{75}=\$0.316=32$ cents

The second rack is the least expensive
per CD space.

Chapter 2 Review Problems

1. $7x - 3 = -5x - 18$
$12x = -15$
$x = -1.25$

2. $8 - 2(x + 3) = 24 - (x - 6)$
$8 - 2x - 6 = 24 - x + 6$
$2 - 2x = 30 - x$
$-x = 28$
$x = -28$

3. $4(x - 1) + 2 = 3x + 8 - 2x$
$4x - 4 + 2 = x + 8$
$4x - 2 = x + 8$
$3x = 10$
$x = \dfrac{10}{3}$

4. $x - \dfrac{7}{5} = \dfrac{1}{3}x + \dfrac{7}{15}$
$15x - 21 = 5x + 7$
$10x = 28$
$x = \dfrac{14}{5}$

5. $\dfrac{1}{9}x - 1 = \dfrac{1}{2}\left(x + \dfrac{1}{3}\right)$
$2x - 18 = 9x + 3$
$-7x = 21$
$x = -3$

6. $\dfrac{x - 4}{2} - \dfrac{1}{5} = \dfrac{7x + 1}{20}$
$10x - 40 - 4 = 7x + 1$
$10x - 44 = 7x + 1$
$3x = 45$
$x = 15$

7. $4.6x = 2(1.6x - 2.8)$
$4.6x = 3.2x - 5.6$
$1.4x = -5.6$
$x = -4$

8. $0.6 - 0.2(x - 3) + 0.5 = 1.5$
$6 - 2(x - 3) + 5 = 15$
$6 - 2x + 6 + 5 = 15$
$17 - 2x = 15$
$-2x = -2$
$x = 1$

9. $2(3ax - 2y) - 6ax = -3(ax + 2y)$
$6ax - 4y - 6ax = -3ax - 6y$
$3ax = -2y$
$a = \dfrac{-2y}{3x}$

10. $4x - 8y = 5$
$-8y = 5 - 4x$
$8y = 4x - 5$
$y = \dfrac{4x - 5}{8}$

11. (a) $C = \dfrac{5F - 160}{9}$

$9C = 5F - 160$

$5F = 9C + 160$

$F = \dfrac{9C + 160}{5}$

(b) $F = \dfrac{9(10) + 160}{5}$

$F = 50°$ when $C = 10°$.

12. (a) $\quad P = 2W + 2L$

$2W = P - 2L$

$W = \dfrac{P - 2L}{2}$

(b) $W = \dfrac{100 - 2(20.5)}{2} = 29.5$ meters

13. $|x + 1| = 8$

$x + 1 = 8$ or $x + 1 = -8$

$\quad x = 7 \qquad\qquad x = -9$

14. $|x - 3| = 12$

$x - 3 = 12$ or $x - 3 = -12$

$\quad x = 15 \qquad\qquad x = -9$

15. $|4x - 5| = 7$

$4x - 5 = 7$ or $4x - 5 = -7$

$\quad 4x = 12 \qquad\qquad 4x = -2$

$\quad x = 3 \qquad\qquad x = -\dfrac{1}{2}$

16. $|3x + 2| = 20$

$3x + 2 = 20$ or $3x + 2 = -20$

$\quad 3x = 18 \qquad\qquad 3x = -22$

$\quad x = 6 \qquad\qquad x = -\dfrac{22}{3}$

17. $|3 - x| = |5 - 2x|$

$3 - x = 5 - 2x$ or $3 - x = -5 + 2x$

$\quad x = 2 \qquad\qquad -3x = -8$

$\qquad\qquad\qquad\qquad x = \dfrac{8}{3}$

18. $|x + 8| = |2x - 4|$

$x + 8 = 2x - 4$ or $x + 8 = -2x + 4$

$\quad -x = -12 \qquad\qquad 3x = -4$

$\quad x = 12 \qquad\qquad x = -\dfrac{4}{3}$

19. $\left|\dfrac{1}{4}x - 3\right| = 8$

$\dfrac{1}{4}x - 3 = 8$ or $\dfrac{1}{4}x - 3 = -8$

$x - 12 = 32 \qquad x - 12 = -32$

$\quad x = 44 \qquad\qquad x = -20$

20. $|4 - 7x| = 25$

$4 - 7x = 25$ or $4 - 7x = -25$

$\quad -7x = 21 \qquad\qquad -7x = -29$

$\quad x = -3 \qquad\qquad x = \dfrac{29}{7}$

21. $|2x - 8| + 7 = 12$

$|2x - 8| = 5$

$2x - 8 = 5$ or $2x - 8 = -5$

$\quad 2x = 13 \qquad\qquad 2x = 3$

$\quad x = \dfrac{13}{x} \qquad\qquad x = \dfrac{3}{2}$

22. Let $x =$ the number.

$$\frac{5}{8}x = 290$$

$$5x = 2320$$

$$x = 464$$

The number is 464.

23. Let $x =$ the number of women.

$$2x - 200 + x = 280$$

$$3x - 200 = 280$$

$$3x = 480$$

$$x = 160$$

$$2x - 200 = 120$$

There are 160 women and 120 men attending Western Tech.

24. Let $x =$ how many miles she drove.

$$30(2) + 0.12x = 102$$

$$60 + 0.12x = 102$$

$$0.12x = 42$$

$$x = 350$$

She drove 350 miles.

25. Let $x =$ number of miles from airport to hotel.

$$2.50 + 0.35\left(\frac{x - \frac{1}{5}}{\frac{1}{5}}\right) = 14.75$$

$$1.75\left(x - \frac{1}{5}\right) = 12.25$$

$$x - \frac{1}{5} = 7$$

$$x = 7.2$$

It is 7.2 miles from the airport to the hotel.

26. Let $x =$ the amount withheld for retirement.

$$x + x + 13 + 3(x + 13) = 102$$

$$2x + 13 + 3x + 39 = 102$$

$$5x + 52 = 102$$

$$5x = 50$$

$$x = 10$$

$$x + 13 = 23$$

$$3(x + 13) = 69$$

$10 is withheld for retirement, $23 for state tax, and $69 for federal tax.

27. Let $x =$ the width.

$$P = 2L + 2W$$

$$88 = 2(3x + 8) + 2x$$

$$88 = 6x + 16 + 2x$$

$$72 = 8x$$

$$x = 9$$

$$3x + 8 = 35$$

The rectangle has a width of 9 mm and a length of 35 mm.

28. Let $x =$ the number of students enrolled last year.

$$0.88x = 2332$$

$$x = 2650$$

2650 students were enrolled last year.

29. Let $x =$ the number of two-door sedans

$$3x + x = 260,000$$

$$4x = 260,000$$

$$x = 65,000$$

$$3x = 195,000$$

They should manufacture 65,000 two-door sedans and 195,000 four-door sedans.

30. Let x = amount invested at 12%.
$$0.08x + 0.12(7000 - x) = 740$$
$$0.08x + 840 - 0.12x = 740$$
$$-0.04x = -100$$
$$x = 2500$$
$$7000 - x = 4500$$
She should invest $2500 at 8% and $4500 at 12%.

31. Let x = the number of liters of 2% acid.
$$0.02x + 0.05(24 - x) = 0.04(24)$$
$$0.02x + 1.2 - 0.05x = 0.96$$
$$-0.03x = -0.24$$
$$x = 8$$
$$24 - x = 16$$
He should use 8 liters of the 2% acid and 16 liters of the 5% acid.

32. Let x = the number of pounds of the $4.25 a pound coffee.
$$4.25x + 4.50(30 - x) = 4.40(30)$$
$$4.25x + 135 - 4.5x = 132$$
$$-0.25x = -3$$
$$x = 12$$
$$30 - x = 18$$
12 pounds at $4.25 and 18 pounds at $4.50 should be used.

33. Let x = current full-time students.
$$\frac{1}{2}x + \frac{1}{3}(890 - x) = 380$$
$$3x + 1780 - 2x = 2280$$
$$x = 500$$
$$890 - 500 = 390$$
The present number of students is 500 full-time and 390 part-time.

34. $7x + 8 < 5x$
$$2x < -8$$
$$x < -4$$

35. $9x + 3 < 12x$
$$-3x < -3$$
$$x > 1$$

36. $4x - 1 < 3(x + 2)$
$$4x - 1 < 3x + 6$$
$$x < 7$$

37. $3(3x - 2) < 4x - 16$
$$9x - 6 < 4x - 16$$
$$5x < -10$$
$$x < -2$$

38. $(x + 6) - (2x + 7) \le 3x - 9$
$$x + 6 - 2x - 7 \le 3x - 9$$
$$-4x \le -8$$
$$x \ge 2$$

39. $(4x - 3) - (2x + 7) \le 5 - x$
$$4x - 3 - 2x - 7 \le 5 - x$$
$$3x - 10 \le 5$$
$$3x \le 15$$
$$x \le 5$$

40. $\frac{1}{9}x + \frac{2}{9} > \frac{1}{3}$
$$x + 2 > 3$$
$$x > 1$$

41. $\dfrac{7}{4} - 2x \geq -\dfrac{3}{2}x - \dfrac{5}{4}$

$\quad 7 - 8x \geq -6x - 5$

$\quad\quad -2x \geq -12$

$\quad\quad\quad x \leq 6$

42. $\dfrac{1}{3}(x-2) < \dfrac{1}{4}(x+5) - \dfrac{5}{3}$

$\quad 4x - 8 < 3x + 15 - 20$

$\quad\quad x < 3$

43. $\dfrac{1}{3}(x+2) > 3x - 5(x-2)$

$\quad\quad x + 2 > 9x - 15x + 30$

$\quad\quad\quad 7x > 28$

$\quad\quad\quad\quad x > 4$

44. $7x - 6 \leq \dfrac{1}{3}(-2x + 5)$

$\quad 21x - 18 \leq -2x + 5$

$\quad\quad 23x \leq 23$

$\quad\quad\quad x \leq 1$

45. $-3 \leq x < 2$

46. $-4 < x \leq 5$

47. $-8 \leq x \leq -4$

48. $-9 \leq x \leq -6$

49. $x < -2 \ \text{ or } x \geq 5$

50. $x < -3 \text{ or } x \geq 6$

51. $x > -5 \text{ and } x < -1$

52. $x > -8 \text{ and } x < -3$

53. $x + 3 > 8 \text{ or } x + 2 < 6$

$\quad\quad x > 5 \quad\quad\quad x < 4$

54. $x - 2 > 7 \text{ or } x + 3 < 2$

$\quad\quad x > 9 \quad\quad\quad x < -1$

55. $x + 3 > 8 \text{ and } x - 4 < -2$

$\quad\quad x > 5 \quad\quad\quad x < 2$

No solution.

56. $-1 < x + 5 < 8$

$\quad -6 < x < 3$

57. $\ 0 \leq 5 - 3x \leq 17$

$\quad -5 \leq -3x \leq 12$

$\quad \dfrac{5}{3} \geq x \geq -4$

$\quad -4 \leq x \leq \dfrac{5}{3}$

58. $2x - 7 < 3 \ and \ 5x - 1 \geq 8$

$\qquad 2x < 10 \qquad\quad 5x \geq 9$

$\qquad\quad x < 5 \qquad\qquad x \geq \dfrac{9}{5}$

$\qquad \dfrac{9}{5} \leq x \leq 5$

59. $4x - 2 < 8 \ or \ 3x + 1 > 4$

$\qquad 4x < 10 \qquad\quad 3x > 3$

$\qquad\quad x < \dfrac{5}{2} \qquad\quad x > 1$

The solution is all real numbers.

60. $|x + 7| < 15$

$\qquad -15 < x + 7 < 15$

$\qquad -22 < x < 8$

61. $|x + 9| < 18$

$\qquad -18 < x + 9 < 18$

$\qquad -27 < x < 9$

62. $\left| \dfrac{1}{2}x + 2 \right| < \dfrac{7}{4}$

$\qquad -\dfrac{7}{4} < \dfrac{1}{2}x + 2 < \dfrac{7}{4}$

$\qquad -7 < 2x + 8 < 7$

$\qquad -15 < 2x < -1$

$\qquad -\dfrac{15}{2} < x < -\dfrac{1}{2}$

63. $\left| \dfrac{1}{5}x + 3 \right| < \dfrac{11}{5}$

$\qquad -\dfrac{11}{5} < \dfrac{1}{5}x + 3 < \dfrac{11}{5}$

$\qquad -11 < x + 15 < 11$

$\qquad -26 < x < -4$

64. $|2x - 1| \geq 9$

$\qquad 2x - 1 \leq -9 \ or \ 2x - 1 \geq 9$

$\qquad 2x \leq -8 \qquad\quad 2x \geq 10$

$\qquad\quad x \leq -4 \qquad\qquad x \geq 5$

65. $|3x - 1| \geq 2$

$\qquad 3x - 1 \leq -2 \ or \ 3x - 1 \geq 2$

$\qquad 3x \leq -1 \qquad\quad 3x \geq 3$

$\qquad\quad x \leq -\dfrac{1}{3} \qquad\quad x \geq 1$

66. $|3(x - 1)| \geq 5$

$\qquad 3(x - 1) \leq -5 \ or \ 3(x - 1) \geq 5$

$\qquad 3x - 3 \leq -5 \qquad\quad 3x - 3 \geq 5$

$\qquad 3x \leq -2 \qquad\qquad 3x \geq 8$

$\qquad\quad x \leq -\dfrac{2}{3} \qquad\qquad x \geq \dfrac{8}{3}$

67. $|2(x - 3)| \geq 4$

$\qquad 2(x - 3) \leq -4 \ or \ 2(x - 3) \geq 4$

$\qquad 2x - 6 \leq -4 \qquad\quad 2x - 6 \geq 4$

$\qquad 2x \leq 2 \qquad\qquad 2x \geq 10$

$\qquad\quad x \leq 1 \qquad\qquad\quad x \geq 5$

68. Let $x =$ the number of minutes he talks.

$\qquad 3.95 + 0.64(x - 1) \leq 13.05$

$\qquad 3.95 + 0.64x - 0.64 \leq 13.05$

$\qquad\qquad\qquad 0.64x \leq 9.74$

$\qquad\qquad\qquad\qquad x \leq 15.21875$

He can talk for a maximum of 15 minutes.

69. Let $x =$ the number of packages.
$$170 + 200 + 77.5x \leq 1765$$
$$77.5x \leq 1395$$
$$x \leq 18$$
A maximum of eighteen packages can be carried.

70. Let $x =$ the number of desks.
$$135 + 165 + 79x \leq 1090$$
$$79x \leq 790$$
$$x \leq 10$$
They may take a maximum of ten desks on the elevator.

71. Let $x =$ the weight of the package.
$$0.33 + 0.22(x - 1) \leq 3.50$$
$$0.22x - 0.22 \leq 3.17$$
$$0.22x \leq 3.39$$
$$x \leq 15.\overline{409}$$
The package could weigh a maximum of 15 ounces.

72. $|d - s| \leq 0.03$
$$-0.03 \leq d - s \leq 0.03$$
$$-0.03 \leq d - 856.45 \leq 0.03$$
$$856.42 \leq d \leq 856.48$$

73. $1.04(2,312,000) \leq x \leq 1.06(2,854,000)$
$$2,404,480 \leq x \leq 3,025,240$$

Chapter 2 Test

1. $5x - 8 = -6x - 10$
$$11x = -2$$
$$x = -\frac{2}{11}$$

2. $3(7 - 2x) = 14 - 8(x - 1)$
$$21 - 6x = 14 - 8x + 8$$
$$2x = 1$$
$$x = \frac{1}{2}$$

3. $\frac{1}{3}(-x + 1) + 4 = 4(3x - 2)$
$$-x + 1 + 12 = 36x - 24$$
$$-37x = -37$$
$$x = 1$$

4. $0.5x + 1.2 = 4x - 3.05$
$$-3.5x = -4.25$$
$$x = \frac{-4.24}{-3.5} = \frac{17}{14}$$
$$x = 1\frac{3}{14}$$

5. $$L = a + d(n - 1)$$
$$L = a + dn - d$$
$$L - a + d = dn$$
$$n = \frac{L - a + d}{d}$$

6. $F = \frac{9}{5}C + 32$
$$5F = 9C + 160$$
$$9C = 5F - 160$$
$$C = \frac{5F - 160}{9}$$

7. $C = \frac{5F - 160}{9}$
$$C = \frac{5(-40) - 160}{9}$$
$$C = -40°$$

8. $H = \dfrac{1}{2}r + 3b - \dfrac{1}{4}$

$4H = 2r + 12b - 1$

$2r = 4H - 12b + 1$

$r = \dfrac{4H - 12b + 1}{2}$

9. $|5x - 2| = 37$

$5x - 2 = 37 \text{ or } 5x - 2 = -37$

$5x = 39 \qquad\qquad 5x = -35$

$x = \dfrac{39}{5} \qquad\qquad x = -7$

10. $\left|\dfrac{1}{2}x + 3\right| - 2 = 4$

$\left|\dfrac{1}{2}x + 3\right| = 6$

$\dfrac{1}{2}x + 3 = 6 \text{ or } \dfrac{1}{2}x + 3 = -6$

$x + 6 = 12 \qquad x + 6 = -12$

$x = 6 \qquad\qquad x = -18$

11. Let $x =$ the length of first side

$x + 2x + x + 5 = 69$

$4x = 64$

$x = 16$

$2x = 32$

$x + 5 = 21$

The first side is 16 meters, the second side is 32 meters, and the third side is 21 meters.

12. Let $x =$ the number of hours the computer was actually used.

$200 + 280(12) + 10x = 12{,}560$

$10x = 9000, \; x = 900$

The computer was used 900 hours.

13. Let $x =$ gallons of 50% antifreeze

$0.50x + 0.90(10 - x) = 0.60(10)$

$0.5x + 9 - 0.9x = 6$

$-0.4x = -3$

$x = 7.5$

$10 - 7.5 = 2.5$

She should use 2.5 gallons of 90% and 7.5 gallons of 50%.

14. Let $x =$ amount invested at 6%.

$0.06x + 0.10(5000 - x) = 428$

$0.06x + 500 - 0.1x = 428$

$-0.04x = -72$

$x = 1800$

$5000 - x = 3200$

$1800 was invested at 6% and $3200 was invested at 10%.

15. $5 - 6x < 2x + 21$

$-8x < 16$

$x > -2$

16. $-\dfrac{1}{2} + \dfrac{1}{3}(2 - 3x) \geq \dfrac{1}{2}x + \dfrac{5}{3}$

$-3 + 4 - 6x \geq 3x + 10$

$-9x \geq 9$

$x \leq -1$

17. $-16 < 2x + 4 < -6$

$-20 < 2x < -10$

$-10 < x < -5$

18. $x - 4 \le -6$ *or* $2x + 1 \ge 3$

$x \le -2 \qquad 2x \ge 2$

$x \ge 1$

19. $|7x - 3| \le 18$

$-18 \le 7x - 3 \le 18$

$-15 \le 7x \le 21$

$-\dfrac{15}{7} \le x \le 3$

20. $|3x + 1| > 7$

$3x + 1 < -7$ *or* $3x + 1 > 7$

$3x < -8 \qquad 3x > 6$

$x < -\dfrac{8}{3} \qquad x > 2$

Cumulative Test for Chapters 1-2

1. $-12, -3, 0, \dfrac{1}{4}, 2.16, 2.333..., -\dfrac{5}{8}, 3$

2. Associative property of addition.

3. $\sqrt{49} + 3(2 - 6)^2 - (-3)$

$= 7 + 3(-4)^2 + 3$

$= 7 + 3(16) + 3$

$= 7 + 48 + 3$

$= 55 + 3$

$= 58$

4. $\left(-2x^{-3}y^4\right)^{-2} = (-2)^{-2} x^{-3(-2)} y^{4(-2)}$

$= \dfrac{1}{(-2)^2} x^6 y^{-8}$

$= \dfrac{x^6}{4y^8}$

5. $\dfrac{7ab^3}{-14a^5b^{-2}} = \dfrac{b^5}{-2a^4}$

6. $P = 2(9) + 2(18) = 54$ cm

7. $A = \pi r^2 = 3.14(7)^2 = 153.86$ sq. in.

8. $2x(3x - 4) - 5x^2(2 - 6x)$

$= 6x^2 - 8x - 10x^2 + 30x^3$

$= 30x^3 - 4x^2 - 8x$

9. $\dfrac{1}{4}(x + 5) - \dfrac{5}{3} = \dfrac{1}{3}(x - 2)$

$3x + 15 - 20 = 4x - 8$

$-x = -3$

$x = 3$

10. $h = \dfrac{2}{3}(b + d)$

$3h = 2b + 2d$

$2b = 3h - 2d$

$b = \dfrac{3h - 2d}{2}$

11. Let $x = $ length of first side.

$x + x + 10 + 2x - 5 = 105$

$4x + 5 = 105$

$4x = 100$

$x = 25$

$x + 10 = 35$

$2x - 5 = 45$

The first side is 25 meters, the second side is 35 meters, and the third side is 45 meters.

12. Let x = the number of miles he drove.

$$19(4) + 0.23x = 154.20$$
$$76 + 0.23x = 154.2$$
$$0.23x = 78.2$$
$$x = 340$$

He drove 340 miles.

13. Let x = the number of gallons at 50%.

$$0.50x + 0.80(9-x) = 0.70(9)$$
$$0.5x + 7.2 - 0.8x = 6.3$$
$$-0.3x = -0.9$$
$$x = 3$$
$$9 - x = 6$$

He should use 3 gallons at 50% and 6 gallons at 80%.

14. Let x = amount invested at 12%.

$$0.12x + 0.10(6500 - x) = 690$$
$$0.12x + 650 - 0.1x = 690$$
$$0.02x = 40$$
$$x = 2000$$
$$6500 - x = 4500$$

She invested $2000 at 12% and $4500 at 10%.

15. $-4 - 3x < -2x + 6$

$$-x < 10$$
$$x > -10$$

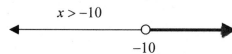

16. $\dfrac{1}{3}(x+2) \le \dfrac{1}{5}(x+6)$

$$5x + 10 \le 2x + 18$$
$$2x \le 8$$
$$x \le 4$$

17. $-7 < 5x + 3 < 18$

$$-10 < 5x < 15$$
$$-2 < x < 3$$

18. $x + 5 \le -4$ or $2 - 7x \le 16$

$$x \le -9 \qquad -7x \le 14$$
$$x \ge -2$$

19. $\left|\dfrac{1}{2}x + 2\right| \le 8$

$$-8 \le \dfrac{1}{2}x + 2 \le 8$$
$$-16 \le x + 4 \le 16$$
$$-20 \le x \le 12$$

20. $|3x - 4| > 11$

$$3x - 4 < -11 \text{ or } 3x - 4 > 11$$
$$3x < -7 \qquad\qquad 3x > 15$$
$$x < -\dfrac{7}{3} \qquad\qquad x > 5$$

Chapter 3

1. $5x + 2y = -12$

$5a + 2(6) = -12$

$5a + 12 = -12$

$5a = -24$

$a = -\dfrac{24}{5}$

2. $y = -\dfrac{1}{2}x + 5$

x	y
0	5
2	4
4	3

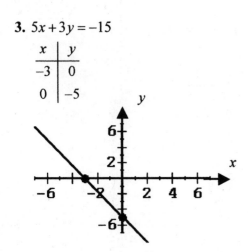

3. $5x + 3y = -15$

x	y
−3	0
0	−5

4. $m = \dfrac{y_2 - y_1}{x_2 - x_1}$

$m = \dfrac{3 - (-6)}{-2 - (-1)} = -9$

5. $m = \dfrac{y_2 - y_1}{x_2 - x_1} = \dfrac{0 - (-4)}{5 - (-13)} = \dfrac{2}{9}$

$m_\perp = -\dfrac{1}{m}$

$m_\perp = -\dfrac{1}{\frac{2}{9}} = -\dfrac{9}{2}$

6. $y - y_1 = m(x - x_1)$

$y - (-3) = -2(x - 7)$

$y + 3 = -2x + 14$

$y = -2x + 11$

7. $3x - 5y = 10$

$5y = 3x - 10$

$y = \dfrac{3}{5}x - 2,$

$m = \dfrac{3}{5},\ m_\perp = -\dfrac{5}{3}$

$y - y_1 = m_\perp(x - x_1)$

$y - (-2) = -\dfrac{5}{3}(x - (-1))$

$y + 2 = -\dfrac{5}{3}x - \dfrac{5}{3}$

$3y + 6 = -5x - 5$

$5x + 3y = -11$

8. $y > -\dfrac{1}{2}x + 3$, Test point: $(0,0)$

$0 > -\dfrac{1}{2}(0) + 3$

$0 > 3$

False. Shade the half-plane not containing $(0,0)$

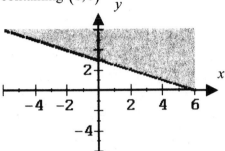

9. $2x - 3y \le -15$: Test point $(0,0)$.

$2(0) - 3(0) \le -15$. False, shade the half-plane not containing $(0,0)$.

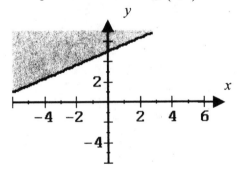

10. Domain $= \{0, \ 3, \ 4, \ 5\}$

Range $= \{1, \ 2, \ 4, \ 6, \ 7\}$. Since two ordered pairs, $(3,4)$ and $(3,2)$, have the same first coordinate, the relation is not a function.

11. Function, no two ordered pairs have the same first coordinate.

12. Not a function, at least two pairs have the same first coordinate.

13. $f(x) = -2x^2 + x - 3$

$f(-3) = -2(-3)^2 + (-3) - 3 = -24$

14. $g(x) = |2x - 3|$

$g(-1) = |2(-1) - 3| = |-5| = 5$

15. $p(x) = x^2 + 4$

x	$p(x)$
-2	8
-1	5
0	4
1	5
2	8

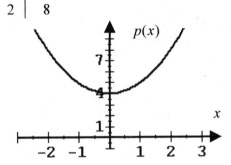

16. $h(x) = |x - 2|$

x	$h(x)$
-1	3
0	2
1	1
2	0
3	1

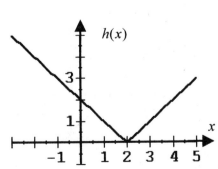

17.

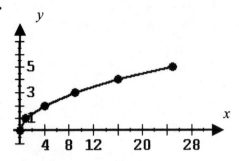

3.1 Exercises

1. Graphs are used to show the relationships among the <u>variables</u> in an equation.

3. To locate the point (a,b), assuming that $a,b > 0$, we move a units to the right and b units up. If $a \neq b$ the graphs of the points will be different. Thus, the order of the numbers in (a,b) matters. $(1,3)$ is not the same as $(3,1)$.

5. $y = 3x - 7$

$y = 3(-2) - 7 = -13$. $(-2,13)$ is a solution of $y = 3x - 7$.

7. $7x + 14y = -21$

$$7x + 14\left(\frac{1}{2}\right) = -21$$

$$14x + 14 = -42$$

$$14x = -56$$

$$x = -4$$

$\left(-4, \frac{1}{2}\right)$ is a solution of $7x + 14y = -21$.

9. $y = 5x - 2$

x	y
0	−2
1	3
2	8

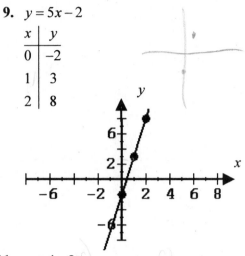

11. $y = 4 - 2x$

x	y
−1	6
0	4
1	2

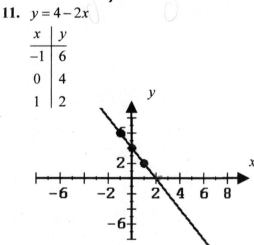

13. $y = \frac{2}{3}x - 4$

x	y
−3	−6
0	−4
3	−2

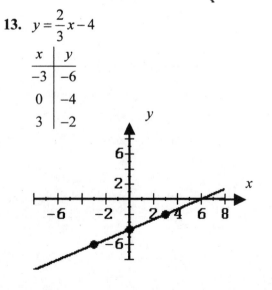

15. $2y - 3x = 6$

x	y
-2	0
0	3
2	6

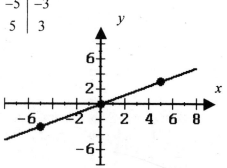

17. $2x - 3y = -9$

x	y
-4.5	0
0	3
3	5

19. $5y - 4 = 3x - 4$
$3x - 5y = 0$

x	y
0	0
-5	-3
5	3

21. $y = -x$

x	y
-2	2
0	0
2	-2

23. $2y + 8 = 0$, $2y = -8$, $y = -4$
The graph is a horizontal line.

x	y
0	-4
1	-4
2	-4

25. $2x - 3 = 3$ $x = -3$

x	y
-3	2
-3	1
-3	0

27. $2.5x - 5y = -20$

x	y
-4	2
-2	3
0	4

29. $2x - (y+3) = x + 3y + 1$

$$2x - y - 3 = x + 3y + 1$$
$$x - 4y = 4$$

x	y
-4	-2
0	-1
4	0

31. $y = 0.06x - 0.04$

x	y
0	-0.04
1	0.02
2	$.08$

33. From the table the greatest increase occurred between 1995 and 2000.

5 year period	increase
1980 to 1985	$277 - 269 = 8$
1985 to 1990	$346 - 277 = 69$
1990 to 1995	$406 - 346 = 60$
1995 to 2000	$537 - 406 = 131$

35. From the table the median weekly earnings were closest in 2000.

year	difference
1980	$380 - 269 = 111$
1985	$406 - 277 = 129$
1990	$481 - 346 = 135$
1995	$538 - 406 = 132$
2000	$627 - 537 = 90$

37. $(627 - 380) \div (380) = 0.65 = 65\%$

39. (a) $G(m) = 900 - 15m$

m	$G(m) = 900 - 15m$
0	$900 - 15(0) = 900$
10	$900 - 15(10) = 750$
20	$900 - 15(20) = 600$
30	$900 - 15(30) = 450$
60	$900 - 15(60) = 0$

(b)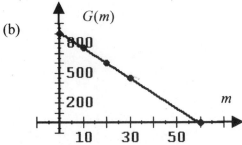

(c) $m = 61$ has no meaning since the tank is empty after $m = 60$ minutes.

41. $h = 5600 - 190t$

t	$h(t) = 5600 - 190t$
0	5600
5	4650
10	3700
20	1800
29	90

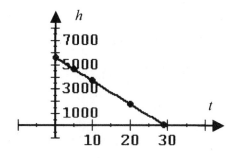

43. $y = 1.36x - 1.83$

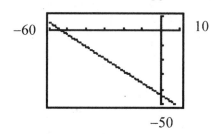

45. $y = -0.810x - 43.82$

47. $3(x-6) + 2 \le 4(x+2) - 21$

$3x - 18 + 2 \le 4x + 8 - 21$

$-x \le 3$

$x \ge -3$

49. Let $x =$ sales price of house.

$9100 = 0.07(100,000) + 0.03(x - 100,000)$

$9100 = 7000 + 0.03x - 3000$

$0.03x = 5100$

$x = \$170,000$ was the sales price.

3.2 Exercises

1. Slope measures *vertical* change (rise) versus *horizontal* change (run).

3. The slope of a horizontal line is 0.

5. $m = \dfrac{y_2 - y_1}{x_2 - x_1} = \dfrac{-7 - 5}{-3 - (-3)} = \dfrac{-12}{0}$. The line passing through $(-3, -7)$ and $(-3, 5)$ does not have a slope because division by 0 is undefined.

7. $m = \dfrac{y_2 - y_1}{x_2 - x_1} = \dfrac{-2 - (-6)}{-2 - 6} = -\dfrac{1}{2}$

9. $m = \dfrac{y_2 - y_1}{x_2 - x_1} = \dfrac{4 - 0}{\dfrac{1}{2} - 2} = -\dfrac{8}{3}$

11. $m = \dfrac{y_2 - y_1}{x_2 - x_1} = \dfrac{-3.5 - (-2.8)}{2.9 - 1.5} = -\dfrac{1}{2}$

13. $m = \dfrac{y_2 - y_1}{x_2 - x_1} = \dfrac{-2 - \dfrac{1}{4}}{\dfrac{3}{2} - \dfrac{3}{2}}$; no slope.

15. $m = \dfrac{\text{rise}}{\text{run}} = \dfrac{33.6}{56} = 0.6$

17. $m = \dfrac{\text{rise}}{\text{run}} = \dfrac{35.7}{142.8} = 0.25$

19. $m = \dfrac{\text{rise}}{\text{run}}$

$0.16 = \dfrac{\text{rise}}{500}$; rise(fall) $= 80$ ft

21. $m = \dfrac{y_2 - y_1}{x_2 - x_1} = \dfrac{1-2}{7-6.5} = -2$

$m_{\parallel} = -2$

23. $m = \dfrac{y_2 - y_1}{x_2 - x_1} = \dfrac{\frac{1}{2} - 5}{-9 - (-6)} = \dfrac{3}{2}$

$m_{\parallel} = \dfrac{3}{2}$

25. $m = \dfrac{y_2 - y_1}{x_2 - x_1} = \dfrac{12-9}{8-3} = \dfrac{3}{5}$

$m_{\perp} = -\dfrac{5}{3}$

27. $m = \dfrac{y_2 - y_1}{x_2 - x_1} = \dfrac{-\frac{1}{2} - \frac{5}{2}}{2-1} = -3$

$m_{\perp} = \dfrac{1}{3}$

29. $m = \dfrac{y_2 - y_1}{x_2 - x_1} = \dfrac{0 - 4.2}{-8.4 - 0} = \dfrac{1}{2}$

$m_{\perp} = -2$

31. $m_k = \dfrac{-9-11}{-3-1} = 5, m_h = \dfrac{-13-7}{-2-2} = 5$.

Lines k and h are parallel because they have the same slope.

33. To be a parallelogram, AD must be parallel to BC and AB must be parallel to DC.

$m_{AD} = \dfrac{1-2}{2-(-4)} = -\dfrac{1}{6}$

$m_{BC} = \dfrac{-2-(-1)}{-1-(-7)} = -\dfrac{1}{6}$

$m_{AB} = \dfrac{1-(-2)}{2-(-1)} = 1$

$m_{CD} = \dfrac{-1-2}{-7-(-4)} = 1$.

Since $m_{AD} = m_{BC}$ and $m_{AB} = m_{CD}$ the opposite sides are parallel and $ABCD$ is a parallelogram.

35. (a) $m = \dfrac{\text{rise}}{\text{run}} = \dfrac{5}{60} = \dfrac{1}{12}$

(b) $\dfrac{1}{12} = \dfrac{\text{rise}}{24}$, rise $= \dfrac{24}{12} = 2$ ft

(c) $\dfrac{1}{12} = \dfrac{1.7}{\text{run}}$, run $= 20.4$ ft

The ramp will have a horizontal distance of 20.4 ft.

37. $2(3-6)^3 + 20 \div (-10) = 2(-3)^3 + (-2)$

$\qquad = 2(-27) + (-2)$

$\qquad = -54 + (-2)$

$\qquad = -56$

39. $8x(x-1) - 2(x+y)$

$\qquad = 8x^2 - 8x - 2x - 2y$

$\qquad = 8x^2 - 10x - 2y$

3.3 Exercises

1. First determine the slope from the coordinates of the points. Then substitute the slope and the coordinates of one of the points into the point-slope form of the equation of a line. This may be rewritten in standard form or slope-intercept form.

3. $y = mx + b, \ y = \dfrac{3}{4}x - 9$

5. $y = mx + b, \ y = \dfrac{3}{4}x + \dfrac{1}{2}$
$$4y = 3x + 2$$
$$3x - 4y = -2$$

7. From the graph, $m = \dfrac{1}{2}, \ b = 3$.
$$y = \dfrac{1}{2}x + 3$$

9. From the graph, $m = 2, \ b = 0$.
$$y = 2x$$

11. From the graph, $m = -\dfrac{5}{2}, \ b = 3$.
$$y = -\dfrac{5}{2}x + 3$$

13. $2x - 3y = -8$
$$3y = 2x + 8$$
$$y = \dfrac{2}{3}x + \dfrac{8}{3}$$
$$m = \dfrac{2}{3}, \ b = \dfrac{8}{3}$$

15. $\dfrac{1}{2}x + 4y = 5$
$$x + 8y = 10$$
$$8y = -x + 10$$
$$y = -\dfrac{1}{8}x + \dfrac{5}{4}, \ m = -\dfrac{1}{8}, \ b = -\dfrac{1}{8}$$

17. $y = 3x + 4, \ m = 3, \ b = 4$

x	y
0	4
-1	1

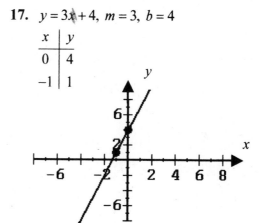

19. $5x - 4y = -20$. The intercepts are

x	y
0	5
-4	0

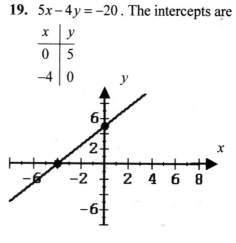

21. $y - y_1 = m(x - x_1)$
$$y - (-2) = 5(x - (-7))$$
$$y + 2 = 5x + 35$$
$$y = 5x + 33$$

23. $y - y_1 = m(x - x_1)$

$$y - (-1) = -\frac{5}{3}(x - 0)$$

$$3y + 3 = -5x$$

$$3y = -5x - 3$$

$$y = -\frac{5}{3}x - 1$$

25. $m = \dfrac{y_2 - y_1}{x_2 - x_1}$

$$m = \frac{-2 - (-3)}{7 - (-1)} = \frac{1}{8}$$

$$y - y_1 = m(x - x_1)$$

$$y - (-2) = \frac{1}{8}(x - 7)$$

$$y + 2 = \frac{1}{8}x - \frac{7}{8}$$

$$y = \frac{1}{8}x - \frac{23}{8}$$

27. $m = \dfrac{y_2 - y_1}{x_2 - x_1}$

$$m = \frac{1 - 0}{\frac{7}{6} - \left(-\frac{1}{3}\right)}$$

$$m = \frac{2}{3}$$

$$y - y_1 = m(x - x_1)$$

$$y - 1 = \frac{2}{3}\left(x - \frac{7}{6}\right)$$

$$3y - 3 = 2x - \frac{7}{3}$$

$$3y = 2x + \frac{2}{3}$$

$$y = \frac{2}{3}x + \frac{2}{9}$$

29. $3x - y = -5$

$$y = 3x + 5, \ m = 3, \ m(\| \text{ line}) = 3$$

$$y - 0 = 3(x - (-1))$$

$$y = 3x + 3$$

$$3x - y = -3$$

31. $2y + x = 7$

$$2y = -x + 7, \ y = -\frac{1}{2}x + \frac{7}{2}, \ m = -\frac{1}{2}$$

$$m(\| \text{ line}) = -\frac{1}{2}$$

$$y - (-4) = -\frac{1}{2}(x - (-5))$$

$$2y + 8 = -x - 5$$

$$x + 2y = -13$$

33. $y = 5x, \ m = 5, \ m(\perp \text{ line}) = -\dfrac{1}{5}$

$$y - (-2) = -\frac{1}{5}(x - 4)$$

$$5y + 10 = -x + 4$$

$$x + 5y = -6$$

35. $x - 4y = 2, \ y = \dfrac{1}{4}x - 2, \ m = \dfrac{1}{4}$

$$m(\perp \text{ line}) = -4$$

$$y - (-1) = -4(x - 3)$$

$$y + 1 = -4x + 12$$

$$4x + y = 11$$

37. $5x - 6y = 19, \ y = \dfrac{5}{6}x - \dfrac{19}{6}, \ m_1 = \dfrac{5}{6}$

$$6x + 5y = -30, \ y = -\frac{6}{5}x - 6, \ m_2 = -\frac{6}{5}$$

$$m_1 m_2 = \frac{5}{6}\left(-\frac{6}{5}\right) = -1$$

The lines are perpendicular.

39. $y = \dfrac{2}{3}x + 6$, $m_1 = \dfrac{2}{3}$

$-2x - 3y = -12$

$y = -\dfrac{2}{3}x + 4$, $m_2 = -\dfrac{2}{3}$

$m_1 \neq m_2$, $m_1 m_2 = -\dfrac{4}{9} \neq -1$

The lines are neither parallel or perpendicular.

41. $y = \dfrac{3}{7}x - \dfrac{1}{14}$, $m_1 = \dfrac{3}{7}$

$14y + 6x = 3$

$y = -\dfrac{3}{7}x + \dfrac{1}{2}$, $m_2 = -\dfrac{3}{7}$

$m_1 \neq m_2$

$m_1 m_2 = \dfrac{3}{7}\left(-\dfrac{3}{7}\right) = -\dfrac{9}{49} \neq -1$

The lines are neither parallel or perpendicular.

43. $y = -2.39x + 2.04$, $y = -2.39x - 0.87$

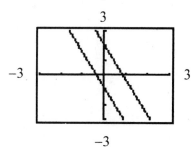

45. $m = \dfrac{167.9 - 68.7}{18 - 0}$

$m = 5.51$

$y = mx + b$

$y = 5.51x + 68.7$

47. $2016 - 1980 = 36$

x	$y = 5.51x + 68.7$
0	68.7
18	167.9
36	267.06

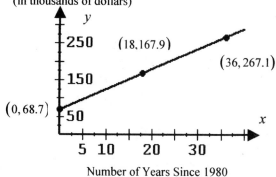

Cost of Homes
(in thousands of dollars)

Number of Years Since 1980

49. $m = \dfrac{117.3 - 87.7}{18 - 0} = 1.64$

$y = 1.64x + 87.7$

51. $2016 - 1980 = 36$

x	$y = 1.64x + 87.7$
0	87.7
18	117.3
36	146.7

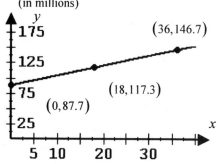

Number of Housing Units
(in millions)

Number of Years Since 1980

Cumulative Review Problems

53. $11 - (x + 2) = 7(3x + 6)$
$11 - x - 2 = 21x + 42$
$22x = -33$
$$x = -\frac{3}{2}$$

55. $70 + 70(0.01x) + 3 = 82.10$
$70 + 0.7x + 3 = 82.1$
$0.7x = 9.1$
$x = 13$

Putting Your Skills To Work

1.

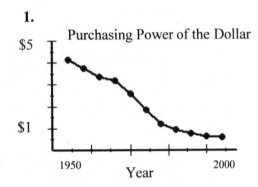

2. From the table, the 5-year period during which the purchasing power of the dollar showed the greatest decline, 0.71, was from 1970-1975.

Year	Power	Decline	Year	Power	Decline
1950	4.15	–	1980	1.22	0.64
1955	3.73	0.42	1985	0.93	0.29
1960	3.37	0.36	1990	0.77	0.16
1965	3.17	0.20	1995	0.66	0.11
1970	2.57	0.60	2000	0.59	0.07
1975	1.86	0.71			

3. The constant dollar value of a $20,000 home in 1955 is $20,000 \times 3.73 = \$74,600$ while the constant dollar value of a $110,000 home in 2000 is $110,000 \times 0.59 = \$64,900$. The 1955 home was more expensive.

4. The constant dollar value of the $1500 tuition in 1960 was $1500 \times 3.37 = \$5055$ while the constant dollar value of the $11,500 tuition in 1995 was $11,500 \times 0.66 = \$7590$. The tuition in 1995 was more expensive.

3.4 Exercises

1. $y > -2x + 4$, Test point: $0, 0$
$0 > -2(0) + 4$, $0 > 4$ False

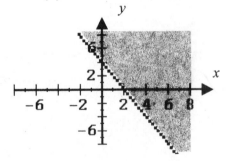

3. $y < \frac{2}{3}x - 2$, Test point: $0, 0$

$0 < \frac{2}{3}(0) - 2$, $0 < -2$ False

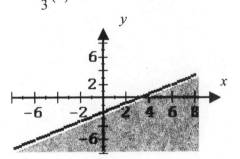

5. $y \geq -\dfrac{5}{3}x + 3$

Test point: $(0,0)$

$0 \geq -\dfrac{5}{3}(0) + 3$

$0 \geq 3$ False

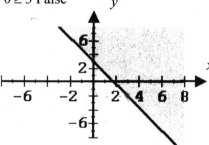

7. $5y - x \leq 15$

Test point: $(0,0)$

$5(0) - 0 \leq 15$

$0 \leq 15$ True

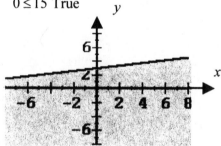

9. $x - y > -2$

Test point: $(0,0)$

$0 - 0 > -2$

$0 > -2$ True

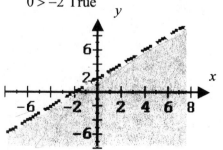

11. $3x - 2y \geq 0$

Test point: $(3,0)$

$3(3) - 2(0) \geq 0$

$9 \geq 0$ True

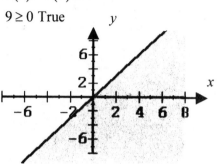

13. $(0,0)$ can't be used as a test point because it is on the line. Using $(-4,2)$ as a test point gives

$3(-4) - 2(2) \geq 0$

$-12 + 4 \geq 0$

$-8 \geq 0$ False

15. $x > -4$

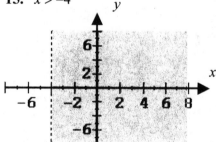

17. $y \leq -1$

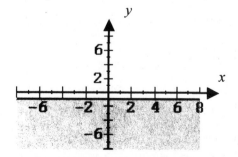

19. $-8x \le -12$, $x \ge \dfrac{3}{2}$

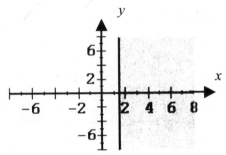

21. $4y \ge 2$, $y \ge \dfrac{1}{2}$

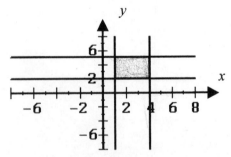

23. $x \ge 1$, $y \ge 2$, $x \le 4$, $y \le 5$

$1 \le x \le 4$, $2 \le y \le 5$

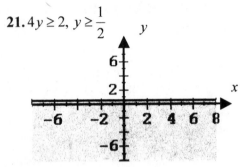

25. $200x + 60y \le 1200$, Test point: $(0,0)$

$200(0) + 60(0) \le 1200$, $0 \le 1200$ True

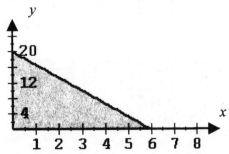

Cumulative Review Exercises

27. $A = \dfrac{1}{2}a(b+c)$

$A = \dfrac{1}{2}(6.0)(2.5 + 5.5) = 24$

3.5 Exercises

1. A relation is any set of ordered pairs. A function is a set of ordered pairs in which no two different ordered pairs have the same first coordinate.

3. A function may be described as a set of ordered pairs, as an equation, and as a graph.

5. Domain $= \left\{ \dfrac{1}{2}, -3, \dfrac{3}{2} \right\}$

Range $= \{5,\ 7,\ -1\}$

Relation is *not* a function.

7. Domain $= \{40, -18, 38, 57\}$

Range $= \{10, 27, -15\}$

Relation is a function.

9. Domain $= \{6, 8, 10, 12, 14\}$

Range $= \{38, 40, 42, 44, 46\}$

Relation is a function.

11. Domain $= \{$Jan., Feb., Mar., April, May, June, July, Aug., Sept., Oct., Nov., Dec.$\}$

Range $= \{81, 80, 79\}$

Relation is a function.

13. Domain = {Chicago, New York}

Range = {1454, 1350, 1250,

1136, 1127, 1046}

Relation is *not* a function.

15. Domain = {32, 41, 50, 59, 68, 95}

Range = {0, 5, 10, 15, 20, 35}

17. Function

19. Not a function.

21. Function

23. Not a function.

25. $g(x) = 2x - 5$

$g(-3) = 2(-3) - 5$

$g(-3) = -11$

27. $g(x) = 2x - 5$

$g\left(\frac{1}{3}\right) = 2\left(\frac{1}{3}\right) - 5$

$g\left(\frac{1}{3}\right) = -4\frac{1}{3}$

29. $h(x) = \frac{2}{3}x + 2$

$h(-3) = \frac{2}{3}(-3) + 2 = 0$

31. $h(x) = \frac{2}{3}x + 2$

$h\left(\frac{1}{2}\right) = \frac{2}{3} \cdot \frac{1}{2} + 2$

$h\left(\frac{1}{2}\right) = 2\frac{1}{3}$

33. $r(x) = 2x^2 - 4x + 1$

$r(1) = 2(1)^2 - 4(1) + 1$

$r(1) = -1$

35. $r(x) = 2x^2 - 4x + 1$

$r(0.1) = 2(0.1)^2 - 4(0.1) + 1$

$r(0.1) = 0.62$

37. $t(x) = x^3 - 3x^2 + 2x - 3$

$t(10) = 10^3 - 3(10)^3 + 2(10) - 3$

$t(10) = 717$

39. $t(x) = x^3 - 3x^2 + 2x - 3$

$t(-1) = (-1)^3 - 3(-1)^2 + 2(-1) - 3$

$t(-1) = -9$

41. $f(x) = x + 3$

x	$f(x) = x + 3$
-2	1
-1	2
0	3
1	4
2	5

Range = {1, 2, 3, 4, 5}

43. $h(x) = \frac{2}{3}x - 4$

$\frac{2}{3}x - 4 = y,\ x = \frac{3}{2}(y+4)$	y
6	0
9	2
11	$\frac{10}{3}$
12	4

Domain = {6, 9, 11, 12}

Cumulative Review Problems

45. $5(3-x)-2(3y+x)$
$= 15-5x-6y-2x$
$= -7x-6y+15$

47. starting bal. $-$ checks $+$ deposit $= 100$

$$\text{checks} = \begin{cases} 78.91 \\ 280.55 \\ 116.01 \\ 196.69 \\ +424.98 \\ \hline 1097.14 \end{cases}$$

$763.21\text{-}1097.14+ \text{deposit} = 100$

$\text{deposit} = \$433.93$

3.6 Exercises

1. $f(x) = \dfrac{3}{4}x+2$

x	$f(x)$
-4	-1
0	2
4	5

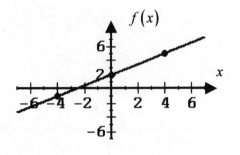

3. $f(x) = -\dfrac{1}{3}x+6$

x	y
0	6
3	5
6	4

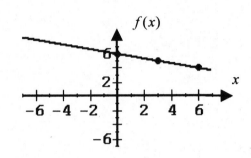

5. $c(x) = 0.10x+35$

x	$c(x)$
0	35
100	45
200	55
250	60

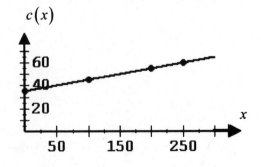

7. $p(x) = -1500x + 45{,}000$

x	$p(x)$
0	45,000
10	30,000
30	0

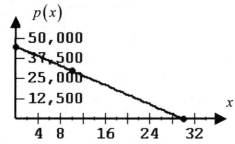

9. $g(x) = |x - 3|$

x	0	1	2	3	4	5
$g(x)$	3	2	1	0	1	2

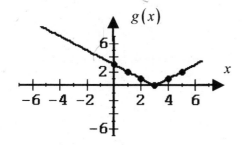

11. $f(x) = |x| + 2$

x	-2	-1	0	1	2
$f(x)$	4	3	2	3	4

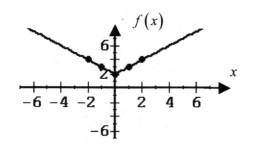

13. $g(x) = x^2 - 4$

x	-3	-2	-1	0	1	2	3
$g(x)$	5	0	-3	-4	-3	0	5

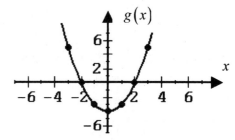

15. $f(x) = (x - 3)^2$

x	$f(x)$
1	4
2	1
3	0
4	4

17. $f(x) = x^3 + 2$

x	$f(x)$
-1	1
0	2
1	3

19. $p(x) = -x^3$

x	$p(x)$
-1	1
0	0
1	-2

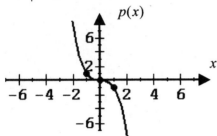

21. $f(x) = \dfrac{2}{x}$

x	-4	-2	-1	$-\dfrac{1}{2}$	$\dfrac{1}{2}$	1	2	4
$f(x)$	$-\dfrac{1}{2}$	-1	-2	-4	4	2	1	$\dfrac{1}{2}$

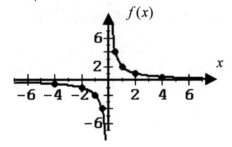

23. $h(x) = -\dfrac{4}{x}$

x	-4	-2	-1	1	2	4
$h(x)$	1	2	4	-4	-2	-1

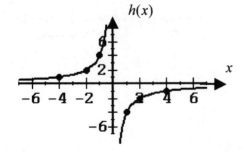

25. $y_1 = x^2$, $y_2 = 0.4x^2$, $y_3 = 2.6x^2$

From the graphs in the next column we see that the larger the coefficient of x^2 the faster the graph rises.

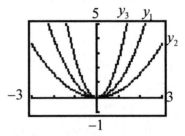

27. From the graph $f(2) = -4$.

x	$f(x)$
-1	2
1	-2
3	-6

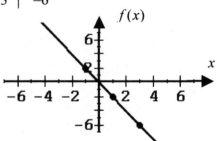

29. From the graph $f(2) = -6$.

x	$f(x)$
-3	-1
-2	2
-1	3
0	2
1	-1

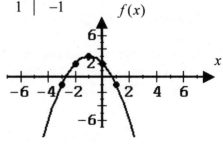

31. From the graph $f(2) = 5$

x	$f(x)$
-2	5
-1	4
0	3
1	4
3	6

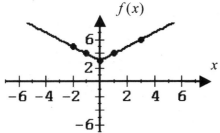

33. (a)

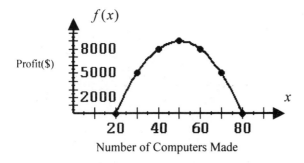

Number of Computers Made

(b) From the graph, 50 computers should be made each day for maximum profit.

(c) To earn a profit of $8000 or more each day the company should manufacture between 40 and 60 computers.

(d) If the company manufactures 82 computers per day they will operate at a loss.

(e) If the company manufactures forty five computers per day the profit will be approximately $8700.

Cumulative Review Problems

35. $2(3ax - 4y) = 5(ax + 3)$

$$6ax - 8y = 5ax + 15$$
$$ax = 8y + 15$$
$$x = \frac{8y + 15}{a}$$

37. $\dfrac{1}{3}x + \dfrac{1}{5}y = \dfrac{1}{10} + \dfrac{1}{2}x$

$$10x + 6y = 3 + 15x$$
$$5x = 6y - 3$$
$$x = \frac{6y - 3}{5}$$

39. $0.8(3.9 \times 10^6 \text{ mi}^2)\left(\dfrac{5280^2 \text{ ft}^2}{\text{mi}^2}\right)$

$$= 8.698 \times 10^{13} \text{ ft}^2$$

Chapter 3 Review Problems

1. $y = -\dfrac{1}{4}x - 1$

x	y
-4	0
0	-1
4	-2

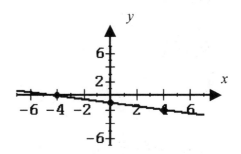

2. $y = -\dfrac{3}{2}x + 5$

x	y
0	5
2	2
4	−1

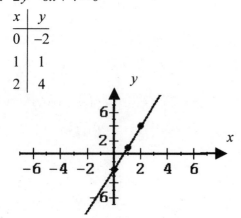

3. $2y - 6x + 4 = 0$

x	y
0	−2
1	1
2	4

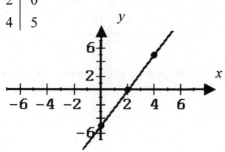

4. $5(x - y) = 10 - 3y$

$5x - 5y = 10 - 3y$

$5x - 2y = 10$

x	y
0	−5
2	0
4	5

5. $-3y + 2 = -8y - 13$

$\qquad 5y = -15$

$\qquad y = -3$

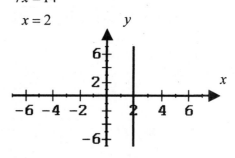

6. $5x - 6 = -2x + 8$

$\qquad 7x = 14$

$\qquad x = 2$

7. $P = 140x - 2000 = 0$

$140x = 2000$

$x = 14\dfrac{2}{7}$

The company must sell 15 or more microcomputers each day to make a profit.

8. $m = \dfrac{y_2 - y_1}{x_2 - x_1} = \dfrac{-4 - (-3)}{-8 - 2} = \dfrac{1}{10}$

9. $m = \dfrac{y_2 - y_1}{x_2 - x_1} = \dfrac{\dfrac{1}{3} - 2}{\dfrac{2}{3} - 4} = \dfrac{1}{2}$

$m_\perp = -\dfrac{1}{m} = -\dfrac{1}{\dfrac{1}{2}} = -2$

10.
$$y = mx + b$$
$$y = \frac{2}{3}x + (-4)$$
$$3y = 2x - 12$$
$$2x - 3y = 12$$

11. $y - y_1 = m(x - x_1)$
$$y - (-2) = -4\left(x - \frac{1}{2}\right)$$
$$y + 2 = -4x + 2$$
$$4x + y = 0$$

12. $y - y_1 = m(x - x_1)$
$$y - 1 = 0(x - (-3)) = 0$$
$$y - 1 = 0$$
$$y = 1$$

13. $m = \dfrac{y_2 - y_1}{x_2 - x_1} = \dfrac{6 - \frac{-1}{2}}{5 - (-1)} = \dfrac{13}{12}$
$$y - y_1 = m(x - x_1)$$
$$y - 6 = \frac{13}{12}(x - 5)$$
$$12y - 72 = 13x - 65$$
$$13x - 12y = -7$$

14. A line with undefined slope is a vertical line, $x = -6$.

15. $7x + 8y - 12 = 0$
$$8y = -7x + 12$$
$$y = -\frac{7}{8}x + \frac{3}{2}$$
$$m = -\frac{7}{8}, \ m_\perp = -\frac{1}{m} = -\frac{1}{-\frac{7}{8}} = \frac{8}{7}$$

(continued)

15. (continued)
$$y - y_1 = m_\perp(x - x_1)$$
$$y - 5 = \frac{8}{7}(x - (-2))$$
$$7y - 35 = 8x + 16$$
$$8x - 7y = -51$$

16. $3x - 2y = 8$
$$2y = 3x - 8$$
$$y = \frac{3}{2}x + 4, \ m = \frac{3}{2}, \ m_{\parallel} = \frac{3}{2}$$
$$y - y_1 = m(x - x_1)$$
$$y - 1 = \frac{3}{2}(x - 5)$$
$$2y - 2 = 3x - 15$$
$$3x - 2y = -13$$

17. $y < 2x + 4$, Test point: $(0,0)$
$$0 < 2(0) + 4, \ 0 < 4 \text{ True}$$

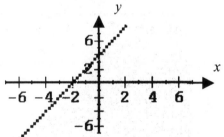

18. $y < 3x + 1$

Test point: $(0,0)$, $0 < 3(0) + 1$,

$0 < 1$, True

19. $y > -\dfrac{1}{2}x + 3$

Test point: $(0,0)$

$0 > -\dfrac{1}{2}(0,0) + 3$

$0 > 3$ False

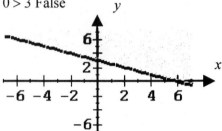

20. $y > -\dfrac{2}{3}x + 0$

Test point: $(0,0)$

$0 > -\dfrac{2}{3}(0) + 1$

$0 > 1$ False

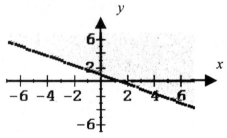

21. $3x + 4y \le -12$

Test point: $(0,0)$

$3(0) + 4(0) \le -12$

$0 \le -12$ False

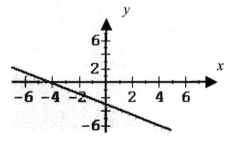

22. $5x + 3y \le -15$

Test point: $(0,0)$

$5(0) + 3(0) \le -15$

$0 \le -15$ False

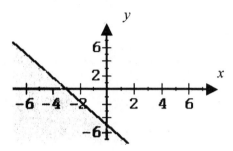

23. $x \le 3y$

Test point: $(0,3)$

$0 \le 3(3)$

$0 \le 9$ True

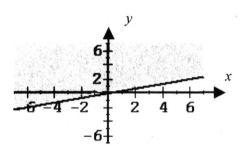

24. $3x - 5 < 7$

$3x < 12$

$x < 4$

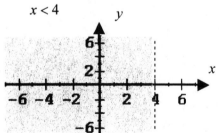

25. $5y - 2 > 3y - 10$

$\qquad 2y > -8$

$\qquad\quad y > -4$

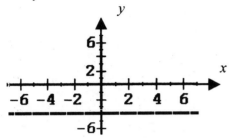

26. Domain: $\{0,1,2,3\}$

Range: $\{0,1,4,9,16\}$

Relation is not a function.

27. Domain $= \{-20,-18,-16,-12\}$

Range $= \{14,16,18\}$

Relation is a function.

28. Function, no two ordered pairs have the same first coordinate.

29. Function, no two ordered pairs have the same first coordinate.

30. Not a function, at least two ordered pairs have the same first coordinate.

31. $f(x) = 3x - 8$

$f(-2) = 3(-2) - 8 = -14$

32. $g(x) = 2x^2 - 3x - 5$

$g(-3) = 2(-3)^2 - 3(-3) - 5 = 22$

33. $h(x) = x^3 + 2x^2 - 5x + 8$

$h(-1) = (-1)^3 + 2(-1)^2 - 5(-1) + 8$

$h(-1) = 14$

34. $p(x) = |-6x - 3|$

$p(3) = |-6(3) - 3| = |-21| = 21$

35. $f(x) = 2|x - 1|$

x	$f(x) = 2|x-1|$
-2	4
0	2
1	0
2	2
3	4

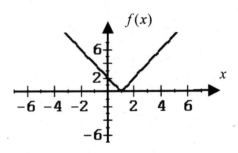

36. $g(x) = x^2 - 3$

x	$g(x) = x^2 - 3$
-2	1
-1	-2
0	-3
1	-2
2	1

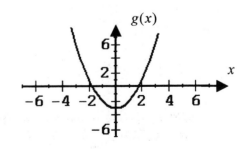

37. $h(x) = x^3 + 2$

x	$g(x) = x^3 + 2$
-2	-6
-1	1
0	2
1	3
2	10

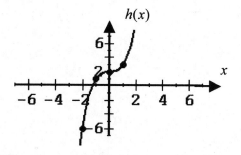

38. From the graph $f(-2) = 3$.

x	$f(x)$
-3	4
-1	2
0	-1
1	0
2	-1
3	0
4	1

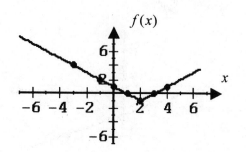

39. From the graph $f(-2) = 0$.

x	-1	-3	-4	-5	-6	-7
$f(x)$	5	-3	-4	-3	0	5

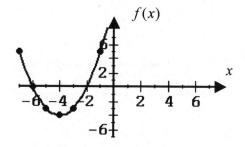

40. $f(x) = -\dfrac{4}{5}x + 3$

x	$f(x) = -\dfrac{4}{5}x + 3$
-5	7
0	3
10	-5

41. $f(x) = 2x^2 - 3x + 4$

x	$f(x) = 2x^2 - 3x + 4$
-3	31
0	4
4	24

42. $f(x) = \dfrac{7}{2x+3}$

x	$f(x) = \dfrac{7}{2x+3}$
-2	-7
0	$\dfrac{7}{3}$
2	1

43. $f(x) = 3x^3 - 4$

x	$f(x) = 3x^3 - 4$
-1	-7
0	-4
2	20

44. $m = \dfrac{y_2 - y_1}{x_2 - x_1} = \dfrac{-5 - (-6)}{3 - (-5)} = \dfrac{1}{8}$

45. $m = \dfrac{y_2 - y_1}{x_2 - x_1} = \dfrac{2 - 10}{-4 - 1} = \dfrac{8}{5}$

$m_n = -\dfrac{1}{m} = -\dfrac{1}{\dfrac{8}{5}} = -\dfrac{5}{8}$

46. $m = \dfrac{y_2 - y_1}{x_2 - x_1} = \dfrac{8 - (-6)}{4.5 - 2.5} = 7$

$m_p = m = 7$

47. $y = mx + b$

$y = \dfrac{5}{6}x + (-5)$

$6y = 5x - 30$

$5x - 6y = 30$

48. $m = \dfrac{y_2 - y_1}{x_2 - x_1} = \dfrac{6 - 3}{5 - (-7)} = \dfrac{1}{4}$

$y - y_1 = m(x - x_1)$

$y - 6 = \dfrac{1}{4}(x - 5)$

$y - 6 = \dfrac{1}{4}x - \dfrac{5}{4}$

$y = \dfrac{1}{4}x + \dfrac{19}{4}$

49. $y = 5x - 2, \ m = 5, \ m_\parallel = 5$

$y - y_1 = m_\parallel(x - x_1)$

$y - 10 = 5(x - 4)$

$y - 10 = 5x - 20$

$y = 5x - 10$

50. $3x - 6y = 9$

$6y = 3x - 9$

$y = \dfrac{1}{2}x - \dfrac{3}{2}, \ m = \dfrac{1}{2}, \ m_\perp = -2$

$y - y_1 = m(x - x_1)$

$y - (-1) = -2(x - (-2))$

$y + 1 = -2x - 4$

$2x + y = -5$

51. $f(x) = 15,000 + 500x$

52. $f(x) = 35 + 0.15x$

53. $f(x) = 18,000 - 65x$

Chapter 3 Test

1. $2x + 3y = -10$

x	y
-2	-2
1	-4
8	-6

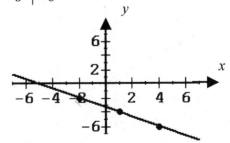

2. $m = \dfrac{y_2 - y_1}{x_2 - x_1}$

$m = \dfrac{-3 - (-6)}{2 - \dfrac{1}{2}} = 2$

3. $9x + 7y = 13$

$7y = -9x + 13$

$y = -\dfrac{9}{7}x + \dfrac{13}{7}$

$m = -\dfrac{9}{7}$

4. $6x - 7y - 1 = 0$

$7y = 6x - 1$

$y = \dfrac{6}{7}x - \dfrac{1}{7}$

$m = \dfrac{6}{7}, \ m_\perp = -\dfrac{7}{6}$

$y - y_1 = m_\perp (x - x_1)$

$y - (-2) = -\dfrac{7}{6}(x - 0)$

$y + 2 = -\dfrac{7}{6}x$

$6y + 12 = -7x$

$7x + 6y = -12$

5. $\quad m = \dfrac{y_2 - y_1}{x_2 - x_1}$

$m = \dfrac{-2 - (-1)}{5 - (-3)} = -\dfrac{1}{8}$

$y - y_1 = m(x - x_1)$

$y - (-2) = -\dfrac{1}{8}(x - 5)$

$8y + 16 = -x + 5$

$x + 8y = -11$

6. $y = 2$

7. $y \geq -4x$

Test point: $(0,2)$

$2 \geq -4(2)$

$2 \geq -8$ True

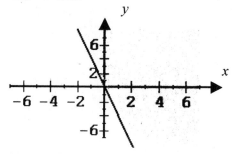

8. $4x - 2y < -6$

Test point: $(0,0)$

$4(0) - 2(0) < -6$

$0 < -6$ False

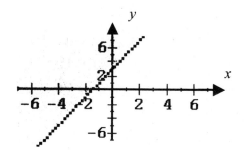

9. Domain $= \{0, 1, 2\}$

Range $= \{-4, -1, 0, 1, 4\}$

10. $f(x) = 2x - 3$

$f\left(\dfrac{3}{4}\right) = 2\left(\dfrac{3}{4}\right) - 3$

$f\left(\dfrac{3}{4}\right) = -1\dfrac{1}{2}$

11. $g(x) = \dfrac{1}{2}x^2 + 3$

$g(-4) = \dfrac{1}{2}(-4)^2 + 3$

$g(-4) = 11$

12. $h(x) = \left| -\dfrac{2}{3}x + 4 \right|$

$h(-9) = \left| -\dfrac{2}{3}(-9) + 4 \right|$

$h(-9) = |10|$

$h(-9) = 10$

13. $p(x) = -2x^3 + 3x^2 + x - 4$

$p(-2) = -2(-2)^3 + 3(-2)^2 + (-2) - 4$

$p(-2) = 22$

14. $g(x) = 5 - x^2$

x	$g(x) = 5 - x^2$
-2	1
-1	4
0	5
1	4
2	1

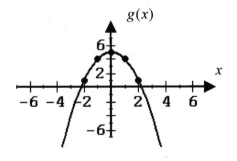

15. $h(x) = x^3 - 2$

x	$h(x) = x^3 - 2$
-1	-3
0	-2
1	-1

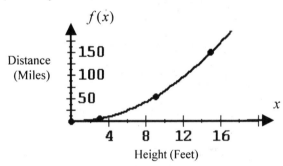

16. From the graph you can see 10 miles if you are 4 feet above the water.

Cumulative Test for Chapters 1-3

1. Inverse property of addition.

2. $3(4-6)^2 + \sqrt{16} + 12 \div (-3)$

$= 3(-2)^2 + 4 + (-4)$

$= 3(4)$

$= 12$

3. $\left(3x^2 y^{-3}\right)^{-4} = 3^{-4} x^{-8} y^{12} = \dfrac{y^{12}}{81x^8}$

4. $5x(2x-3y)-3(x^2+4)$

$=10x^2-14xy-3x^2-12$

$=7x^2-14xy-12$

5. $|3x-2|=8$

$3x-2=8$ or $3x-2=-8$

$3x=10$ $3x=-6$

$x=\dfrac{10}{3}$ $x=-2$

6. $3(x-2)>6$ or $5-3(x+1)>8$

$3x-6>6$ $5-3x-3>8$

$3x>12$ $-3x>6$

$x>4$ $x<-2$

```
<-----------○-------○----------->
           -2       4
```

7. $3b=\dfrac{1}{2}(3x+2y)$

$6b=3x+2y$

$3x=6b-2y$

$x=\dfrac{6b-2y}{3}$

8. Let $w=$ the width.

$P=2l+2w$

$92=2(2w+1)+2w$

$46=2w+1+w$

$3w=45$

$w=15$

$2w+1=31$

The width of the insulator is 15 cm and the length of the insulator is 31 cm.

9. Let $x=$ the amount invested at 5%.

$0.05x+0.08(3000-x)=189$

$0.05x+240-0.08x=189$

$-0.03x=-51$

$x=1700$

$3000-1700=1300$

Sharim invested $1700 at 5% and $1700 at 8%.

10. $A=\dfrac{1}{2}\pi r^2$

$A=\dfrac{1}{2}(3.14)(3)^2=14.13$ sq. in.

11. $4x-6y=10$

x	y
-2	-3
1	-1
4	1

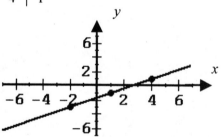

12. $m=\dfrac{y_2-y_1}{x_2-x_1}=\dfrac{5-1}{6-(-2)}=\dfrac{1}{2}$

13. $m=\dfrac{y_2-y_1}{x_2-x_1}=\dfrac{1-3}{5-4}=-2$

$y-y_1=m(x-x_1)$

$y-1=-2(x-5)$

$y-1=-2x+10$

$2x+y=11$

14.
$$y = \frac{2}{3}x - 4$$

$$m = \frac{2}{3},\ m_\perp = -\frac{3}{2}$$

$$y - y_1 = m_\perp (x - x_1)$$

$$y - (-3) = -\frac{3}{2}(x - (-2))$$

$$2y + 6 = -3x - 6$$

$$3x + 2y = -12$$

15. $D = \left\{ \frac{1}{2}, 2, 3, 5 \right\}$

$R = \{-1, 2, 7, 8\}$

Yes, the relation is a function.

16. $f(x) = -2x^2 - 4x + 1$

$$f(-3) = -2(-3)^2 - 4(-3) + 1$$

$$f(-3) = -5$$

17. $p(x) = -\frac{1}{3}x + 2$

x	$p(x) = -\frac{1}{3}x + 2$
-3	3
0	2
3	1

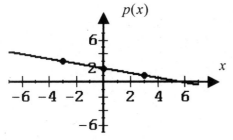

18. $h(x) = |x - 2|$

x	$h(x) = |x - 2|$
0	2
1	1
2	0
3	1
4	2

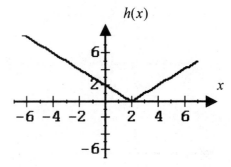

19. $r(x) = \frac{3}{x}$

x	$r(x) = \frac{3}{x}$	x	$r(x) = \frac{3}{x}$
-3	-1	1	3
-2	$-\frac{3}{2}$	$\frac{3}{2}$	2
$-\frac{3}{2}$	-2	2	$\frac{3}{2}$
-1	-3	3	1

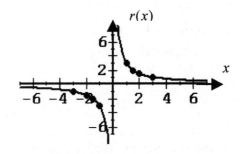

20. $f(x) = x^2 - 3$

x	$f(x) = x^2 - 3$
-2	1
-1	-2
0	-3
1	-2
2	1

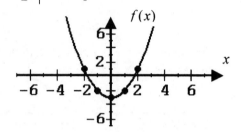

21. $y \leq -\dfrac{3}{2}x + 3$

Test point: $(0,0)$

$0 \leq -\dfrac{3}{2}(0) + 3$

$0 \leq 3$ True

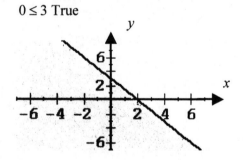

22. $f(x) = -2x^3 + 4$

x	$f(x) = -2x^3 + 4$
-2	20
0	4
3	-50

23. $f(x) = 32,500 - 1400x$

24. $2005 - 1990 = 15$

$$f(15) = 32,500 - 1400(15)$$
$$f(15) = 11,500$$

11,500 cars will travel each weekday on Route 1 in the year 2005.

Chapter 4

1. $5x - 2y = 27$ **(1)**

   $3x - 5y = -18$ **(2)**

   Multiply **(1)** by 5, and **(2)** by -2 and add

   $25x - 10y = 135$

   $\underline{-6x + 10y = 36}$

   $19x = 171$

   $x = 9$

   Substitute $x = 9$ into **(1)**

   $5(9) - 2y = 27$

   $45 - 2y = 27$

   $2y = 18$

   $y = 9$

   The solution is $(9, 9)$.

2. $7x + 3y = 15$ **(1)**

   $\dfrac{1}{3}x - \dfrac{1}{2}y = 2$ **(2)**

   Multiply **(2)** by 6 to clear fractions

   $2x - 3y = 12$ **(3)**

   Add **(1)** and **(3)**

   $7x + 3y = 15$

   $\underline{2x - 3y = 12}$

   $9x = 27$

   $x = 3$

   Substitute $x = 3$ into **(1)**

   $7(3) + 3y = 15$

   $3y = -6$

   $y = -2$

   The solution is $(3, -2)$.

3. $2x = 3 + y$ **(1)**

   $3y = 6x - 9$ **(2)**

   Solve **(1)** for y

   $y = 2x - 3$ and substitute into **(2)**

   $3(2x - 3) = 6x - 9$

   $6x - 9 = 6x - 9$

   $0 = 0$

   The equations are dependent and the system has an infinite number of solutions.

4. $2x + 7y = -10$ **(1)**

   $5x + 6y = -2$ **(2)**

   Multiply **(1)** by 5 and **(2)** by -2 and add

   $10x + 35y = -50$

   $\underline{-10x - 12y = 4}$

   $23y = -46$

   $y = -2$

   Substitute $y = -2$ into **(1)**

   $2x + 7(-2) = -10$

   $2x - 14 = -10$

   $2x = 4$

   $x = 2$

   The solution is $(2, -2)$.

5. $5x - 2y + z = -1$ **(1)**

   $3x + y - 2z = 6$ **(2)**

   $-2x + 3y - 5z = 7$ **(3)**

   Multiply **(1)** by 2 and add to **(2)**

   $10x - 4y + 2z = -2$

   $\underline{3x + y - 2z = 6}$

   $13x - 3y = 4$ **(4)**

   (continued)

5. (continued)
Multiply **(1)** by 5 and add to **(3)**
$$25x - 10y + 5z = -5$$
$$\underline{-2x + 3y - 5z = 7}$$
$$23x - 7y = 2 \qquad \textbf{(5)}$$
Multiply **(4)** by 7 and **(5)** by -3 and
add
$$91x - 21y = 28$$
$$\underline{-69x + 21y = -6}$$
$$22x = 22$$
$$x = 1$$
Substitute $x = 1$ into **(5)**
$$23(1) - 7y = 2$$
$$7y = -21$$
$$y = -3$$
Substitute $x = 1, \ y = 3$ into **(1)**
$$5(1) - 2(3) + z = -1$$
$$5 - 6 + z = -1$$
$$z = 0$$
The solution is $(1, 3, 0)$.

6.
$$x + y + 2z = 9 \qquad \textbf{(1)}$$
$$3x + 2y + 4z = 16 \qquad \textbf{(2)}$$
$$2y + z = 10 \qquad \textbf{(3)}$$

Solve **(3)** for z and substitute into **(1)**
and **(2)**
$$z = 10 - y$$
$$x + y + 2(10 - 2y) = 9$$
$$x + y + 20 - 4y = 9$$
$$x - 3y = -11 \qquad \textbf{(4)}$$
$$3x + 2y + 4(10 - 2y) = 16$$
$$3x + 2y + 40 - 8y = 16$$
$$3x - 6y = -24 \qquad \textbf{(5)}$$
Multiply (4) by -2 and (5) and add to
(5)

(continued)

6. (continued)
$$-2x + 6y = 22$$
$$\underline{3x + 6y = -24}$$
$$x = -2$$

Substitute $x = -2$ into **(5)**
$$3(-2) - 6y = -24$$
$$-6y = -18$$
$$y = 3$$
Substitute $x = -2, \ y = 3$ into $\quad$ **(1)**
$$-2 + 3 + 2z = 9$$
$$2z = 8$$
$$z = 4$$
The solution is $(-2, 3, 4)$.

7. $x - 2z = -5 \qquad \qquad \textbf{(1)}$
$y - 3z = -3 \qquad \qquad \textbf{(2)}$
$2x - z = -4 \qquad \qquad \textbf{(3)}$

Solve **(3)** for z and substitute into **(1)**
$$z = 2x + 4$$
$$x - 2(2x + 4) = -5$$
$$x - 4x - 8 = -5$$
$$-3x = 3$$
$$x = -1$$
Substitute $x = -1$ into **(3)**
$$2(-1) - z = -4$$
$$-z = -2$$
$$z = 2$$
Substitute $z = 2$ into **(2)**
$$y - 3(2) = -3$$
$$y - 6 = -3$$
$$y = 3$$
The solution is $(-1, 3, 2)$.

8. Let $x =$ cost of shirt and $y =$ cost of pants.

$$2x + 3y = 75 \qquad \textbf{(1)}$$
$$3x + 5y = 121 \qquad \textbf{(2)}$$

Multiply **(1)** by 3 and **(2)** by -2 and add

$$6x + 9y = 225$$
$$\underline{-6x - 10y = -242}$$
$$-y = -17$$
$$y = 17$$

Substitute $y = 17$ into **(1)**

$$2x + 3(17) = 75$$
$$2x = 24$$
$$x = 12$$

The shirts cost \$12 and the pants cost \$17.

9. Let $x =$ the number of A packets, $y =$ the number of B packets, and $z =$ the number of C packets.

$$3x + 2y + 2z = 21 \qquad \textbf{(1)}$$
$$2x + 4y + 2z = 22 \qquad \textbf{(2)}$$
$$4x + 5y + z = 26 \qquad \textbf{(3)}$$

Solve **(3)** z and substitute into **(1)** and **(2)**

$$z = 26 - 4x - 5y$$
$$3x + 2y + 2(26 - 4x - 5y) = 21$$
$$3x + 2y + 52 - 8x - 10y = 21$$
$$5x + 8y = 31 \quad \textbf{(4)}$$
$$2x + 4y + 2(26 - 4x - 5y) = 22$$
$$2x + 4y + 52 - 8x - 1y = 22$$
$$6x + 6y = 30$$
$$x + y = 5$$
$$y = 5 - x \ \textbf{(5)}$$

Substitute $y = 5 - x$ into **(4)**

(continued)

9. (continued)
$$5x + 8(5 - x) = 31$$
$$-3x = -9$$
$$x = 3$$

Substitute $x = 3$ into **(5)**

$y = 5 - 3 = 2$. Substitute $x = 3$, $y = 2$ into **(3)**

$$4(3) + 5(2) + z = 26$$
$$z = 4$$

She should use 3 of packet A, 2 of packet B, and 4 of packet C.

10. Let $x =$ the number of easy exercises and $y =$ the number of hard exercises.

$$x + y = 10 \qquad \textbf{(1)}$$
$$7x + 10y = 82 \qquad \textbf{(2)}$$

Solve **(1)** for y and substitute into **(2)**

$$y = 10 - x$$
$$7x + 10(10 - x) = 82$$
$$-3x = -18$$
$$x = 6$$

Substitute $x = 6$ into **(1)**

$$6 + y = 10$$
$$y = 4$$

You can solve four hard exercises and 6 easy exercises.

11. $2x + 2y \geq -4$

Test point: $(0, 0)$

$$2(0) + 2(0) \geq -4$$
$$0 \geq -4 \text{ True}$$
$$-3x + y \leq 2$$

Test point: $(0, 0)$

$$-3(0) + 0 \leq 2$$
$$0 \leq 2 \text{ True}$$

(continued)

11. (continued)

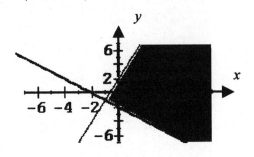

12. $x - 3y < -6$

Test point: $(0,0)$

$$0 - 3(0) < -6$$

$$0 < -6 \text{ False}$$

$x < 3$

Test point: $(0,0)$

$$0 < 3 \text{ True}$$

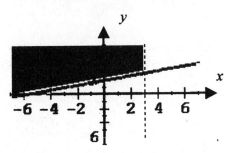

4.1 Exercises

1. There is no solution. There is no point (x, y) that satisfies both equations.

The graph of such a system yields two parallel lines.

3. $4x + 1 = 6 - y$

LHS: $4\left(\dfrac{3}{2}\right) + 1 = 6 + 1 = 7$

RHS: $6 - (-1) = 6 + 1 = 7$

$2x - 5y = 8$

$2\left(\dfrac{3}{2}\right) - 5(-1) = 3 + 5 = 8$

$\left(\dfrac{3}{2}, -1\right)$ is a solution.

5. $3x + y = 2$
$2x - y = 3$

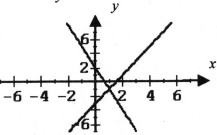

$(1, -1)$ is the solution.

7. $2x + 3y = 6$
$2x + y = -2$

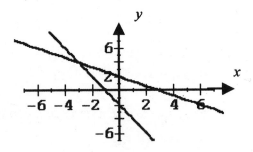

$(-3, 4)$ is the solution.

9. $y = -x + 3$
$3x + 3y = -2$

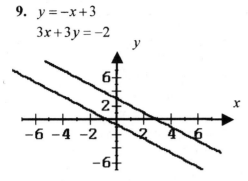

The system has no solution.

11. $3x + 2y = -17$ **(1)**
    $2x + y = 3$ **(2)**

Solve **(2)** for y and substitute into **(1)**
$$y = 3 - 2x$$
$$3x + 2(3 - 2x) = -17$$
$$3x + 6 - 4x = -17$$
$$x = 23$$
Substitute $x = 23$ into **(2)** and solve for y
$$2(23) + y = 3$$
$$46 + y = 3$$
$$y = -43$$
$$x = 23, \ y = -43$$
is the solution.

13. $-x + 3y = -8$ **(1)**
    $2x - y = 6$ **(2)**

Solve **(2)** for y and substitute into **(1)**
$$y = 2x - 6$$
$$-x + 3(2x - 6) = -8$$
$$-x + 6x - 18 = -8$$
$$5x = 10$$
$$x = 2$$
Substitute $x = 2$ into **(1)**
(continued)

13. (continued)
$$-2 + 3y = -8$$
$$3y = -6$$
$$y = -2$$
$x = 2, \ y = -2$ is the solution.
Check:

(1) $-2 + 3(-2) \overset{?}{=} -8$
$$-2 - 6 \overset{?}{=} -8$$
$$-8 = -8$$

(2) $2(2) - (-2) \overset{?}{=} 6$
$$4 + 2 \overset{?}{=} 6$$
$$6 = 6$$

15. $\dfrac{1}{5}x - \dfrac{1}{2}y = 1$ **(1)**

    $\dfrac{1}{5}x - 3y = -9$ **(2)**

Multiply **(2)** by 5 to clear fractions, solve for x, then substitute the result into **(1)**
$$x - 15y = -45$$
$$x = 15y - 45$$
$$\frac{1}{5}(15y - 45) - \frac{1}{2}y = 1$$
$$3y - 9 - \frac{1}{2}y = 1$$
$$6y - 18 - y = 2$$
$$5y = 20$$
$$y = 4$$
Substitute $y = 4$ into **(1)**
$$\frac{1}{5}x - \frac{1}{2}(4) = 1$$
(continued)

15. (continued)

$$\frac{1}{5}x - 2 = 1$$

$$\frac{1}{5}x = 3$$

$$x = 15$$

$x = 15,\ y = 4$ is the solution.

Check:

(1) $\frac{1}{5}(15) - \frac{1}{2}(4) \overset{?}{=} 1$

$$3 - 2 \overset{?}{=} 1$$

$$1 = 1$$

(2) $\frac{1}{5}(15) - 3(4) \overset{?}{=} -9$

$$3 - 12 \overset{?}{=} -9$$

$$-9 = -9$$

17. $3(x-1) + 2(y+2) = 6$ **(1)**

$4(x-2) + y = 2$ **(2)**

Solve **(2)** for y and substitute into **(1)**

$$4x - 8 + y = 2$$

$$y = 10 - 4x$$

$$3x - 3 + 2(10 - 4x + 2) = 6$$

$$3x - 3 + 20 - 8x + 4 = 6$$

$$-5x = -15$$

$$x = 3$$

Substitute $x = 3$ into **(2)**

$$4(3-2) + y = 2$$

$$4 + y = 2$$

$$y = -2$$

$x = 3,\ y = -2$ is the solution.

19. $9x + 2y = 2$ **(1)**

$3x + 5y = 5$ **(2)**

(continued)

19. (continued)

Multiply **(2)** by -3 and add to **(1)**

$$9x + 2y = 2$$

$$-9x - 15y = -15$$

$$\overline{-13y = -13}$$

$$y = 1$$

Substitute $y = 1$ into **(1)**

$$9x + 2(1) = 2$$

$$9x + 2 = 2$$

$$9x = 0$$

$$x = 0$$

$x = 0,\ y = 1$ is the solution.

21. $6s - 3t = 1$ **(1)**

$5s + 6t = 15$ **(2)**

Multiply **(1)** by 2 and add to **(2)**

$$12s - 6t = 2$$

$$5s + 6t = 15$$

$$\overline{17s = 17}$$

$$s = 1$$

Substitute $s = 1$ into **(2)**

$$5(1) + 6t = 15$$

$$6t = 10$$

$$t = \frac{3}{5}$$

$s = 1,\ t = \dfrac{3}{5}$ is the solution.

23. $\dfrac{7}{2}x + \dfrac{5}{2}y = -4$ **(1)**

$3x + \dfrac{2}{3}y = 1$ **(2)**

Clear fractions

$$7x + 5y = -8 \quad \textbf{(1)}$$

$$9x + 2y = 3 \quad \textbf{(2)}$$

(continued)

23. (continued)

Multiply **(1)** by 2 and **(2)** by -5 and add

$$14x + 10y = -16$$
$$\underline{-45x - 10y = -15}$$
$$-31x = -31$$
$$x = 1$$

Substitute $x = 1$ into **(2)**

$$9(1) + 2y = 3$$
$$2y = -6$$
$$y = -3$$

$x = 1,\ y = -3$ is the solution.

25.
$$1.6x + 1.5y = 1.8 \qquad \textbf{(1)}$$
$$0.4x + 0.3y = 0.6 \qquad \textbf{(2)}$$

Multiply **(2)** by -0.5 and add to **(1)**

$$1.6x + 1.5y = 1.8$$
$$\underline{-2.0x - 1.5y = -3.0}$$
$$-0.4x = -1.2$$
$$x = 3$$

Substitute $x = 3$ into **(1)**

$$1.6(3) + 1.5y = 1.8$$
$$4.8 + 1.5y = 1.8$$
$$1.5y = -3.0$$
$$y = -2$$

$x = 3,\ y = -2$ is the solution.

Check:

(1) $1.6(3) + 1.5(-2) \overset{?}{=} 1.8$

$$4.8 - 3.0 \overset{?}{=} 1.8$$
$$1.8 = 1.8$$

(2) $0.4(3) + 0.3(-2) \overset{?}{=} 0.6$

$$1.2 - 0.6 \overset{?}{=} 0.6$$
$$0.6 = 0.6$$

27.
$$2x + y = 4 \qquad \textbf{(1)}$$
$$\frac{2}{3}x + \frac{1}{4}y = 2 \qquad \textbf{(2)}$$

Solve **(1)** for y and substitute into **(2)**

$$y = 4 - 2x$$
$$\frac{2}{3}x + \frac{1}{4}(4 - 2x) = 2$$
$$8x + 12 - 6x = 24$$
$$2x = 12$$
$$x = 6$$

Substitute $x = 6$ into **(1)**

$$2(6) + y = 4$$
$$12 + y = 4$$
$$y = -8$$

Check:

(1) $2(6) + (-8) \overset{?}{=} 4$

$$12 - 8 \overset{?}{=} 4$$
$$4 = 4$$

(2) $\frac{2}{3}(6) + \frac{1}{4}(-8) \overset{?}{=} 2$

$$4 - 2 \overset{?}{=} 2$$
$$2 = 2$$

29.
$$0.2x = 0.1y - 1.2 \qquad \textbf{(1)}$$
$$2x - y = 6 \qquad \textbf{(2)}$$

Solve **(2)** for y and substitute into **(1)**

$$y = 2x - 6$$
$$0.2x = 0.1(2x - 6) - 1.2$$
$$0.2x = 0.2x - 0.6 - 1.2$$
$$0 = -1.8$$

This is an inconsistent system of equations and has no solution.

31. $5x - 7y = 12$ **(1)**
      $-10x + 14y = -24$ **(2)**

Multiply **(1)** by 2 and add to **(2)**
$10x - 14y = 24$
$-10x + 14y = -24$
$\overline{0 = 0}$

This is a dependent system of equations and has an infinite number of solutions.

33. Multiply both equations by 10 to clear decimals
$8x + 9y = 13$ **(1)**
$6x - 5y = 45$ **(2)**

Multiply **(1)** 5 and **(2)** by 9 and add
$40x + 45y = 65$
$\underline{54x - 45y = 405}$
$94x = 4760$

$x = 5$

Substitute $x = 5$ into **(1)**
$8(5) + 9y = 13$

$40 + 9y = 13$

$9y = -27$

$y = -3$

$x = 5,\ y = -3$ is the solution.

35. Solve the first equation for a and substitute into the second equation.
$$a = \frac{4}{5}b - \frac{1}{5}$$

$$15\left(\frac{4}{5}b - \frac{1}{5}\right) - 12b = 4$$

$$12b - 3 - 12b = 4$$

$$-3 = 4$$

This is an inconsistent system of equations and has no solution.

37. Solve the first equation for y and substitute into the second equation.
$$y = 14 - \frac{3}{8}x$$

$$2x - \frac{7}{4}\left(14 - \frac{3}{8}x\right) = 18$$

$$8x - 98 + \frac{21}{8}x = 72$$

$$\frac{85}{8}x = 170$$

$$x = 16$$

Substitute $x = 16$ into the first equation
$$\frac{3}{8}(16) + y = 14$$

$$6 + y = 14$$

$$y = 8$$

$x = 16,\ y = 8$ is the solution.

39. (a) $y = 200 + 50x$ for Old World Tile
     $y = 300 + 30x$ for Modern Bath

(b)

x	$y = 300 + 30x$		x	$y = 200 + 50x$
0	300		0	200
4	420		4	400
8	540		8	600

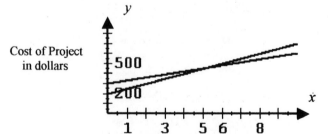

Cost of Project in dollars

Number of Hours of Labor

(continued)

39. (continued)

(c) From the graph the cost will be the same for 5 hours of installing new tile.

(d) From the graph Modern Bathroom Headquarters will cost less to remove old time and install new tile if the time needed to install the new tile is 6 hrs.

41. $y_1 = -1.7x + 3.8$

    $y_2 = 0.7x - 2.1$

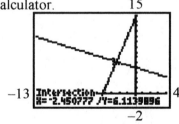

To the nearest hundredth , the point of intersection is $(2.46, -0.38)$.

43. Write $0.5x + 1.1y = 5.5$ as

$y_1 = (5.5 - 0.5x)/1.1$ and

$-3.1x + 0.9y = 13.1$ as

$y_1 = (13.1 + 3.1x)/0.9$ for input to the calculator.

The point of intersection is $(2.45, 6.11)$.

Cumulative Review Problems

45. $\dfrac{\$200,000,000}{9,000,000 \text{ tons}} \left(\dfrac{\text{ton}}{2000 \text{ pounds}} \right)$

$\approx \$0.01$ per pound

Putting Your Skills to Work

1. $\dfrac{100,000 - 12,000}{2000 - 1900} = 880$ lost per year

$c(t) = 100,000 - 880t$ where t is the number of years since 1900.

2. $\dfrac{70,000 - 10,500}{2000 - 1970} \approx 1980$ lost per year

$r(t) = 70,000 - 1980t$ where t is the where t is the number of years since 1900.

3. $c(90) = 100,000 - 880(90) = 20,800$

$c(t) = 20,800 - 880t$ where now t is the number of years since 1990.

4. $r(20) = 70,000 - 1980(20) = 30,400$

$r(t) = 30,400 - 1980t$ where now t is the number of years since 1990.

5.

t	$c(t)$		t	$r(t)$
0	20,800		0	30,400
5	16,400		5	20,500
10	12,000		10	10,600

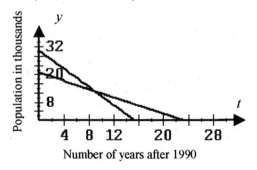

6. The slope of the rhino line is -1980, a decrease of 1980/yr. The slope of the cheetah line is -880, a decrease of 880/yr. Rhinos are decreasing faster.

7. $30,400 - 1980t = 20,800 - 880t$

$$1100t = 9600$$

$$t = 8.7$$

The two populations will be equal 8.7 years after 1990. In the year 1999 the two populations were equal.

8. The cheetah will become extinct when
$20,800 - 880t = 0$

$$880t = 20,800$$

$$t \approx 23.6$$

$1990 + 23.6 = 2013.6$. The cheetah will become extinct sometime between 2013 and 2014. The rhino will become extinct when
$30,400 - 1980t = 0$

$$1980t = 30,400$$

$$t \approx 15.4$$

$1990 + 15.4 = 2005.4$. The rhino will become extinct sometime between 2005 and 2006.

4.2 Exercises

1. $2x - 3y + 2z = -7$ **(1)**
$x + 4y - z = 10$ **(2)**
$3x + 2y + z = 4$ **(3)**

(1)

$$2(2) - 3(1) + 2(-4) \overset{?}{=} -7$$

$$4 - 3 - 8 \overset{?}{=} -7$$

$$-7 = -7$$

(2)

$$2 + 4(1) - (-4) \overset{?}{=} 10$$

$$2 + 4 + 4 \overset{?}{=} 10$$

$$10 = 10$$

(continued)

1. (continued)

(3)

$$3(2) + 2(1) + (-4) \overset{?}{=} 4$$

$$6 + 2 - 4 \overset{?}{=} 4$$

$$4 = 4$$

$(2, 1, -4)$ is a solution.

3. $x + y + 2z = 0$ **(1)**
$2x - y - z = 1$ **(2)**
$x + 2y + 3z = 1$ **(3)**
Add **(1)** and **(2)**
$3x + z = 1$ **(4)**
Add $2 \cdot$ **(2)** and **(3)**
$5x + z = 3$ **(5)**
Subtract **(5)** from **(4)**
$-2x = -2$

$$x = 1$$

Substitute $x = 1$ into **(5)**
$5(1) + z = 3$

$$z = -2$$

Substitute $x = 1$, $z = -2$ into **(1)**
$1 + y + 2(-2) = 0$

$$1 + y - 4 = 0$$

$$y = 3$$

$x = 1$, $y = 3$, $z = -3$ is the solution.

5. $x + 2y - 3z = -11$ **(1)**
$-2x + y - z = -11$ **(2)**
$x + y + z = 6$ **(3)**
Add **(2)** and **(3)**
$-x + 2y = -5$ **(4)**
Add $3 \cdot$ **(3)** and **(1)**
$4x + 5y = 7$ **(5)**
Add $4 \cdot$ **(4)** and **(5)**
$13y = -13$

$$y = -1$$

(continued)

5. (continued)

Substitute $y = -1$ into **(5)**

$$4x + 5(-1) = 7$$
$$4x - 5 = 7$$
$$4x = 12$$
$$x = 3$$

Substitute $x = 3$, $y = -1$ into **(3)**

$$3 + (-1) + z = 6$$
$$2 + z = 6$$
$$z = 4$$

$x = 3$, $y = -1$, $z = 4$ is the solution.

7.
$$\begin{array}{ll} -4x + y - 3z = 2 & \textbf{(1)} \\ 5x - 3y + 4z = 1 & \textbf{(2)} \\ 3x - 2y + 5z = 1 & \textbf{(3)} \end{array}$$

Add $3 \cdot$ **(1)** and **(2)**
$$-7x - 5z = 7 \qquad \textbf{(4)}$$

Add $2 \cdot$ **(1)** and **(3)**
$$-5x - z = 5 \qquad \textbf{(5)}$$

Add $-5 \cdot$ **(5)** and **(4)**
$$18x = -18$$
$$x = -1$$

Substitute $x = -1$ into **(5)**
$$-5(-1) - z = 5$$
$$5 - z = 5$$
$$-z = 0$$
$$z = 0$$

Substitute $x = -1$ and $z = 0$ into **(1)**
$$-4(-1) + y - 3(0) = 2$$
$$4 + y = 2$$
$$y = -2$$

$x = -1$, $y = -2$, $z = 0$ is the solution.

9.
$$\begin{array}{ll} x + 4y - z = -5 & \textbf{(1)} \\ -2x - 3y + z = 5 & \textbf{(2)} \\ x - \dfrac{2}{3}y + z = \dfrac{11}{3} & \textbf{(3)}\text{(continued)} \end{array}$$

9. (continued)

Add **(1)** and **(3)**
$$2x + \frac{10}{3}y = -\frac{4}{3} \qquad \textbf{(4)}$$

Add $2 \cdot$ **(1)** and **(2)**
$$5y = -5$$
$$y = -1$$

Substitute $y = -1$ into **(4)**
$$2x + \frac{10}{3}(-1) = -\frac{4}{3}$$
$$6x - 10 = -4$$
$$6x = 6$$
$$x = 1$$

Substitute $x = 1$ and $y = -1$ into **(3)**
$$1 - \frac{2}{3}(-1) + z = \frac{11}{3}$$
$$3 + 2 + 3z = 11$$
$$3z = 6$$
$$z = 2$$

$x = 1$, $y = -1$, $z = 2$ is the solution.

11.
$$\begin{array}{ll} 2x - 3y + 2z = -7 & \textbf{(1)} \\ \dfrac{3}{2}x + y + \dfrac{1}{2}z = 2 & \textbf{(2)} \\ x + 4y - z = 10 & \textbf{(3)} \end{array}$$

Add **(1)** and $2 \cdot$ **(3)**
$$4x + 5y = 13 \qquad \textbf{(4)}$$

Add **(3)** and $2 \cdot$ **(2)**
$$4x + 6y = 14 \qquad \textbf{(5)}$$

Subtract **(5)** from **(4)**
$$-y = -1$$
$$y = 1$$

Substitute $y = 1$ into **(5)**
$$4x + 6(1) = 14$$
$$4x = 8$$
$$x = 2$$

(continued)

11. (continued)

Substitute $x = 2$, $y = 1$ into **(3)**

$$2 + 4(1) - z = 10$$
$$6 - z = 10$$
$$-z = 4$$
$$z = -4$$

$x = 2$, $y = 1$, $z = -4$ is the solution.

13.
$$a = 8 + 3b - 2c \qquad \textbf{(1)}$$
$$4a + 2b - 3c = 10 \qquad \textbf{(2)}$$
$$c = 10 + b - 2a \qquad \textbf{(3)}$$

Substitute a from **(1)** into **(2)**

$$4(8 + 3b - 2c) + 2b - 3c = 10$$
$$32 + 12b - 8c + 2b - 3c = 10$$
$$14b - 11c = -22 \quad \textbf{(4)}$$

Substitute a from **(1)** into **(3)**

$$c = 10 + b - 2(8 + 3b - 2c)$$
$$c = 10 + b - 16 - 6b + 4c$$
$$5b - 3c = -6 \qquad \textbf{(5)}$$

Add $5 \cdot$ **(4)** and $-14 \cdot$ **(5)**

$$-13c = -26$$
$$c = 2$$

Substitute $c = 2$ into **(5)**

$$5b - 3(2) = -6$$
$$5b - 6 = -6$$
$$5b = 0$$
$$b = 0$$

Substitute $b = 0$, $c = 2$ into **(1)**

$$a = 8 + 3(0) - 2(2)$$
$$a = 8 - 4$$
$$a = 4$$

$a = 4$, $b = 0$, $c = 2$ is the solution.

15. Multiply all three equations by 10 to clear decimals.

$$2a + b + 2c = 1 \qquad \textbf{(1)}$$
$$3a + 2b + 4c = -1 \qquad \textbf{(2)}$$
$$6a + 11b + 2c = 3 \qquad \textbf{(3)} \text{ (con't)}$$

15. (continued)

Subtract **(3)** from **(1)**

$$-4a - 10b = -2$$
$$2a + 5b = 1 \qquad \textbf{(4)}$$

Add $-2 \cdot$ **(1)** and **(2)**

$$-a = -3$$
$$a = 3$$

Substitute $a = 3$ into **(4)**

$$2(3) + 5b = 1$$
$$6 + 5b = 1$$
$$5b = -5$$
$$b = -1$$

Substitute $a = 3$, $b = -2$ into **(1)**

$$2(3) + (-1) + 2c = 1$$
$$6 - 1 + 2c = 1$$
$$2c = -4$$
$$c = -2$$

$a = 3$, $b = -1$, $c = -2$ is the solution.

17.
$$-x + 3y - 2z = 11 \qquad \textbf{(1)}$$
$$2x - 4y + 3z = -15 \qquad \textbf{(2)}$$
$$3x - 5y - 4z = 5 \qquad \textbf{(3)}$$

Add $3 \cdot$ **(1)** and **(3)**

$$4y - 10z = 38 \qquad \textbf{(4)}$$

Add $2 \cdot$ **(1)** and **(2)**

$$2y - z = 7 \qquad \textbf{(5)}$$

Add **(4)** and $2 \cdot$ **(5)**

$$-8z = 24$$
$$z = -3$$

Substitute $z = -3$ into **(5)**

$$2y - (-3) = 7$$
$$2y + 3 = 7$$
$$2y = 4$$
$$y = 2$$

Substitute $z = -3$, $y = 2$ into **(1)**

$$-x + 3(2) - 2(-3) = 11$$

(continued)

17. (continued)

$$-x + 6 + 6 = 11$$

$$-x + 12 = 11$$

$$-x = -1$$

$$x = 1$$

$x = 1$, $y = 2$, $z = -3$ is the solution.

19. When a calculator is used it is convenient to keep all three equations together as the operations are performed.

$$x - 4y + 4z = -3.72186$$

$$-x + 3y - z = 5.98115$$

$$2x - y + 5z = 7.93645$$

Now perform two operations on the system; first, add the first equation to the second and add -2 times the first to the third. Note that this *does not* change the first equation but the second and third.

$$x - 4y + 4z = -3.72186$$

$$-y + 3z = 2.25929$$

$$7y - 3z = 15.38017$$

Add the second equation to the third

$$x - 4y + 4z = -3.72186$$

$$-y + 3z = 2.25929$$

$$6y = 17.63946$$

From the third equation
$y = 2.93991$ which may be substituted into the second equation to give
$z = 1.73307$. Substituting these values for x and y into the first equation gives
$x = 1.10551$.
The solution, to five decimal places, is
$x = 1.10551$, $y = 2.93991$, $z = 1.73307$

21.
$$\begin{aligned} x + y &= 1 & \textbf{(1)} \\ y - z &= -3 & \textbf{(2)} \\ 2x + 3y + z &= 1 & \textbf{(3)} \end{aligned}$$

Solve **(1)** for y
$$y = 1 - x \qquad \textbf{(4)}$$

Solve **(2)** for z
$$z = y + 3 \qquad \textbf{(5)}$$

Substitute y from **(4)** into **(5)**
$$z = 1 - x + 3 = 4 - x \qquad \textbf{(6)}$$

Substitute y from **(4)** and z from **(6)** into **(3)**
$$2x + 3(1 - x) + 4 - x = 1$$

$$2x + 3 - 3x + 4 - x = 1$$

$$-2x + 7 = 1$$

$$-2x = -6$$

$$x = 3$$

Substitute $x = 3$ into **(6)**
$$z = 4 - 3$$

$$z = 1$$

Substitute $x = 3$ into **(4)**
$$y = 1 - 3$$

$$y = -2$$

Substitute $y = -2$, $z = 1$ into **(3)**
$$2x + 3(-2) + 1 = 1$$

$$2x - 6 + 1 = 1$$

$$2x = 6$$

$$x = 3$$

$x = 3$, $y = -2$, $z = 1$ is the solution.

23.
$$\begin{aligned} -y + 2z &= 1 & \textbf{(1)} \\ x + y + z &= 2 & \textbf{(2)} \\ -x + 3z &= 2 & \textbf{(3)} \end{aligned}$$

Add **(2)** and **(3)**
$$y + 4z = 4 \qquad \textbf{(4)}$$

Add **(1)** and **(4)**
$$6z = 5$$

$$z = \frac{5}{6} \qquad \text{(continued)}$$

23. (continued)

Substitute $z = \dfrac{5}{6}$ into **(4)**

$$y + 4\left(\dfrac{5}{6}\right) = 4$$

$$6y + 20 = 24$$

$$6y = 4$$

$$y = \dfrac{2}{3}$$

Substitute $z = \dfrac{5}{6}$, $y = \dfrac{2}{3}$ into **(2)**

$$x + \dfrac{2}{3} + \dfrac{5}{6} = 2$$

$$x = \dfrac{1}{2}$$

$x = \dfrac{1}{2}$, $y = \dfrac{2}{3}$, $z = \dfrac{5}{6}$ is the solution.

25.
$$\begin{array}{ll} x - 2y + z = 0 & \textbf{(1)} \\ -3x - y = -6 & \textbf{(2)} \\ y - 2z = -7 & \textbf{(3)} \end{array}$$

Multiply **(1)** by 2 and add to **(3)**
$$2x - 3y = -7 \qquad \textbf{(4)}$$

Multiply **(2)** by -3 and add to **(4)**
$$11x = 11$$
$$x = 1$$

Substitute $x = 1$ into **(2)**
$$-3 - y = -6$$
$$y = 3$$

Substitute $x = 1$, $y = 3$ into **(1)**
$$1 - 6 + z = 0$$
$$z = 5$$

$x = 1$, $y = 3$, $z = 5$ is the solution.

27.
$$\begin{array}{ll} \dfrac{a}{2} - b + c = 8 & \textbf{(1)} \\[2mm] \dfrac{3}{2}a + b + 2c = 0 & \textbf{(2)} \\[2mm] a + c = 2 & \textbf{(3)} \end{array}$$

Add **(1)** and **(2)**
$$2a + 3c = 8 \qquad \textbf{(4)}$$
Subtract $3 \cdot$ **(3)** from **(4)**
$$-a = 2$$
$$a = -2$$

Substitute $a = -2$ into **(3)**
$$-2 + c = 2$$
$$c = 4$$

Substitute $a = 2$, $c = -4$ into **(1)**
$$\dfrac{a}{2} - (-5) + 4 = 8$$
$$a + 10 + 8 = 16$$
$$a = -2$$
$a = -2$, $b = -5$, $c = 4$ is the solution.

29.
$$\begin{array}{ll} 2x + y = -3 & \textbf{(1)} \\ 2y + 16z = -18 & \textbf{(2)} \\ -7x - 3y + 4z = 6 & \textbf{(3)} \end{array}$$

Solve **(1)** for x, $x = \dfrac{-3 - y}{2}$ and **(2)** for

z, $z = \dfrac{-18 - 2y}{16}$ and substitute into **(3)**

$$-7\left(\dfrac{-3 - y}{2}\right) - 3y + 4\left(\dfrac{-18 - 2y}{16}\right) = 6$$

$$\dfrac{21}{2} + \dfrac{7y}{2} - 3y - \dfrac{18}{4} - \dfrac{y}{2} = 6$$

$$6 = 6$$

The system of equations is a dependent system and has an infinite number of solutions.

31.
$$3x + 3y - 3z = -1 \quad \textbf{(1)}$$
$$4x + y - 2z = 1 \quad \textbf{(2)}$$
$$-2x + 4y - 2z = -8 \quad \textbf{(3)}$$
Subtract **(3)** from **(2)**
$$6x - 3y = 9 \quad \textbf{(4)}$$
Multiply **(1)** by -2 and add to 3 times **(2)**
$$6x - 3y = 5 \quad \textbf{(5)}$$
Comparing **(4)** and **(5)** gives $5 = 4$ which is false. This is an inconsistent system of equations and has no solution.

Cumulative Review Problems

33. $m = \dfrac{y_2 - y_1}{x_2 - x_1} = \dfrac{4 - 3}{1 - (-2)} = \dfrac{1}{3}$

$$y - y_1 = m(x - x_1)$$

$$y - 4 = \frac{1}{3}(x - 1)$$

$$3y - 12 = x - 1$$

$$x - 3y = -11$$

35. Let $c = $ the number of cattle purchased, $h = $ the number of horses purchased, and $s = $ the number of sheep purchased, then
$$c + 601 = 1.8(h + 346)$$
$$s + 545 = 1.74(h + 346)$$
where h can be any number. Let $h = 0$, which gives
$$c + 601 = 1.8(346) = 622.8$$
$$c = 21.8$$
$$s + 545 = 1.74(346) = 602.04$$
$$s = 57.04$$
The rancher should by no horses, 22 cattle, and 57 sheep. After the purchase he will have 346 horses, 623 cattle and 602 sheep.

4.3 Exercises

1. Let $x = $ number of heavy equipment operators and $y = $ number of laborers.
$$x + y = 35, \ y = 35 - x$$
$$140x + 90y = 3950$$
$$140x + 90(35 - x) = 3950$$
$$140x + 3150 - 90x = 3950$$
$$50x = 800$$
$$x = 16$$
$$y = 35 - x = 35 - 16 = 19$$

16 heavy equipment operators and 19 laborers were employed.

3. $x = $ number of tickets for regular coach seats
$y = $ number of tickets for sleeper car seats
$$x + y = 98 \quad \textbf{(1)}$$
$$120x + 290y = 19,750 \quad \textbf{(2)}$$
Solve **(1)** for y and substitute into **(2)**.
$$y = 98 - x$$
$$120x + 290(98 - x) = 19,750$$
$$120x + 28,420 - 290x = 19,750$$
$$-170x = -8670$$
$$x = 51$$
Substitute 51 for x in **(1)**
$$51 + y = 98$$
$$y = 47$$
Number of coach tickets = 51
Number of sleeper tickets = 47

5. $x = $ number of managers with computer experience
$y = $ number of managers without computer experience

(continued)

5. (continued)

$$2x + 5y = 190, \quad y = \frac{190 - 2x}{5}$$

$$3x + 8y = 295$$

$$3x + 8\left(\frac{190 - 2x}{5}\right) = 295$$

$$15x + 1520 - 16x = 1475$$

$$-x = -45$$

$$x = 45$$

$$y = \frac{190 - 2(45)}{5} = 20$$

The company should train 45 managers with computer experience and 20 managers without computer experience.

7. $x = $ number of packages of old fertilizer
$y = $ number of packages of new fertilizer

$$50x + 65y = 3125, \quad y = \frac{3125 - 50x}{65}$$

$$60x + 45y = 2925$$

$$60x + 45\left(\frac{3125 - 50x}{65}\right) = 2925$$

$$3900x + 140,625 - 2250x = 190,125$$

$$1650x = 49,500$$

$$x = 30$$

$$y = \frac{3125 - 50(30)}{65} = 25$$

The farmer should use 30 packages of the old fertilizer and 25 packages of the new fertilizer.

9. $x = $ cost of one doughnut
$y = $ cost of one large coffee
$$3x + 4y = 4.91 \quad \textbf{(1)} \quad \text{(continued)}$$

9. (continued)
$$5x + 6y = 7.59 \quad\quad \textbf{(2)}$$
Multiply **(1)** by 5 and add to -3 times **(2)**

$$15x + 20y = 24.55$$
$$\underline{-15x - 18y = -22.77}$$
$$2y = 1.78$$
$$y = 0.89$$

Substitute 0.89 for y in **(1)** and solve for x
$$3x + 4(0.89) = 4.91$$
$$3x = 1.35$$
$$x = 0.45$$

The cost of one doughnut is $0.45. The cost of one large coffee is $0.89.

11. $x = $ speed of plane in still air
$y = $ speed of wind

$$(x - y)\left(\frac{7}{6}\right) = 210$$

$$(x + y)\left(\frac{5}{6}\right) = 210$$

$$x - y = 180$$
$$\underline{x + y = 252}$$
$$2x = 432$$
$$x = 216$$
$$216 + y = 252$$
$$y = 36$$
plane: 216 mph; wind: 36 mph

13. $x = $ speed in still air
$y = $ speed of wind
$$\frac{7}{2}(x - y) = 630$$
$$3(x + y) = 630$$
(continued)

13. (continued)

$$x - y = 180$$
$$\underline{x + y = 210}$$
$$2x = 390$$
$$x = 195$$
$$195 + y = 210$$
$$y = 15$$

Speed of plane in still air = 195mph
Speed of wind = 15 mph

15. x = number of free throws
y = number of regular shots

$$x + y = 21, \ y = 21 - x$$
$$x + 2y = 32$$
$$x + 2(21 - x) = 32$$
$$x + 42 - 2x = 32$$
$$-x = -10$$
$$x = 10$$
$$y = 21 - 10 = 11$$

Number of free throws = 10
Number of regular shots = 11

17. x = number of highway miles
y = number of city miles

$$x + y = 432, \ y = 432 - x$$
$$\frac{x}{32} + \frac{y}{24} = 16$$
$$\frac{x}{32} + \frac{432 - x}{24} = 16$$
$$24x + 32(432 - x) = 12,288$$
$$24x + 13,824 - 32x = 12,288$$
$$-8x = -1536$$
$$x = 192$$
$$y = 432 - x = 240$$

Carlos drove 192 highway miles and 240 city miles.

19. x = cost of truck, y = cost of car

$$256x + 183y = 5,791,948,$$
$$y = \frac{5,791,948 - 256x}{183}$$
$$64x + 107y = 2,507,612$$
$$64x + 107\left(\frac{5,791,948 - 256x}{183}\right)$$
$$= 2,507,612$$
$$11,712x + 619,738,436 - 27,392x$$
$$= 458,892,996$$
$$-15,680x = -160,845,440$$
$$x = 10,258$$
$$y = \frac{5,791,948 - 256(10,258)}{183} = 17,300$$

Cars cost \$10,258 and trucks cost \$17,300.

21. x = number of 15-passenger vehicles
y = number of 7-passenger vehicles
z = number of 5-passenger vehicles

$$15x + 7y + 5z = 98 \qquad \textbf{(1)}$$
$$x + y + z = 14 \qquad \textbf{(2)}$$
$$x + y = 6 \qquad \textbf{(3)}$$

Subtract **(3)** from **(2)**
$z = 8$. Substitute $z = 8$ into **(1)**
$$15x + 7y + 5(8) = 98$$
$$15x + 7y = 58 \qquad \textbf{(4)}$$
Substitute $y = 6 - x$ from **(3)** into **(4)**
$$15x + 7(6 - x) = 58$$
$$15x + 42 - 7x = 58$$
$$8x = 16$$
$$x = 2, \ y = 6 - x = 4$$

Number of 15-passenger vehicles is 2, the number of 7-passenger vehicles is 4, and the number of 5-passenger vehicles is 8.

23. $x =$ number of adults
$y =$ number of high school students
$z =$ number of children
$$5x + 3y + 2z = 1010 \qquad \textbf{(1)}$$
$$7x + 4y + 3z = 1390 \qquad \textbf{(2)}$$
$$x + y + z = 300 \qquad \textbf{(3)}$$
Multiply -7 times **(3)** and add to **(2)**
$$-3y - 4z = -710, \; 3y + 4z = 710 \; \textbf{(4)}$$
Multiply -5 times **(3)** and add to **(1)**
$$-2y - 3z = -490, \; 2y + 3z = 490 \; \textbf{(5)}$$
Add $2 \cdot$ **(4)** and $-3 \cdot$ **(5)**
$$-z = -50$$
$$z = 50$$
Substitute $z = 50$ into **(5)**
$$2y + 3(50) = 490$$
$$2y = 340$$
$$y = 170$$
Substitute $y = 170$, $z = 50$ into **(3)**
$$x + 170 + 50 = 300$$
$$x = 80$$

The number of adults attending was 80, the number of high school students was 170, and the number of children was 50.

25. $x =$ number of children under 12
$y =$ number of adults
$z =$ number of senior citizens
$$x + y + z = 12,000 \qquad \textbf{(1)}$$
$$0.25x + y + 0.5z = 10,700 \qquad \textbf{(2)}$$
$$0.35x + 1.5y + 0.5z = 15,820 \qquad \textbf{(3)}$$
Add -0.25 times **(1)** to **(2)**
$$-0.25x - 0.25y - 0.25z = -3000$$
$$\underline{0.25x + y + 0.5z = 10,700}$$
$$0.75y + 0.25z = 7700 \qquad \textbf{(4)}$$
Add -0.35 times **(1)** to **(3)**
(continued)

25. (continued)
$$-0.35x - 0.35y - 0.35z = -4200$$
$$\underline{0.35x + 1.5y + 0.5z = 15,820}$$
$$0.75y + 0.25z = 7700 \qquad \textbf{(5)}$$
Add -1.15 times **(4)** to 0.75 times **(5)**
$$-.08626y - 0.2875z = -8855$$
$$\underline{0.8625y + 0.1125z = 8715}$$
$$-0.175z = -140$$
$$z = 800$$
Substitute 800 for z in **(1)**
$$x + y = 11,200$$
Add -1 times **(3)** to **(2)**
$$0.25x + y + 0.5z = 10,700$$
$$\underline{-0.35x - 1.5y - 0.5z = -15,820}$$
$$-0.1x - 0.5y = -5120 \qquad \textbf{(7)}$$
Add 0.5 times **(6)** to **(7)**
$$0.5x + 0.5y = 5600$$
$$\underline{-0.1x - 0.5y = -5120}$$
$$0.4x = 480$$
$$x = 1200$$
Substitute 1200 for x in **(6)**
$$1200 + y = 11,200$$
$$y = 10,000$$

children under 12 $= 1200$
adults $= 10,000$
senior citizens $= 800$

27. $x =$ number of medium sandwiches
$y =$ number of large sandwiches
$z =$ number of extra large sandwiches
$$x + y + z = 24 \qquad \textbf{(1)}$$
$$6x + 10y + 14z = 15.5(16) = 248 \qquad \textbf{(2)}$$
$$2.5x + 3y + 4.5z = 82 \qquad \textbf{(3)}$$
Add -6 times (1) to (2) and -2.5 times (1) to (3)
(continued)

27. (continued)

$$x + y + z = 24 \qquad \textbf{(1)}$$
$$4y + 8z = 104 \qquad \textbf{(2)}$$
$$0.5y + 2z = 22 \qquad \textbf{(3)}$$

Add $-\dfrac{1}{8}$ times **(2)** to **(3)**

$$x + y + z = 24 \qquad \textbf{(1)}$$
$$4y + 8z = 104 \qquad \textbf{(2)}$$
$$z = 9 \qquad \textbf{(3)}$$

Substitute $z = 9$ back into **(2)**

$$4y + 8(9) = 104$$
$$4y = 32$$
$$y = 8$$

Substitute $y = 8$, $z = 9$ back into **(1)**

$$x + 8 + 9 = 24$$
$$x = 7$$

Nick sold 7 medium sandwiches, 8 large sandwiches, and 9 extra large sandwiches.

29. $x =$ number of Box A
$y =$ number of Box B
$z =$ number of Box C

$$10x + 5y + 4z = 51 \qquad \textbf{(1)}$$
$$3x + 2y + z = 16 \qquad \textbf{(2)}$$
$$3x + 3y + 2z = 23 \qquad \textbf{(3)}$$

Add $-\dfrac{3}{10}$ times **(1)** to **(2)** and **(3)**

$$10x + 5y + 4z = 51 \qquad \textbf{(1)}$$
$$0.5y - 0.2z = 0.7 \qquad \textbf{(2)}$$
$$1.5y + 0.8z = 7.7 \qquad \textbf{(3)}$$

Add -3 times **(2)** to **(3)**

$$10x + 5y + 4z = 51 \qquad \textbf{(1)}$$
$$0.5y - 0.2z = 0.7 \qquad \textbf{(2)}$$
$$1.4z = 5.6 \qquad \textbf{(3)}$$
$$z = 4$$

Substitute $z = 4$ back into **(2)**

(continued)

29. (continued)

$$0.5y - 0.2(4) = 0.7 \qquad \textbf{(2)}$$
$$0.5y = 1.5$$
$$y = 3$$

Substitute $z = 4$, $y = 3$ back into **(1)**

$$10x + 5(3) + 4(4) = 51$$
$$10x = 20$$
$$x = 2$$

The shipping manager can prepare 2 of the A Boxes, 3 of the B Boxes, and 4 of the C Boxes.

31. $x =$ measure of first angle
$y =$ measure of second angle
$z =$ measure of second angle

$$x + y + z = 180$$
$$x + y = \frac{1}{2}z$$
$$8x = y + z$$

Clear fractions and write variables on LHS

$$x + y + z = 180 \qquad \textbf{(1)}$$
$$2x + 2y - z = 0 \qquad \textbf{(2)}$$
$$8x - y - z = 0 \qquad \textbf{(3)}$$

Add **(1)** and **(2)**

$$3x + 3y = 180$$
$$x + y = 60 \qquad \textbf{(4)}$$

Add **(1)** and **(3)**

$$9x = 180$$
$$x = 20$$

Substitute $x = 20$ into **(4)**

$$20 + y = 60$$
$$y = 40$$

Substitute $x = 20$, $y = 40$ into **(1)**

$$20 + 40 + z = 180, \; z = 120$$

The angles are $20°$, $40°$, $120°$.

33. x, y, z, q = number of A, B, C, and D packages respectively

$$42x \quad +20y \quad +0z \quad +10q \quad =134$$
$$20x \quad +10y \quad +20z \quad +0q \quad =150$$
$$34x \quad +0y \quad +10z \quad +20q \quad =178$$
$$50x \quad +35y \quad +30z \quad +40q \quad =405$$

Add $-\dfrac{20}{42} \times$ first eqn. to second eqn.

Add $-\dfrac{30}{42} \times$ first eqn. to third eqn.

Add $-\dfrac{50}{42} \times$ first eqn. to fourth eqn.

$$42x \quad +20y \quad +0z \quad +10q \quad =134$$
$$0x \quad +\dfrac{10}{21}y \quad +20z \quad -\dfrac{100}{21}q \quad =\dfrac{1810}{21}$$
$$0x \quad -\dfrac{340}{21}y \quad +10z \quad +\dfrac{250}{21}q \quad =\dfrac{1460}{21}$$
$$0x \quad +\dfrac{235}{21}y \quad +30z \quad +\dfrac{590}{21}q \quad =\dfrac{5155}{21}$$

Add $34 \times$ second eqn. to third eqn.
Add $-23.5 \times$ second eqn. to fourth eqn.

$$42x \quad +20y \quad +0z \quad +10q \quad =134$$
$$0x \quad +\dfrac{10}{21}y \quad +20z \quad -\dfrac{100}{21}q \quad =\dfrac{1810}{21}$$
$$0x \quad +0y \quad +690z \quad -150q \quad =3000$$
$$0x \quad +0y \quad -440z \quad +140q \quad =-1780$$

Add $\dfrac{44}{69} \times$ third eqn. to fourth eqn.

(continued)

33. (continued)

$$42x \quad +20y \quad +0z \quad +10q \quad =134$$
$$0x \quad +\dfrac{10}{21}y \quad +20z \quad -\dfrac{100}{21}q \quad =\dfrac{1810}{21}$$
$$0x \quad +0y \quad +690z \quad -150q \quad =3000$$
$$0x \quad +0y \quad +0z \quad +\dfrac{1023}{23}q \quad =\dfrac{3060}{23}$$

$\dfrac{1020}{23}q = \dfrac{3060}{23}$, $q=3$

Substitute $q=3$ into third equation
$690z - 150(3) = 3000$, $z=5$
Substitute $z=5$, $q=3$ into second eqn
$\dfrac{10}{21}y + 20(5) - \dfrac{10}{21}(3) = \dfrac{1810}{21}$, $y=1$
Substitute $y=1$, $z=5$, $q=3$ into the first equation
$42x + 20(5) + 0(5) + 10(3) = 134$
$x=2$

The scientist should use 2 A packages, 1 B package, 5 C packages, and 3 D packages.

Cumulative Review Problems

35. $0.06x + 0.15(0.5 - x) = 0.04$
$0.06x + 0.075 - 0.15x = 0.04$
$-0.09x = -0.035$
$x = \dfrac{7}{18}$

4.4 Exercises

1. $y \geq 2x - 1$
Test point: $(0, 0)$
$0 \geq 2(0) - 1$
$0 \geq -1$ True

(continued)

1. (continued)

$x + y \leq 6$

Test point: (0,0)

$0 + 0 \leq 6$

$0 \leq 6$ True

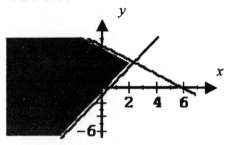

3. $y \geq -4x$

Test Point: (0,2)

$2 \geq -4(0)$

$2 \geq 0$ True

$y \geq 3x - 2$

Test point: (0,0)

$0 \geq 3(0) - 2$

$0 \geq -2$ True

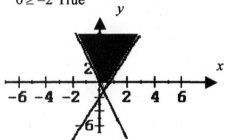

5. $y \geq 2x - 3$

Test point: (0,0)

$0 \geq 2(0) - 3$

$0 \geq -3$ True

$y \leq \frac{2}{3}x$

(continued)

5. (continued)

Test point: (0,2)

$2 \leq \frac{2}{3}(0)$

$2 \leq 0$ False

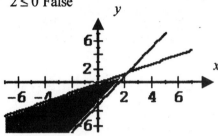

7. $x - y \geq 1$

Test point: (0,0)

$0 - 0 \geq -1$

$0 \geq -1$ True

$-3x - y \leq 4$

Test point: (0,0)

$-3(0) - 0 \leq 4$

$0 \leq 4$ True

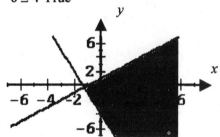

9. $x + 2y < 6$

Test point: (0,0)

$0 + 2(0) < 6$

$0 < 6$ True

$y < 3$

Test point: (0,0)

$0 < 3$ True

(continued)

9. (continued)

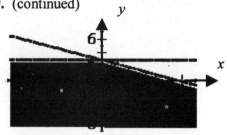

11. $y < 4,\ x > -2$

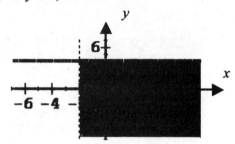

13. $3x + 2y < 6$

Test point: $(0,0)$

$3(0) + 2(0) < 6$

$0 < 6$ True

$3x + 2y > -6$

Test point: $(0,0)$

$3(0) + 2(0) > -6$

$0 > -6$ True

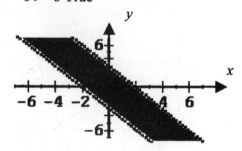

15. $x - 4y \geq 4$

Test point: $(0,0)$

$0 - 4(0) \geq 4$

$0 \geq 4$ False

$3x + y \leq 3$

Test point: $(0,0)$

$3(0) + 0 \leq 3$

$0 \leq 3$ True

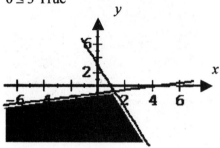

17. $x + y \leq 5$

Test point: $(0,0)$

$0 + 0 \leq 5$

$0 \leq 5$ True

$2x - y \geq 1$

Test point: $(0,0)$

$2(0) - 0 \geq 1$

$0 \geq 1$ False

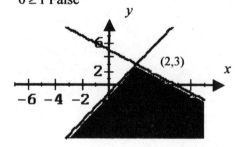

19. $x + 3y \leq 12$

Test point: $(0,0)$

$0 + 3(0) \leq 12$

$0 \leq 12$ True

$y < x$

Test point: $(2,0)$

$0 < 2$ True

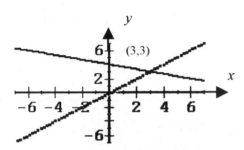

21. $x + y \geq 1$

Test point: $(0,0)$

$0 + 0 \geq 1$

$0 \geq 1$ False

$x - y \geq 1$

Test point: $(0,0)$

$0 - 0 \geq 1$

$0 \geq 1$ False

$x \geq 3$

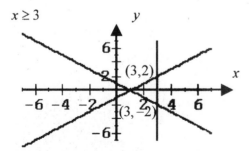

23. $y \leq 3x + 6$

Test point: $(0,0)$

$0 \leq 3(0) + 6$

$0 \leq 6$ True

(continued)

23. (continued)

$4x + 3y \leq 3$

Test point: $(0,0)$

$4(0) + 3(0) \leq 3$

$0 \leq 3$ True

$x \geq -2, \; y \geq -3$

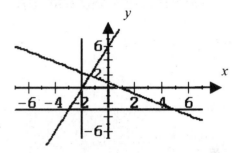

25. (a)

$N \leq 2D$

Test point: $(2,2)$

$2 \leq 2(2)$

$2 \leq 4$ True

$4N + 3D \leq 20$

Test point: $(2,2)$

$4(2) + 3(2) \leq 20$

$8 + 6 \leq 20$

$14 \leq 20,$ True

$N \geq 0, \; D \geq 0$

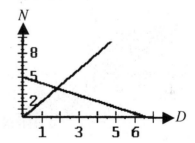

(b) Yes, $(3,2)$ is in the shaded region.

(c) No, $(1,4)$ is not in the shaded region.

Cumulative Review

27. $m = \dfrac{y_2 - y_1}{x_2 - x_1} = \dfrac{-4 - (-2)}{3 - (-1)} = -\dfrac{1}{2}$

29. $m = \dfrac{y_2 - y_1}{x_2 - x_1} = \dfrac{6 - 1}{2 - (-2)} = \dfrac{5}{4}$

$y - y_1 = m(x - x_1)$

$y - 6 = \dfrac{5}{4}(x - 2)$

$4y - 24 = 5x - 10$

$5x - 4y = -14$

31. $x =$ money taken in or rainy day
$y =$ money taken in on sunny day

$2x + 5y = 23,400 \Rightarrow y = \dfrac{23,400 - 2x}{5}$

$4x + 3y = 25,800$

$4x + 3\left(\dfrac{23,400 - 2x}{5}\right) = 25,800$

$20x + 70,200 - 6x = 129,000$

$14x = 58,800$

$x = 4200$

$y = \dfrac{23,400 - 2x}{5} = \dfrac{23,400 - 2(4200)}{5}$

$y = 3000$

The Cape Cod Cinema takes in $4200 on a rainy day and $3000 on a sunny day.

33. $x =$ sales, $y =$ number of weeks

Hector: $0.05x + 200y = 7400$

$y = \dfrac{7400 - 0.05x}{200}$

Fernando: $0.08x + 100y = 9200$

(continued)

31. (continued)

$0.08x + 100\dfrac{7400 - 0.05x}{200} = 9200$

$16x + 740,000 - 5x = 1,840,000$

$11x = 1,100,000$

$x = 100,000$

$y = \dfrac{7400 - 0.05x}{200}$

$y = \dfrac{7400 - 0.05(100,000)}{200} = 12$

They had each worked 12 weeks and had $100,000 in sales.

Chapter 4 Review Problems

1. $x + 2y = 8$
$x - y = 2$

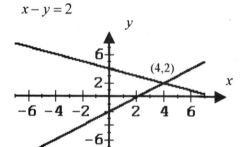

2. $x + y = 2$
$3x - y = 6$

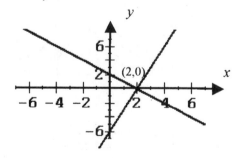

The solution is (2,0).

3. $2x + y = 6$
$3x + 4y = 4$

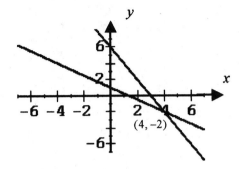

The solution is $(4, -2)$.

4. $3x - 2y = -9$, $y = \dfrac{9 + 3x}{2}$

$2x + y = 1$

$2x + \dfrac{9 + 3x}{2} = 1$

$4x + 9 + 3x = 2$

$7x = -7$

$x = -1$

$y = \dfrac{9 + 3x}{2}$

$y = \dfrac{9 + 3(-1)}{2}$

$y = 3$

5. $-6x - y = 1$, $y = -6x - 1$

$3x - 4y = 31$

$3x - 4(-6x - 1) = 31$

$3x + 24x + 4 = 31$

$27x = 27$

$x = 1$

$y = -6x - 1$

$y = -6(1) - 1$

$y = -7$

6. $4x + 3y = 10$, $y = \dfrac{10 - 4x}{3}$

$5x - y = 3$

$5x - \dfrac{10 - 4x}{3} = 3$

$15x - 10 + 4x = 9$

$19x = 19$, $x = 1$

$y = \dfrac{10 - 4x}{3} = \dfrac{10 - 4(1)}{3} = 2$

7. $-7x + y = -4$, $y = 7x - 4$

$5x + 2y = 11$

$5x + 2(7x - 4) = 11$

$5x + 14x - 8 = 11$

$19x = 19$

$x = 1$

$y = 7x - 4 = 7(1) - 4 = 3$

8. $-2x + 5y = -12$, $-6x + 15y = -36$

$3x + y = 1$, $\underline{6x + 2y = 2}$

                        $17y = -34$

                         $y = -2$

$3x + y = 1$

$3x - 2 = 1$

$3x = 3$

$x = 3$

9. $-4x + 2y = -16$, $-12x + 6y = -48$

$3x + 5y = -1$, $\underline{12x + 20y = -4}$

                        $26y = -52$

                        $y = -2$

$3x + 5y = -1$

$3x + 5(-2) = -1$

$3x = -9$

$x = -3$

10. $7x-4y=2,$ $-35x+20y=-10$

    $6x-5y=-3,$ $\underline{24x-20y=-12}$

$$-11x=-22$$
$$x=2$$

$6x-5y=-3$

$6(2)-5y=-3$

$-5y=-15$

$y=3$

11. $\quad 3x+4y=-6,$ $6x+8y=-12$

  $-2x+7y=-25,$ $\underline{-6x+21y=-75}$

$$29y=-87$$
$$y=-3$$

$3x+4y=-6$

$3x+4(-3)=-6$

$3x=6$

$x=2$

12. $2x+4y=9,\ y=\dfrac{9-2x}{4}$

$3x+6y=8$

$3x+6\dfrac{9-2x}{4}=8$

$12x+54-12x=32$

$54=32$

No solution. Inconsistent system.

13. $x+5y=10,\ x=10-5y$

$y=2-\dfrac{1}{5}x=2-\dfrac{1}{5}(10-5y)$

$5y=10-10+5y$

$0=0$

Infinite number of solutions.
Dependent system of equations.

14. $7x+6y=-10$

$2x+y=0,\ y=-2x$

$7x+6(-2x)=-10$

$7x-12x=-10$

$-5x=-10$

$x=2$

$y=-2x=-2(2)=-4$

15. $7x+6y=-10,\ y=\dfrac{-10-7x}{6}$

$2x+y=0$

$2x+\dfrac{-10-7x}{6}=0$

$2x-10-7x=0$

$-5x=10$

$x=-2$

$2x+y=0$

$2(-2)+y=0$

$y=4$

16. $x+\dfrac{1}{3}y=1,\ x=1-\dfrac{1}{3}y$

$\dfrac{1}{4}x-\dfrac{3}{4}y=-\dfrac{9}{4}$

$\dfrac{1}{4}\left(1-\dfrac{1}{3}y\right)-\dfrac{3}{4}y=-\dfrac{9}{4}$

$1-\dfrac{1}{3}y-3y=-9$

$-\dfrac{10}{3}y=-10$

$y=3$

$x=1-\dfrac{1}{3}y=1-\dfrac{1}{3}(3)=1-1$

$x=0$

17. $\dfrac{2}{3}x + y = \dfrac{14}{3}$

$\underline{\dfrac{2}{3}x - y = -\dfrac{22}{3}}$

$\dfrac{4}{3}x = -\dfrac{8}{3}$

$x = -2$

$\dfrac{2}{3}x + y = \dfrac{14}{3}$

$\dfrac{2}{3}(-2) + y = \dfrac{14}{3}$

$-4 + 3y = 14$

$3y = 18$

$y = 6$

18. $9a + 10b = 7, \quad 36a + 40b = 28$

$6a - 4b = 10, \quad \underline{60a - 40b = 100}$

$96a = 128$

$a = \dfrac{4}{3}$

$9a + 10b = 7$

$9\left(\dfrac{4}{3}\right) + 10b = 7$

$12 + 10b = 7$

$10b = -5$

$b = -\dfrac{1}{2}$

19. $3a + 5b = 8, \quad 6a + 10b = 16$

$2a + 4b = 3, \quad \underline{-6a - 12b = -9}$

$-2b = -7$

$b = -\dfrac{7}{2}$

(continued)

19. (continued)

$3a + 5b = 8$

$3a + 5\left(-\dfrac{7}{2}\right) = 8$

$6a - 35 = 16$

$6a = 51$

$a = \dfrac{17}{2}$

20. $x + 3 = 3y + 1, \quad x = 3y - 2$

$1 - 2(x - 2) = 6y + 1$

$1 - 2(3y - 2 - 2) = 6y + 1$

$1 - 6y + 8 = 6y + 1$

$-12y = -8$

$y = \dfrac{2}{3}$

$x = 3y - 2 = 3\left(\dfrac{2}{3}\right) - 2 = 2 - 2$

$x = 0$

21. $10(x + 1) - 13 = -8y, \quad y = \dfrac{10x - 3}{-8}$

$4(2 - y) = 5(x + 1)$

$8 - 4y = 5x + 4$

$8 - 4\left(\dfrac{10x - 3}{-8}\right) = 5x + 4$

$16 + 10x - 3 = 10x + 8$

$13 = 8$

No solution. Inconsistent system of equations.

22. $0.3x - 0.2y = 0.7, \quad 0.6x - 0.4y = 1.4$

$-0.6x + 0.4y = 0.3, \quad \underline{-0.6x + 0.4y = 0.3}$

$0 = 1.7$

No solution. Inconsistent system of equations.

23. $0.2x - 0.1y = 0.8,$ $0.6x - 0.3y = 2.4$
$0.1x + 0.3y = 1.1,$ $\underline{0.1x + 0.3y = 1.1}$
$0.7x = 3.5$
$x = 5$

$0.2x - 0.1y = 0.8$
$0.2(5) - 0.1y = 0.8$
$-0.1y = -0.2$
$y = 2$

24. $3x - 2y - z = 3$ (1)
$2x + y + z = 1$ (2)
$-x - y + z = -4$ (3)
Add (1) and (2)
$5x - y = 4$ (4)
Add (1) and (3)
$2x - 3y = -1$ (5)

Substitute y from (4) into (5)
$2x - 3(5x - 4) = -1$
$2x - 15x + 12 = -1$
$-13x = -13$
$x = 1$
Substitute into (4)
$5(1) - y = 4$
$-y = -1$
$y = 1$

Substitute $x = 1$, $y = 1$ into (3)
$-1 - 1 + z = -4$
$z = -2$

$x = 1$, $y = 1$, $z = -2$ is the solution to the system of equations.

25. $-2x + y - z = -7$ (1)
$x - 2y - x = 2$ (2)
$6x + 4y + 2z = 4$ (3)

$\frac{1}{2} \times (1) + (2),$

$3 \times (1) + (3),$

$-2x + y - z = -7$ (1)

$0x - \frac{3}{2}y - \frac{3}{2}z = -\frac{3}{2}$ (2)

$0x + 7y - z = -17$ (3)

$\frac{14}{3} \times (2) + 3,$

$-2x + y - z = -7$ (1)

$0x - \frac{3}{2}y - \frac{3}{2}z = -\frac{3}{2}$ (2)

$0x + 0y - 8z = -24$ (3)

$z = 3$, from (2)

$0x - \frac{3}{2}y - \frac{3}{2}(3) = -\frac{3}{2}$ (2)

$3y + 9 = 3$
$3y = -6$
$y = -2$, from (1)

$-2x + y - z = -7$ (1)

$-2x + (-2) - 3 = -7$
$-2x - 5 = -7$
$-2x = -2$
$x = 1$

$x = 1$, $y = -2$, $z = 3$ is the solution to the system of equations.

26.
$$2x + 5y + 3z = 10 \qquad \textbf{(1)}$$
$$x + y + 5z = 42 \qquad \textbf{(2)}$$
$$2x + y = 7 \qquad \textbf{(3)}$$

Solve **(1)** for z
$z = 3 - 2x - 5y$ and substitute into **(2)**
$$x + y + 5(3 - 2x - 5y) = 42$$
$$x + y + 15 - 10x - 25y = 42$$
$$-9x - 24y = 27 \qquad \textbf{(4)}$$
Solve **(3)** for y, $y = 7 - 2x$ and
substitute into **(4)**
$$-9x - 24(7 - 2x) = 27$$
$$-9x - 168 + 48x = 27$$
$$39x = 195$$
$$x = 5$$
Substitute $x = 5$ into $y = 7 - 2x$
$$y = 7 - 2(5) = 7 - 10 = -3$$
$$y = -3$$
Substitute $x = 5$, $y = -3$ into
$z = 3 - 2x - 5y$
$$z = 3 - 2(5) - 5(-3) = 8$$
$$z = 8$$

$x = 5$, $y = -3$, $z = 8$ is the solution.

27.
$$x + 2y + z = 5 \qquad \textbf{(1)}$$
$$3x - 8y = 17 \qquad \textbf{(2)}$$
$$2y + z = -2 \qquad \textbf{(3)}$$
Solve **(1)** for x, $x = 5 - 2y - z$ and
substitute into **(2)**
$$3(5 - 2y - z) - 8y = 17$$
$$15 - 6y - 3z - 8y = 17$$
$$3z = -14y - 2$$
$$z = \frac{-14y - 2}{3}.$$

Substitute this into **(3)**

(continued)

27. (continued)
$$2y + \frac{-14y - 2}{3} = -2$$
$$6y - 14y - 2 = -6$$
$$-8y = -4$$
$$y = \frac{1}{2}$$
Substitute $y = \frac{1}{2}$ into **(3)**
$$2\left(\frac{1}{2}\right) + z = -2$$
$$1 + z = -2$$
$$z = -3$$
Substitute $y = \frac{1}{2}$ into **(2)**
$$3x - 8\left(\frac{1}{2}\right) = 17$$
$$3x - 4 = 17$$
$$3x = 21$$
$$x = 7$$
$x = 7$, $y = \frac{1}{2}$, $z = -3$ is the solution.

28.
$$2x - 4y + 3z = 0 \qquad \textbf{(1)}$$
$$x - 2y - 5z = 13 \qquad \textbf{(2)}$$
$$5x + 3y - 2z = 19 \qquad \textbf{(3)}$$

Multiply **(2)** by -2 and add to **(1)**
$$13z = -26$$
$$z = -2$$

Multiply **(2)** by -5 and add to **(3)**
$$13y + 23z = -46$$
$$13y + 23(-2) = -46$$
$$13y = 0$$
$$y = 0$$

(continued)

28. (continued)

Substitute $y = 0$, $z = -2$ into **(2)**

$x - 2(0) - 5(-2) = 13$

$x + 10 = 13$

$x = 3$

$x = 3$, $y = 0$, $z = -2$ is the solution.

29. $5x + 2y + 3z = 10$ **(1)**

$6x - 3y + 4z = 24$ **(2)**

$-2x + y + 2z = 2$ **(3)**

Add 3 times **(3)** to **(2)**

$10z = 30$

$z = 3$

Add -2 times **(3)** to **(1)**

$9x - z = 6$

$9x - (3) = 6$

$9x = 9$

$x = 1$

Substitute $x = 1$, $z = 3$ into **(3)**

$-2(1) + y + 2(3) = 2$

$-2 + y + 6 = 2$

$y = -2$

$x = 1$, $y = -2$, $z = 3$ is the solution.

30. $3x + 2y = 7$ **(1)**

$2x + 7z = -26$ **(2)**

$5y + z = 6$ **(3)**

Solve **(3)** for z and substitute into **(2)**

$2x + 7(6 - 5y) = -26$

$2x + 42 - 35y = -26$

$2x = 35y - 68$

$x = \dfrac{35y - 68}{2}$

Substitute this into **(1)**

$3\left(\dfrac{35y - 68}{2}\right) + 2y = 7$

(continued)

30. (continued)

$3\left(\dfrac{35y - 68}{2}\right) + 2y = 7$

$105y - 204 + 4y = 14$

$109y = 218$

$y = 2$

Substitute $y = 2$ into **(3)**

$5(2) + z = 6$

$10 + z = 6$

$z = -4$

Substitute $y = 2$ into **(1)**

$3x + 2(2) = 7$

$3x + 4 = 7$

$3x = 3$

$x = 1$

$x = 1$, $y = 2$, $z = -4$ is the solution.

31. $x - y = 2$ **(1)**

$5x + 7y - 5z = 2$ **(2)**

$3x - 5y + 2z = -2$ **(3)**

Add 2 times **(2)** to 5 times **(3)**

$25x - 11y = -6$ **(4)**

Solve **(1)** for x and substitute into **(4)**

$x = 2 + y$

$25(2 + y) - 11y = -6$

$50 + 25y - 11y = -6$

$14y = -56$

$y = -4$

Substitute $y = -4$ into **(1)**

$x - (-4) = 2$

$x + 4 = 2$

$x = -2$

Substitute $x = -2$, $y = -4$ into **(3)**

(continued)

31. (continued)

$3(-2)-5(-4)+2z=-2$

$-6+20+2z=-2$

$2z=-16$

$z=-8$

$x=-2,\ y=-4,\ z=-8$ is the solution.

32. $v=$ speed of plane in still air

$w=$ speed of wind

$720=(v-w)\cdot 3 \qquad v-w=240$

$720=(v+w)(2.5) \qquad \underline{v+w=288}$

$\qquad\qquad\qquad\qquad\quad 2v=528$

$\qquad\qquad\qquad\qquad\quad\ \ v=264$

$w=288-v=288-264=24$

The speed of the plane in still air is 264 mph and the wind speed is 24 mph.

33. $x=$ number of new employees

$y=$ number of laid-off employees

$25x+10y=275 \qquad$ **(1)**

$8x+3y=86 \qquad$ **(2)**

Add 3 times **(1)** to -10 times **(2)**

$75x+30y=825$

$\underline{-80x-30y=-860}$

$-5x=-35$

$x=7$

Substitute $x=7$ into **(1)**

$25(7)+10y=275$

$10y=100$

$y=10$

The company can train 7 new employees and 10 laid-off employees.

34. $x=$ number of laborers

$y=$ number of mechanics

$70x+90y=1950 \qquad$ **(1)**

(continued)

34. (continued)

$80x+100y=2200 \qquad$ **(2)**

Divide both equations by 10

$7x+9y=195 \qquad$ **(1)**

$8x+10y=220 \qquad$ **(2)**

Add -8 times **(1)** to 7 times **(2)**

$-2y=-20$

$y=10$

Substitute $y=10$ into **(2)**

$8x+10(10)=220$

$8x=120$

$x=15$

The circus hired 15 laborers and 10 mechanics.

35. $x=$ number of children's tickets

$y=$ number of adult tickets

$x+y=590 \qquad$ **(1)**

$6x+11y=4790 \qquad$ **(2)**

Solve **(1)** for y substitute into **(2)**

$6x+11(590-x)=4790$

$6x+6490-11x=4790$

$-5x=-1700$

$x=340$

$y=590-x=590-340=250$

There were 340 children's tickets sold and 250 adult tickets.

36. $x=$ cost of hat

$y=$ cost of shirt

$z=$ cost of pants

$2x+5y+4z=129 \qquad$ **(1)**

$x+y+2z=42 \qquad$ **(2)**

$2x+3y+z=63 \qquad$ **(3)**

Add -2 times **(2)** and **(1)**

$3y=45$

$y=15$

(continued)

36. (continued)

Substitute $y = 15$ into **(2)** and solve for x

$x + 15 + 2z = 42$

$x = 27 - 2z$

Substitute this and $y = 15$ into **(3)**

$2(27 - 2z) + 3(15) + z = 63$

$54 - 4z + 45 + z = 63$

$-3z = -36$

$z = 12$

Substitute $y = 15$, $z = 12$ into **(2)**

$x + 15 + 2(12) = 42$

$x + 15 + 24 = 42$

$x = 3$

The hats cost $3, shirts $15, and pants $12.

37. $x = $ number of A packets

$y = $ number of B packets

$z = $ number of C packets

$2x + 3y + 4z = 29$ **(1)**

$4x + y + 3z = 23$ **(2)**

$3x + y + 2z = 17$ **(3)**

Add 3 times **(1)** to -4 times **(2)**

$-10x + 5y = -5$ **(4)**

Add 2 times **(2)** to -3 times **(3)**

$-x - y = -5$ **(5)**

Solve **(5)** for y and substitute into **(4)**

$-10x + 5(-x + 5) = -5$

$-10x - 5x + 25 = -5$

$-15x = -30$

$x = 2$

Substitute $x = 2$ into **(5)**

$-2 - y = -5$

$-y = -3$

$y = 3$

(continued)

37. (continued)

Substitute $x = 2$, $y = 3$ into **(1)**

$2(2) + 3(3) + 4z = 29$

$4 + 9 + 4z = 29$

$4z = 16$

$z = 4$

She should use 2 of the A packets, 3 of the B packets, and 4 of the C packets.

38. $x = $ cost of jelly

$y = $ cost of peanut butter

$z = $ cost of honey

$4x + 3y + 5z = 9.8$ **(1)**

$2x + 2y + z = 4.20$ **(2)**

$3x + 4y + 2z = 7.70$ **(3)**

Add **(1)** and -5 times **(2)**

$-6x - 7y = -11.2$ **(4)**

Add -2 times **(2)** to **(3)**

$-x = -0.7$

$x = 0.7$

Substitute $x = 0.7$ into   **(4)**

$-6(0.7) - 7y = -11.2$

$-7y = -7$

$y = 1$

Substitute $x = 0.7$, $y = 1$ into **(2)**

$2(0.7) + 2(1) + z = 4.2$

$z = 0.8$

The cost of a jar of jelly is $0.70, the cost of a jar of peanut butter is $1, and the cost of a jar of honey is $0.80.

39. $x = $ number of buses

$y = $ number of station wagons

$z = $ number of sedans

$x + y + z = 9$ **(1)**

$40x + 8y + 5z = 127$ **(2)**

(continued)

108

39. (continued)

$8(3y) + 5(2z) = 126$

$24y + 10z = 126$ **(3)**

Add -40 times **(1)** to **(2)**

$-32y - 35z = -233$ **(4)**

Add 32 times **(3)** to 24 times **(4)**

$-520z = -1560$

$z = 3$

Substitute $z = 3$ into **(3)**

$24y + 10(3) = 126$

$24y = 96$

$y = 3$

Substitute $y = 4$, $z = 3$ into **(1)**

$x + 4 + 3 = 9$

$x = 2$

They should use 2 buses, 4 station wagons, and 3 sedans.

40. $-x - 5z = -5$ **(1)**

$13x + 2z = 2$ **(2)**

Solve **(1)** for x and substitute into **(2)**

$x = 5 - 5z$

$13(5 - 5z) + 2z = 2$

$65 - 65z + 2z = 2$

$-63z = -63$

$z = 1$

Substitute $z = 1$ into **(1)**

$-x - 5(1) = -5$

$-x = 0$

$x = 0$

The solution is $x = 0$, $z = 1$.

41. $x + y = 10$ **(1)**

$6x + 9y = 70$ **(2)**

Solve **(1)** for y and substitute into **(2)**

(continued)

41. (continued)

$y = 10 - x$

$6x + 9(10 - x) = 70$

$6x + 90 - 9x = 70$

$-3x = -20$

$x = \dfrac{20}{3}$

Substitute $x = \dfrac{20}{3}$ into **(1)**

$\dfrac{20}{3} + y = 10$

$y = \dfrac{10}{3}$

The solution is $x = \dfrac{20}{3}$, $y = \dfrac{10}{3}$.

42. $2x + 5y = 4$ **(1)**

$5x - 7y = 4$ **(2)**

Solve **(1)** for y and substitute into **(1)**

$y = \dfrac{4 - 2x}{5}$

$5x - 7\dfrac{4 - 2x}{5} = -29$

$25x - 28 + 14x = -145$

$39x = -117$

$x = -3$

Substitute $x = -3$ into **(1)**

$2(-3) + 5y = 4$

$-6 + 5y = 4$

$5y = 10$

$y = 2$

The solution is $x = -3$, $y = 2$.

43. $\dfrac{x}{2} - 3y = -6$ **(1)**

(continued)

43. (continued)

$$\frac{4}{3}x + 2y = 4 \quad \textbf{(2)}$$

Solve **(1)** for x and substitute into **(2)**

$$x = 6y - 12$$

$$\frac{4}{3}(6y - 12) + 2y = 4$$

$$24y - 48 + 6y = 12$$

$$30y = 60$$

$$y = 2$$

$$x = 6y - 12 = 6(2) - 12 = 0$$

The solution is $x = 0,\ y = 2$.

44. $\dfrac{3}{5}x - y = 6 \quad \textbf{(1)}$

$$x + \frac{y}{3} = 10 \quad \textbf{(2)}$$

Solve **(1)** for y and substitute into **(2)**

$$y = \frac{3}{5}x - 6$$

$$x + \frac{\frac{3}{5}x - 6}{5} = 10$$

$$5x + \frac{3}{5}x - 6 = 50$$

$$25x + 3x - 30 = 250$$

$$28x = 280$$

$$x = 10$$

Substitute $x = 10$ into **(2)**

$$10 + \frac{y}{3} = 10$$

$$y = 0$$

The solution is $x = 10,\ y = 0$.

45. $\dfrac{x+1}{5} = y + 2 \quad \textbf{(1)}$

(continued)

45. (continued)

$$\frac{2y + 7}{3} = x - y \quad \textbf{(2)}$$

Solve **(1)** for y and substitute into **(2)**

$$y = \frac{x+1}{5} - 2$$

$$\frac{2\left(\dfrac{x+1}{5} - 2\right) + 7}{3} = x - \frac{x+1}{5} + 2$$

$$2\left(\frac{x+1}{5} - 2\right) + 7 = 3x - \frac{3x+3}{5} + 6$$

$$2(x + 1 - 10) + 35 = 15x - 3x - 3 + 30$$

$$2x - 18 + 35 = 12x + 27$$

$$10x = -10$$

$$x = -1$$

$$y = \frac{x+1}{5} - 2 = \frac{-1+1}{5} - 2 = -2$$

The solution is $x = -1,\ y = -2$.

46. $3(2 + x) = y + 1$

$$6 + 3x = y + 1$$

$$y = 3x + 5 \quad \textbf{(1)}$$

$$5(x - y) = -7 - 3y$$

$$5x - 5y = -7 - 3y$$

$$5x - 2y = -7 \quad \textbf{(2)}$$

Substitute y from **(1)** into **(2)**

$$5x - 2(3x + 5) = -7$$

$$5x - 6x - 10 = -7$$

$$-x = 3$$

$$x = -3$$

$$y = 3x + 5 = 3(-3) + 5 = -9 + 5 = -4$$

The solution is $x = -3,\ y = -4$.

47. $7(x + 3) = 2y + 25$

$$7x + 21 = 2y + 25$$

(continued)

47. (continued)

$7x - 2y = 4$ **(1)**

$3(x - 6) = -2(y + 1)$

$3x - 18 = -2y - 2$

$3x + 2y = 16$ **(2)**

Add **(1)** and **(2)**

$10x = 20$

$x = 2$

Substitute $x = 2$ into **(2)**

$3(2) + 2y = 16$

$2y = 10$

$y = 5$

The solution is $x = 2$, $y = 5$.

48. Multiply both equations by 10 to clear decimals.

$3x - 4y = 9$ **(1)**

$2x - 3y = 4$ **(2)**

Solve **(1)** for y and substitute into **(2)**

$y = \dfrac{3x - 9}{4}$

$2x - 3\dfrac{3x - 9}{4} = 4$

$8x - 9x + 27 = 16$

$-x = -11$

$x = 11$

$y = \dfrac{3x - 9}{4} = \dfrac{3(11) - 9}{4} = 6$

The solution is $x = 11$, $y = 6$

49. Multiply both equations by 100 to clear decimals.

$5x + 8y = -76$ **(1)**

$4x - 3y = 5$ **(2)**

Solve **(1)** for y and substitute into **(2)**

$y = \dfrac{-76 - 5x}{8}$

(continued)

49. (continued)

$4x - 3\dfrac{-76 - 5x}{8} = 5$

$32x + 228 + 15x = 40$

$47x = -188$

$x = -4$

$y = \dfrac{-76 - 5x}{8} = \dfrac{-76 - 5(-4)}{8} = -7$

The solution is $x = -4$, $y = -7$

50. $x - \dfrac{y}{2} + \dfrac{1}{2}z = -1$ **(1)**

$2x + \dfrac{5}{2}z = -1$ **(2)**

$\dfrac{3}{2}y + 2z = 1$ **(3)**

Solve **(2)** for x and **(3)** for y and substitute into **(1)**

$x = \dfrac{-\dfrac{5}{2}z - 1}{2}$

$y = \dfrac{2(1 - 2z)}{3} = \dfrac{2 - 4z}{3}$

$\dfrac{-\dfrac{5}{2}z - 1}{2} - \dfrac{\dfrac{2 - 4z}{3}}{2} + \dfrac{1}{2}z = -1$

$-\dfrac{5}{2}z - 1 - \dfrac{2 - 4z}{3} + z = -2$

$-15z - 6 - 4 + 8z + 6z = -12$

$-z = -2$

$z = 2$

$x = \dfrac{-\dfrac{5}{2}z - 1}{2} = \dfrac{-\dfrac{5}{2}(2) - 1}{2} = -3$

$y = \dfrac{2 - 4z}{3} = \dfrac{2 - 4(2)}{3} = -2$

The solution is

$x = -3$, $y = -2$, $z = 2$.

51.
$$2x - 3y + 2z = 0 \qquad \textbf{(1)}$$
$$x + 2y - z = 2 \qquad \textbf{(2)}$$
$$2x + y - 3z = -1 \qquad \textbf{(3)}$$
Add **(1)** and 2 times **(2)**
$$4x - y = 4 \qquad \textbf{(4)}$$
Add -3 times**(2)** and **(3)**
$$-x - 5y = 1 \qquad \textbf{(5)}$$
Solve**(4)**for y and substitute into**(5)**
$$y = 4x - 4$$
$$-x - 5(4x - 4) = 1$$
$$-x - 20x + 20 = 1$$
$$-21x = -21$$
$$x = 1$$
$$y = 4x - 4 = 4(1) - 4 = 0$$
Substitute $x = 1,\ y = 0$ into **(1)**
$$2(1) - 3(0) + 2z = 0$$
$$2 + 2z = 0$$
$$2z = -2$$
$$z = -1$$
The solution is
$x = 1,\ y = 0,\ z = -1$.

52.
$$x - 4y + 4z = -1 \qquad \textbf{(1)}$$
$$2x - y + 5z = -3 \qquad \textbf{(2)}$$
$$x - 3y + z = 4 \qquad \textbf{(3)}$$
Add **(1)** and -4 times **(3)**
$$-3x + 8y = -17 \qquad \textbf{(4)}$$
Add **(2)** and -5 times **(3)**
$$-3x + 14y = -23 \qquad \textbf{(5)}$$
Subtract **(5)** from **(4)**
$$-6y = 6$$
$$y = -1$$
Substitute $y = -1$ into **(4)**
$$-3x + 8(-1) = -17$$
$$-3x = -9$$
$$x = 3$$

(continued)

52. (continued)
Substitute $x = 3,\ y = -1$ into **(3)**
$$3 - 3(-1) + z = 4$$
$$z = -2$$
The solution is
$x = 3,\ y = -1,\ z = -2$

53.
$$x - 2y + z = -5 \qquad \textbf{(1)}$$
$$2x + z = -10 \qquad \textbf{(2)}$$
$$y - z = 15 \qquad \textbf{(3)}$$
Solve **(3)** for y
$$y = z + 15 \qquad \textbf{(4)}$$
Solve **(2)** for x
$$x = \frac{-z - 10}{2} \qquad \textbf{(5)}$$
Substitute **(4)** and **(5)** into **(1)**
$$\frac{-z - 10}{2} - 2(z + 15) + z = -5$$
$$-z - 10 - 4z - 60 + 2z = -10$$
$$= 3z = 60$$
$$z = -20$$
Substitute $z = -20$ into **(4)** and **(5)**
$$x = \frac{-(-20) - 10}{2} = 5$$
$$y = -20 + 15 = -5$$
The solution
is $x = 5,\ y = -5,\ z = -20$

54. $x - y \le 3$
Test point: $(0,0)$
$$0 - 0 \le 3$$
$$0 \le 3 \text{ True}$$
$$y \le -\frac{1}{4}x + 2$$
$$0 \le -\frac{1}{4}(0) + 2$$
$$0 \le 2 \text{ True}$$

(continued)

54. (continued)

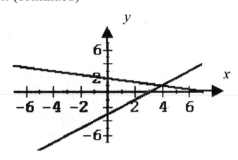

55. $-2x + 3y < 6$

Test point: (0,0)

$-2(0) + 3(0) < 6$

$0 < 6$ True

$y > -2$

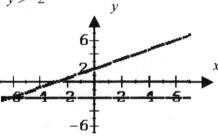

56. $x + y > 1$

Test point: (0,0)

$0 + 0 > 1$

$0 > 1$ False

$2x - y < 5$

Test point: (0,0)

$2(0) - 0 < 5$

$0 < 5$ True

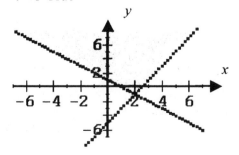

57. $x + y \geq 4$

Test point: (0,0)

$0 + 0 \geq 4$

$0 \geq 4$ False

$y \leq x$

Test point: (2,0)

$0 \leq 2$ True

$x \leq 6$

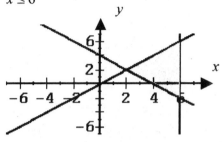

Chapter 4 Test

1. $3x - 2y = -8$ **(1)**

    $x + 6y = 4$ **(2)**

Solve **(2)** for x and substitute into **(1)**

$x = 4 - 6y$

$3(4 - 6y) - 2y = -8$

$12 - 18y - 2y = -8$

$-20y = -20$

$y = 1$

$x = 4 - 6y = 4 - 6(1) = -2$

The solution is $x = -2$, $y = 1$.

2. $6x - 2y = -2$ **(1)**

    $3x + 4y = 14$ **(2)**

Add 2 times **(1)** to **(2)**

$15x = 10$

$x = \dfrac{2}{3}$

           (continued)

2. (continued)

$3x + 4y = 14$

$3\left(\dfrac{2}{3}\right) + 4y = 14$

$4y = 12$

$y = 3$

The solution is $x = \dfrac{2}{3},\ y = 3$.

3. $\dfrac{1}{4}a - \dfrac{3}{4}b = -1$ **(1)**

$\dfrac{1}{3}a + b = \dfrac{5}{3}$ **(2)**

Solve **(2)** for b and substitute into **(1)**

$b = \dfrac{5}{3} - \dfrac{1}{3}a$

$\dfrac{1}{4}a - \dfrac{3}{4}\left(\dfrac{5}{3} - \dfrac{1}{3}a\right) = -1$

$a - 5 + a = -4$

$2a = 1$

$a = \dfrac{1}{2}$

$b = \dfrac{5}{3} - \dfrac{1}{3}a = \dfrac{5}{3} - \dfrac{1}{3} \cdot \dfrac{1}{2} = \dfrac{3}{2}$

The solution is $a = \dfrac{1}{2},\ b = \dfrac{3}{2}$.

4. $7x - 1 = 3(1 + y)$ **(1)**

$7x - 1 = 3 + 3y$

$7x - 3y = 4$ **(3)**

$1 - 6y = -14x - 7$ **(2)**

$14x - 6y = -8$

$7x - 3y = -4$ **(4)**

Comparing **(3)** and **(4)** shows the system is inconsistent and has no solution.

5. $5x - 3y = 3$ **(1)**

$7x + y = 25$ **(2)**

Add 3 times **(2)** to **(1)**

$26x = 78$

$x = 3$

$7x + y = 25$

$7(3) + y = 25$

$y = 4$

The solution is $x = 3,\ y = 4$.

6. $\dfrac{1}{3}x + \dfrac{5}{6}y = 2$ **(1)**

$\dfrac{3}{5}x - y = -\dfrac{7}{5}$ **(2)**

Solve **(2)** for y and substitute into **(1)**

$y = \dfrac{3}{5}x + \dfrac{7}{5}$

$\dfrac{1}{3}x + \dfrac{5}{6}\left(\dfrac{3}{5}x + \dfrac{7}{5}\right) = 2$

$2x + 3x + 7 = 12$

$5x = 5$

$x = 1$

$y = \dfrac{3}{5}x + \dfrac{7}{5}$

$y = \dfrac{3}{5}(1) + \dfrac{7}{5}$

$y = 2$

The solution is $x = 1,\ y = 2$.

7.

$$
\begin{array}{rcrcrcr}
3x & + & 5y & - & 2z & = & -5 \\
2x & + & 3y & - & z & = & -2 \\
2x & + & 4y & + & 6z & = & 18
\end{array}
$$

(continued)

7. (continued)

Multiply first equation by $-\dfrac{2}{3}$ and add to the second and third equation

$$3x + 5y - 2z = -5$$
$$\frac{1}{3}y + \frac{1}{3}z = \frac{4}{3}$$
$$\frac{2}{3}y + \frac{22}{3}z = \frac{64}{3}$$

Multiply second equation by 2 and add to the third equation

$$3x + 5y - 2z = -5$$
$$\frac{1}{3}y + \frac{1}{3}z = \frac{4}{3}$$
$$8z = 24$$

$$8z = 24$$
$$z = 3$$

From second equation

$$-\frac{1}{3}y + \frac{1}{3}(3) = \frac{4}{3}, \ y = -1$$

From the first equation

$$3x + 5(-1) - 2(3) = -5, \ x = 2$$

The solution is $x = 2, \ y = -1, \ z = 3$.

8.

$$3x + 2y = 0$$
$$2x - y + 3z = 8$$
$$5x + 3y + z = 4$$

Multiply the first equation by $-\dfrac{2}{3}$ and add to the second equation and then multiply the first equation by $-\dfrac{5}{3}$ and add to the third equation

(continued)

8. (continued)

$$3x + 2y = 0$$
$$-\frac{7}{3}y + 3z = 8$$
$$-\frac{1}{3}y + z = 4$$

Multiply the second equation by $-\dfrac{1}{7}$ and add to the third equation

$$3x + 2y = 0$$
$$-\frac{7}{3}y + 3z = 8$$
$$+ \frac{4}{7}z = \frac{20}{7}$$

$$\frac{4}{7}z = \frac{20}{7}, \ z = 5$$

From second equation

$$-\frac{7}{3}y + 3(5) = 8, \ y = 3$$

From the first equation

$$3x + 2(3) = 0, \ x = -2$$

The solution is $x = -2, \ y = 3, \ z = 5$.

9.

$$x + 5y + 4z = -3$$
$$x - y - 2z = -3$$
$$x + 2y + 3z = -5$$

Multiply the first equation by -1 and add to the second and third equation

$$x + 5y + 4z = -3$$
$$- 6y - 6z = 0$$
$$- 3y - z = -2$$

Multiply the second equation by $-\dfrac{1}{2}$ and add to the third equation

(continued)

9. (continued)

$$
\begin{array}{rcrcrcr}
x & + & 5y & + & 4z & = & -3 \\
  &   & -6y & - & 6z & = & 0 \\
  &   &   &   & 2z & = & -2
\end{array}
$$

$2z = -2,\ z = -1$

From the second equation

$-6y - 6(-1) = 0,\ y = 1$

From the first equation

$x + 5(1) + 4(-1) = -3,\ x = -4$

The solution is $x = -4,\ y = 1,\ z = -1$.

10. $v =$ speed of plane in still air

$w =$ speed of wind

$$
\begin{array}{ll}
1000 = (v+w) \cdot 2, & v + w = 500 \\
1000 = (v-w) \cdot 2.5 & \underline{v - w = 400} \\
 & 2v = 900 \\
 & v = 450
\end{array}
$$

$450 + w = 500$

$\quad\quad w = 50$

The speed of the plane in still air is 450 mph. The speed of the wind is 50 mph.

11. $x =$ number of station wagons

$y =$ number of two-door sedans

$z =$ number of four-door sedans

$$
\begin{array}{rcrcrcr}
5x & + & 4y & + & 3z & = & 62 \\
4x & + & 3y & + & 3z & = & 52 \\
3x & + & 2y & + & 2z & = & 36
\end{array}
$$

Multiply first equation by $-\dfrac{4}{5}$ and add to the second equation, then multiply the first equation by $-\dfrac{3}{5}$ and add to the third equation.

(continued)

11. (continued)

$$
\begin{array}{rcrcrcr}
5x & + & 4y & + & 3z & = & 62 \\
   &   & -\dfrac{1}{5}y & + & \dfrac{3}{5}z & = & \dfrac{12}{5} \\
   &   & -\dfrac{2}{5}y & + & \dfrac{1}{5}z & = & -\dfrac{6}{5}
\end{array}
$$

Multiply the second equation by -2 and add to the third equation

$$
\begin{array}{rcrcrcr}
5x & + & 4y & + & 3z & = & 62 \\
   &   & -\dfrac{1}{5}y & + & \dfrac{3}{5}z & = & \dfrac{12}{5} \\
   &   &   &   & -z & = & -6
\end{array}
$$

$-z = -6,\ z = 6$

From the second equation

$-\dfrac{1}{5}y + \dfrac{3}{5}(6) = \dfrac{12}{5},\ y = 6$

From the first equation

$5x + 4(6) + 3(6) = 62,\ x = 4$

The line supervisor should send 4 station wagons, 6 two-door sedans, and 6 four-door sedans down the assembly line.

12. $x =$ daily charge

$y =$ mileage charge

$$
\begin{array}{ll}
5x + 150y = 180 & \textbf{(1)} \\
7x + 320y = 274 & \textbf{(2)}
\end{array}
$$

Multiply **(1)** by $-\dfrac{7}{5}$ and add to **(2)**

$110y = 22$

$y = 0.2$

From first equation

$5x + 150(0.2) = 180$

$5x = 150$

$x = 30$

They charge \$30 per day and \$0.20 per mile.

13. $x + 2y \le 6$

Test point: (0,0)

$0 + 2(0) \le 6$

$0 \le 6$ True

$-2x + y \ge -2$

Test point: (0,0)

$-2(0) + 0 \ge -2$

$0 \ge -2$ True

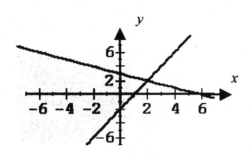

14. $3x + y \ge 8$

Test point: (0,0)

$3(0) + 0 \ge 8$

$0 \ge 8$ False

$x - 2y \ge 5$

Test point: (0,0)

$0 - 2(0) \ge 5$

$0 \ge 5$ False

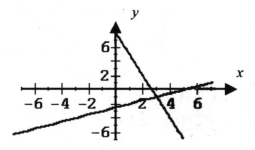

Cumulative Test for Chapters 1-4

1. Identity property of addition.

2. $\sqrt{25} + (2 - 3)^3 + 20 \div (-10)$

$= 5 + (-1)^3 + (-2)$

$= 5 + (-1) + (-2)$

$= 4 + (-2)$

$= 2$

3. $(5x^{-2})(3x^{-4}y^2)$

$= 15x^{-6}y^2$

$= \dfrac{15y^2}{x^6}$

4. $2x - 4[x - 3(2x + 1)]$

$= 2x - 4[x - 6x - 3]$

$= 2x - 4[-5x - 3]$

$= 2x + 20x + 12$

$= 22x + 12$

5. $A = P(3 + 4rt)$

$P = \dfrac{A}{3 + 4rt}$

6. $\dfrac{1}{4}x + 5 = \dfrac{1}{3}(x - 2)$

$3x + 60 = 4x - 8$

$x = 68$

7. $4x - 8y = 10$

x	y
-2	$-\dfrac{9}{4}$
0	$-\dfrac{5}{4}$
4	$\dfrac{3}{4}$

(continued)

7. (continued)

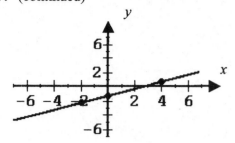

8. $m = \dfrac{y_2 - y_1}{x_2 - x_1} = \dfrac{-1-(-2)}{6-(-4)} = \dfrac{1}{10}$

9. $4x + 3 - 13x - 7 < 2(3-4x)$

$-9x - 4 < 6 - 8x$

$-x < 10$

$x > -10$

-10

10. $\dfrac{2x-1}{3} \le 7$ and $2(x+1) \ge 12$

$2x - 1 \le 21 \qquad 2x + 2 \ge 12$

$2x \le 22 \qquad\quad 2x \ge 10$

$x \le 11 \qquad\quad x \ge 5$

$5 \le x \le 11$

11. $5x + 6y = -2$

$6y = -5x - 2$

$y = -\dfrac{5}{6}x - \dfrac{1}{3}, \; m = -\dfrac{5}{6}, \; m_\perp = \dfrac{6}{5}$

$y - y_1 = m(x - x_1)$

$y - (-3) = \dfrac{6}{5}(x - 2)$

$5y + 15 = 6x - 12$

$6x - 5y = 27$

12. $x =$ length of first side

$x + 7 =$ length of second side

$2x - 6 =$ length of third side

$x + x + 7 + 2x - 6 = 69$

$4x + 1 = 69$

$4x = 68$

$x = 17$

$x + 7 = 24$

$2x - 6 = 28$

The first side is 17 meters, the second side is 24 meters, and the third side is 28 meters.

13. $x =$ amount invested at 7%

$6000 - x =$ amount invested at 9%

$0.07x + 0.09(6000 - x) = 510$

$0.07x + 540 - 0.09x = 510$

$-0.02x = -30$

$x = 1500$

$6000 - x = 4500$

Victor invested $1500 at 7% and $4500 at 9%.

14. $5x + 2y = 2$ 　　**(1)**

$4x + 3y = -4$ 　　**(2)**

Solve **(1)** for y and substitute into **(2)**

$2y = 2 - 5x$

$y = \dfrac{2 - 5x}{2}$

$4x + 3\dfrac{2 - 5x}{2} = -4$

$8x + 6 - 15x = -8$

$-7x = -14$

$x = 2$

$y = \dfrac{2 - 5x}{2} = \dfrac{2 - 5(2)}{2} = -4$

The solution is $x = 2$, $y = -4$.

15.

$$\begin{aligned} 2x + y - z &= 4 \\ x + 2y - 2z &= 2 \\ x - 3y + z &= 4 \end{aligned}$$

Multiply the first equation by $-\dfrac{1}{2}$ and add to the second and third equations

$$\begin{aligned} 2x + y - z &= 4 \\ \tfrac{3}{2}y - \tfrac{3}{2}z &= 0 \\ -\tfrac{7}{2}y + \tfrac{3}{2}z &= 2 \end{aligned}$$

Multiply the second equation by $\dfrac{7}{3}$ and add to the third equation

$$\begin{aligned} 2x + y - z &= 4 \\ \tfrac{3}{2}y - \tfrac{3}{2}z &= 0 \\ -2z &= 2 \end{aligned}$$

$-2z = 2$

$z = -1$

From the second equation

$\dfrac{3}{2}y - \dfrac{3}{2}(-1) = 0$

$y = -1$

From the first equation

$2x + (-1) - (-1) = 4$

$2x = 4$

$x = 2$

The solution is $x = 2,\ y = -1,\ z = -1$.

16. $x = $ cost of shirt

$y = $ cost of slacks

$5x + 8y = 345$ **(1)**

$7x + 3y = 237$ **(2)**

Solve **(1)** for y and substitute into **(2)**

(continued)

16. (continued)

$8y = 345 - 5x$

$y = \dfrac{345 - 5x}{8}$

$7x + 3\dfrac{345 - 5x}{8} = 237$

$56x + 1035 - 15x = 1896$

$41x = 861$

$x = 21$

$y = \dfrac{345 - 5x}{8} = \dfrac{345 - 5(21)}{8} = 30$

The shirts cost \$21 and the slacks cost \$39.

17. $7x - 6y = 17$ **(1)**

$3x + y = 18$ **(2)**

Solve **(2)** for y and substitute into **(1)**

$y = 18 - 3x$

$7x - 6(18 - 3x) = 17$

$7x - 108 + 18x = 17$

$25x = 125$

$x = 5$

$y = 18 - 3x = 18 - 3(5) = 3$

The solution is $x = 5,\ y = 3$.

18.

$$\begin{aligned} x + 3y + z &= 5 \\ 2x - 3y - 2z &= 0 \\ x - 2y + 3z &= -9 \end{aligned}$$

Multiply first equation by -2 and add to the second equation, then multiply the first equation by -1 and add to the third equation.

$$\begin{aligned} x + 3y + z &= 5 \\ -9y - 4z &= -10 \\ -5y + 2z &= -14 \end{aligned}$$

(continued)

18. (continued)

Multiply the second equation by $-\dfrac{5}{9}$ and add to third equation.

$$
\begin{aligned}
x + 3y + z &= 5 \\
- 9y - 4z &= -10 \\
+ \tfrac{38}{9}z &= -\tfrac{76}{9}
\end{aligned}
$$

From third equation

$\dfrac{38}{9}z = -\dfrac{76}{9}$, $z = -2$

From second equation

$-9y - 4(-2) = -10$, $y = 2$

From first equation

$x + 3(2) + (-2) = 5$, $x = 1$

The solution is $x = 1$, $y = 2$, $z = -2$.

19. $-5x + 6y = 2$ **(1)**

$10x - 12y = -4$ **(2)**

Multiplying **(1)** by 2 and adding to **(2)** gives $0 = 0$. This is a dependent system of equations with an infinite number of solutions.

20. $x - y \geq -4$

Test point: $(0,0)$

$0 - 0 \geq -4$

$0 \geq -4$ True

$x + 2y \geq 2$

Test point: $(0,0)$

$0 + 2(0) \geq 2$

$0 \geq 2$ False

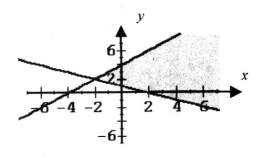

Chapter 5

Pretest Chapter 5

1. $(5x^2 - 3x + 2) + (-3x^2 - 5x - 8)$
$-(x^2 + 3x - 10)$
$= 5x^2 - 3x + 2 - 3x^2 - 5x - 8$
$\quad - x^2 - 3x + 10$
$= x^2 - 11x + 4$

2. $(x^2 - 3x - 4)(2x - 3)$
$= 2x^3 - 3x^2 - 6x^2 + 9x - 8x + 12$
$= 2x^3 - 9x^2 + x + 12$

3. $(5a - 8)(a - 7)$
$= 5a^2 - 35a - 8a + 56$
$= 5a^2 - 43a + 56$

4. $(2y - 3)(2y + 3)$
$= 4y^2 - 3^3$
$= 4y^2 - 9$

5. $(3x^2 + 4)^2$
$= (3x^2)^2 + 2(3x^2)(4) + 4^2$
$= 9x^4 + 24x^2 + 16$

6. $p(x) = 2x^3 - 5x^2 - 6x + 1$
$p(-3) = 2(-3)^3 - 5(-3)^2 - 6(-3) + 1$
$p(-3) = -80$

7. $\dfrac{25x^3y^2 - 30x^2y^3 - 50x^2y^2}{5x^2y^2}$

$= \dfrac{25x^3y^2}{5x^2y^2} - \dfrac{30x^2y^3}{5x^2y^2} - \dfrac{50x^2y^2}{5x^2y^2}$

$= 5x - 6y - 10$

8. $(3y^3 - 5y^2 + 2y - 1) \div (y - 2)$

$$
\begin{array}{r}
3y^2 + y + 4 \\
y-2{\overline{\smash{\big)}\,3y^3 - 5y^2 + 2y - 1}} \\
\underline{3y^3 - 6y^2} \\
y^2 + 2y \\
\underline{y^2 - 2y} \\
4y - 1 \\
\underline{4y - 8} \\
7
\end{array}
$$

$(3y^3 - 5y^2 + 2y - 1) \div (y - 2)$

$= 3y^2 + y + 4 + \dfrac{7}{y - 2}$

9. $(2x^4 + 9x^3 + 8x^2 - 9x - 10) \div (2x + 5)$

$$
\begin{array}{r}
x^3 + 2x^2 - x - 2 \\
2x+5{\overline{\smash{\big)}\,2x^4 + 9x^3 + 8x^2 - 9x - 10}} \\
\underline{2x^4 + 5x^3} \\
4x^3 + 8x^2 \\
\underline{4x^2 + 10x^2} \\
-2x^2 - 9x \\
\underline{-2x^2 - 5x} \\
-4x - 10 \\
\underline{-4x - 10} \\
0
\end{array}
$$

$(2x^4 + 9x^3 + 8x^2 - 9x - 10) \div (2x + 5)$

$= x^3 + 2x^2 - x - 2$

10. $24a^3b^2 + 36a^4b^2 - 60a^3b^3$
$= 12a^3b^2(2 + 3a^2 - 5b)$

121

11. $3x(4x-3y)-2(4x-3y)$
$= (4x-3y)(3x-2)$

12. $10wx+6zx-15yz-25wy$
$= 2x(5w+3z)-5y(3z+5w)$
$= (5x+3z)(2x-5y)$

13. $x^2-7xy+10y^2 = (x-5y)(x-2y)$

14. $4y^2-4y-15 = (2y-5)(2y+3)$

15. $28x^2-19xy+3y^2 = (7x-3y)(4x-y)$

16. $36x^2-60xy+25y^2$
$= (6x-5y)(6x-5y)$
$= (6x-5y)^2$

17. $121x^2-1 = (11x-1)(11x+1)$

18. $8x^3-y^3 = (2x)^3-y^3$
$= (2x-y)((2x)^2+(2x)y+y^2)$
$= (2x-y)(4x^2+2xy+y^2)$

19. $64x^3+27$
$= (4x)^3+(3)^3$
$= (4x+3)((4x)^2-(4x)(3)+3^2)$
$= (4x+3)(16x^2-12x+9)$

20. $x^3y^3-27y^6$
$= (xy)^3-(3y^2)^3$
$= (xy-3y^2)((xy)^2+(xy)(3y^2)+(3y^2)^2)$
$= y(x-3y)(x^2y^2+3xy^3+9y^4)$
$= y(x-3y)(y^2)(x^2+3xy+9y^2)$
$= y^3(x-3y)(x^2+3xy+9y^2)$

21. $2x^3-2x^2-24x = 2x(x^2-x-12)$
$= 2x(x-4)(x-3)$

22. $2x^2+8x-3$, prime

23. $81a^3+126a^2y+49ay^2$
$= a((9a)^2+126ay+(7y)^2)$
$= a((9a)^2+2(9a)(7y)+(7y)^2)$
$= a(9a+7y)^2$

24. $12x^2+x-6=0$
$(3x-2)(4x+3)=0$
$3x-2=0$ or $4x+3=0$
$3x=2$ $\qquad$ $4x=-3$
$x=\dfrac{2}{3}$ $\qquad$ $x=-\dfrac{3}{4}$

25. $3x^2+5x = 7x^2-2x$
$4x^2-7x=0$
$x(4x-7)=0$
$x=0$ or $4x-7=0$
$\qquad\qquad 4x=7$
$\qquad\qquad x=\dfrac{7}{4}$

26. $(x+5)(x-3) = 2x+1$
$x^2+2x-15 = 2x+1$
$x^2-16=0$
$(x-4)(x+4)=0$
$x-4=0$ or $x+4=0$
$x=4$ $\qquad$ $x=4$

27. $A = LW = (3W + 1)W = 52$

$3W^2 + W - 52 = 0$

$(3W + 13)(W - 4) = 0$

$3W + 13 = 0$ or $W - 4 = 0$

$3W + 13 = 0$ $W = 4$

$3W = -13$ $L = 3W + 1 = 13$

$W = -\dfrac{13}{3}$, reject since $W > 0$

The width of the rectangle is 4 m and the length is 13 m.

5.1 Exercises

1. Binomial, 4th degree

3. Monomial, 9th degree

5. Trinomial, 8th degree

7. $(x^2 + 3x - 2) + (-2x^2 - 5x + 1)$
$+ (x^2 - x - 5)$
$= x^2 + 3x - 2 - 2x^2 - 5x + 1 + x^2 - x - 5$
$= -3x - 6$

9. $(4x^3 - 6x^2 - 3x + 5.5)$
$\qquad - (2x^3 + 3x^2 - 5x - 8.3)$
$= 4x^3 - 6x^2 - 3x + 5.5$
$\qquad - 2x^3 - 3x^2 + 5x + 8.3$
$= 2x^3 - 9x^2 + 2x + 13.8$

11. $(5a^3 - 2a^2 - 6a + 8) + (5a + 6)$
$\qquad - (-a^2 - a + 2)$
$= 5a^3 - 2a^2 - 6a + 8 + 5a + 6$
$\qquad\qquad + a^2 + a - 2$
$= 5a^3 - a^2 + 12$

13. $\left(\dfrac{1}{2}x^2 - 7x\right) + \left(\dfrac{1}{3}x^2 + \dfrac{1}{4}x\right)$

$= \dfrac{1}{2}x^2 - 7x + \dfrac{1}{3}x^2 + \dfrac{1}{4}x$

$= \dfrac{5}{6}x^2 - \dfrac{27}{4}x$

$= \dfrac{5}{6}x^2 - 6\dfrac{3}{4}x$

15. $(2.3x^3 - 5.6x^2 - 2) - (5.5x^3 - 7.4x^2 + 2)$

$= 2.3x^3 - 5.6x^2 - 2 - 5.5x^3 + 7.4x^2 - 2$

$= -3.2x^3 + 1.8x^2 - 4$

17. $p(x) = 5x^2 - 9x - 12$

$p(3) = 5(3)^2 - 9(3) - 12 = 6$

19. $g(x) = -2x^3 - 3x^2 + 5x + 8$

$g(-2) = -2(-2)^3 - 3(-2)^2 + 5(-2) + 8$

$g(-2) = 2$

21. $h(x) = 2x^4 - x^3 + 2x^2 - 4x - 3$

$h(-1) = 2(-1)^4 - (-1)^3 + 2(-1)^2 - 4(-1) - 3$

$h(-1) = 6$

23. $2x(3x^2 - 5x + 1) = 6x^2 - 10x^2 + 2x$

25. $-\dfrac{1}{3}xy(2x - 6y + 15)$

$= -\dfrac{2}{3}x^2y + 2xy^2 - 5xy$

27. $(6x + 7)(x + 1) = 6x^2 + 6x + 7x + 7$
$\qquad\qquad\qquad\quad = 6x^2 + 13x + 7$

29. $(5w + 2d)(3a - 4b)$
$= 15aw - 20bw + 3ad - 8bd$

31. $(3x-2y)(-4x+y)$
$$= -12x^2 + 3xy + 8xy - 2y^2$$
$$= -12x^2 + 11xy - 2y^2$$

33. $(7r-s^2)(-4a-11s^2)$
$$= -28ar - 77rs^2 + 4as^2 + 11s^4$$

35. $(2x-3)(x^2-x+1)$
$$= 2x^3 - 2x^2 + 2x - 3x^2 + 3x - 3$$
$$= 2x^3 - 5x^2 + 5x - 3$$

37. $(3x^2-2xy-6y^2)(2x-y)$
$$= 6x^3 - 3x^2y - 4x^2y + 2xy^2 - 12xy^2 + 6y^3$$
$$= 6x^3 - 7x^2y - 10xy^2 + 6y^3$$

39. $\left(\dfrac{3}{2}x^2 - x + 1\right)(x^2 + 2x - 6)$
$$= \frac{3}{2}x^4 + 3x^3 - 9x^2$$
$$\qquad - x^3 - 2x^2 + 6x$$
$$\qquad\qquad + x^2 + 2x - 6$$
$$= \frac{3}{2}x^4 + 2x^3 - 10x^2 + 8x - 6$$

41. $(5a^3 - 3a^2 + 2a - 4)(a-3)$
$$= 5a^4 - 15a^3$$
$$\qquad - 3a^3 + 9a^2$$
$$\qquad\qquad + 2a^2 - 6a$$
$$\qquad\qquad\qquad - 4a + 12$$
$$= 5a^4 - 18a^3 + 11a^2 - 10a + 12$$

43. $(r^2 + 3rs - 2s^2)(3r^2 - 4rs - 2s^2)$
$$= 3r^4 - 4r^3s - 2r^2s^2$$
$$\qquad + 9r^3s - 12r^2s^2 - 6rs^3$$
$$\qquad\qquad - 6r^2s^2 + 8rs^3 + 4s^4$$
$$= 3r^4 + 5r^3s - 20r^2s^2 + 2rs^3 + 4s^4$$

45. $(5x-8y)(5x+8y) = 25x^2 - 64y^2$

47. $(5a-2b)^2 = 25a^2 - 20ab + 4b^2$

49. $(7m-1)^2 = 49m^2 - 14m + 1$

51. $\left(\dfrac{1}{2}x^2 - 1\right)\left(\dfrac{1}{2}x^2 + 1\right) = \dfrac{1}{4}x^2 - 1$

53. $(2a^2b^2 - 3)^2 = 4a^4b^4 - 12a^2b^2 + 9$

55. $(0.6x - 0.5y^2)(0.6x + 0.5y^2)$
$$= 0.36x^2 - 0.25y^4$$

57. $(x+2)(x-3)(2x-5)$
$$= (x^2 - x - 6)(2x-5)$$
$$= 2x^3 - 5x^2 - 2x^2 + 5x - 12x + 30$$
$$= 2x^3 - 7x^2 - 7x + 30$$

59. $(a+3)(2-a)(4-3a)$
$$= (2a - a^2 + 6 - 3a)(4-3a)$$
$$= (-a^2 - a + 6)(4-3a)$$
$$= -4a^2 + 3a^3 - 4a + 3a^2 + 24 - 18a$$
$$= 3a^3 - a^2 - 22a + 24$$

61. $V = (2x^2 + 5x + 8)(3x+5)$
$$= 6x^3 + 10x^2 + 15x^2 + 25x + 24x + 40$$
$$= 6x^3 + 25x^2 + 49x + 40 \text{ cm}^3$$

63. $p(t) = -0.03t^2 + 78$

$p(3) = -0.03(3)^2 + 78$

$p(3) = 77.73$ parts per million

65. $p(t) = -0.03t^2 + 78$

$p(50) = -0.03(50)^2 + 78$

$p(50) = 3$ parts per million

Cumulative Review Problems

67. $\dfrac{3(2)^2 - 6}{5(-2) - (-1)}$

$= \dfrac{3(4) - 6}{-10 + 1} = \dfrac{12 - 6}{-9} = \dfrac{6}{-9} = -\dfrac{2}{3}$

69. $2500 \dfrac{\text{ft}}{\text{min}} \cdot t = (31,000 - 8000) \text{ ft}$

$t = \dfrac{23,000 \text{ ft}}{2500 \dfrac{\text{ft}}{\text{min}}} = 9.2 \text{ min}$

for the jet to reach 8000 ft.

5.2 Exercises

1. $(24x^2 - 8x - 44) \div 4$

$\dfrac{24x^2}{4} - \dfrac{8x}{4} - \dfrac{44}{4} = 6x^2 - 2x - 11$

3. $(27x^4 - 9x^3 + 63x^2) \div 9x$

$\dfrac{27x^4}{9x} - \dfrac{9x^3}{9x} + \dfrac{63x^2}{9x} = 3x^3 - x^2 + 7x$

5. $\dfrac{4x^3 - 2x^2 + 5x}{2x} = \dfrac{4x^3}{2x} - \dfrac{2x^2}{2x} + \dfrac{5x}{2x}$

$= 2x^2 - x + \dfrac{5}{2}$

7. $\dfrac{18a^3b^2 + 12a^2b^2 - 4ab^2}{2ab^2}$

$= \dfrac{18a^3b^2}{2ab^2} + \dfrac{12a^2b^2}{2ab^2} - \dfrac{4ab^2}{2ab^2}$

$= 9a^2 + 6a - 2$

9. $(15x^2 + 23x + 4) \div (5x + 1)$

$$\begin{array}{r} 3x + 4 \\ 5x+1 \overline{) 15x^2 + 23x + 4} \\ \underline{15x^2 + \ 3x} \\ 20x + 4 \\ \underline{20x + 4} \end{array}$$

$(15x^2 + 23x + 4) \div (5x + 1) = 3x + 4$

Check:

$(5x + 1)(3x + 4) = 15x^2 + 23x + 4$

11. $(28x^2 - 29x + 6) \div (4x - 3)$

$$\begin{array}{r} 7x - 2 \\ 4x-3 \overline{) 28x^2 - 29x + 6} \\ \underline{28x^2 - 21x} \\ -8x + 6 \\ \underline{-8x + 6} \end{array}$$

$(28x^2 - 29x + 6) \div (4x - 3) = 7x - 2$

Check:

$(4x - 3)(7x - 2) = 28x^2 - 29x + 6$

13. $(x^3 - x^2 + 11x - 1) \div (x + 1)$

$$\begin{array}{r} x^2 - 2x + 13 \\ x+1\overline{)x^3 - x^2 + 11x - 1} \\ \underline{x^3 + x^2} \\ -2x^2 + 11x \\ \underline{-2x^2 - 2x} \\ 13x - 1 \\ \underline{13x + 13} \\ -14 \end{array}$$

$(x^3 - x^2 + 11x - 1) \div (x + 1)$

$= x^2 - 2x + 13 - \dfrac{14}{x+1}$

Check:

$(x+1)\left(x^2 - 2x + 13 - \dfrac{14}{x+1}\right)$

$= (x+1)(x^2 - 2x + 13) - (x+1)\left(\dfrac{14}{x+1}\right)$

$= x^3 - 2x^2 + 13x + x^2 - 2x + 13 - 14$

$= x^3 - x^2 + 11x - 1$

15. $(2x^3 - x^2 - 7) \div (x - 2)$

$$\begin{array}{r} 2x^2 + 3x + 6 \\ x-2\overline{)2x^3 - x^2 + 0x - 7} \\ \underline{2x^3 - 4x^2} \\ 3x^2 + 0x \\ \underline{3x^2 - 6x} \\ 6x - 7 \\ \underline{6x - 12} \\ 5 \end{array}$$

$(2x^3 - x^2 - 7) \div (x - 2)$

$= 2x^2 + 3x + 6 + \dfrac{5}{x-2}$

17. $\dfrac{8x^3 - 14x^2 - 17x + 5}{2x - 5}$

$$\begin{array}{r} 4x^2 + 3x - 1 \\ 2x-5\overline{)8x^3 - 14x^2 - 17x + 5} \\ \underline{8x^3 - 20x^2} \\ 6x^2 - 17x \\ \underline{6x^2 - 15x} \\ -2x + 5 \\ \underline{-2x + 5} \end{array}$$

$\dfrac{8x^3 - 14x^2 - 17x + 5}{2x - 5} = 4x^2 + 3x - 1$

19. $\dfrac{2x^4 - x^3 + 16x^2 - 4}{2x - 1}$

$$\begin{array}{r} x^3 \qquad + 8x + 4 \\ 2x-1\overline{)2x^4 - x^3 + 16x^2 + 0x - 4} \\ \underline{2x^4 - x^3} \\ 16x^2 + 0x \\ \underline{16x^2 - 8x} \\ 8x - 4 \\ \underline{8x - 4} \end{array}$$

$\dfrac{2x^4 - x^3 + 16x^2 - 4}{2x - 1} = x^3 + 8x + 4$

21. $\dfrac{6t^4 - 5t^3 - 8t^2 + 16t - 8}{3t^2 + 2t - 4}$

$$
\begin{array}{r}
2t^2 - 3t + 2 \\
3t^2 + 2t - 4 \overline{)6t^4 - 5t^3 - 8t^2 + 16t - 8} \\
\underline{6t^4 + 4t^3 - 8t^2} \\
-9t^3 \qquad + 16t \\
\underline{-9t^3 - 6t^2 + 12t} \\
6t^2 + 4t - 8 \\
\underline{6t^2 + 4t - 8}
\end{array}
$$

$\dfrac{6t^4 - 5t^3 - 8t^2 + 16t - 8}{3t^2 + 2t - 4} = 2t^2 - 3t + 2$

23. $A = LW$

$18x^3 - 21x^2 + 11x - 2 = (6x^2 - 5x + 2)W$

$W = \dfrac{18x^3 - 21x^2 + 11x - 2}{6x^2 - 5x + 2}$

$$
\begin{array}{r}
3x - 1 \\
6x^2 - 5x + 2 \overline{)18x^3 - 21x^2 + 11x - 2} \\
\underline{18x^3 - 15x^2 + 6x} \\
-6x^2 + 5x - 2 \\
\underline{6x^2 + 5x - 2}
\end{array}
$$

The width of the solar panel is $3x - 1$ meters.

25. The graphs of $y_1 = \dfrac{2x^2 - x - 10}{2x - 5}$ and $y_2 = x + 2$ coincide.

Cumulative Review Problems

27. $3x + 4(3x - 5) = -x + 12$

$3x + 12x - 20 = -x + 12$

$16x = 32$

$x = 2$

29. $2(x + 5) - 3y = 5x - (2 - y)$

$2x + 10 - 3y = 5x - 2 + y$

$3x = 12 - 4y$

$x = \dfrac{12 - 4y}{3}$

31. From the table, if one person likes Grape, then it is Curt

	Coke	RB	Grape	7-Up	Orange	Ale
Sylvia	yes	no	no			
Curt		yes		yes	no	
Fritz	yes		no	no		yes

5.3 Exercises

1. $(2x^2 - 11x - 8) \div (x - 6)$

$$
\begin{array}{r|rrr}
6 & 2 & -11 & -8 \\
 & & 12 & 6 \\
\hline
 & 2 & 1 & \underline{|-2}
\end{array}
$$

$(2x - 11x - 8) \div (x - 6) =$

$2x + 1 + \dfrac{-2}{x - 6}$

3. $(3x^3 + x^2 - x + 4) \div (x + 1)$

$$
\begin{array}{r|rrrr}
-1 & 3 & 1 & -1 & 4 \\
 & & -3 & 2 & -1 \\
\hline
 & 3 & -2 & 1 & \underline{|3}
\end{array}
$$

$(3x^3 + x^2 - x + 4) \div (x + 1)$

$= 3x^2 - 2x + 1 + \dfrac{3}{x + 1}$

5. $(x^3 + 7x^2 + 17x + 15) \div (x + 3)$

$$\begin{array}{r|rrrr} -3 & 1 & 7 & 17 & 15 \\ & & -3 & -12 & -15 \\ \hline & 1 & 4 & 5 & \underline{0} \end{array}$$

$(x^3 + 7x^2 + 17x + 15) \div (x + 3)$
$= x^2 + 4x + 5$

7. $(8x^3 - 30x^2 - 55x + 27) \div (x - 5)$

$$\begin{array}{r|rrrr} -3 & 8 & -30 & -55 & 27 \\ & & 40 & 50 & -25 \\ \hline & 8 & 10 & -5 & \underline{2} \end{array}$$

$(8x^3 - 30x^2 - 55x + 27) \div (x - 5)$
$= 8x^2 + 10x - 5 + \dfrac{2}{x - 5}$

9. $(x^3 - 2x^2 + 8) \div (x + 2)$

$$\begin{array}{r|rrrr} -2 & 1 & -2 & 0 & 8 \\ & & -2 & 8 & -16 \\ \hline & 1 & -4 & 8 & \underline{-8} \end{array}$$

$(x^3 - 2x^2 + 8) \div (x + 2)$
$= x^2 - 4x + 8 + \dfrac{-8}{x + 2}$

11. $(6x^4 + 13x^3 + 35x - 24) \div (x + 3)$

$$\begin{array}{r|rrrrr} -3 & 6 & 13 & 0 & 35 & -24 \\ & & -18 & 15 & -45 & 30 \\ \hline & 6 & -5 & 15 & -10 & \underline{6} \end{array}$$

(continued)

11. (continued)

$(x^3 - 2x^2 + 8) \div (x + 2)$
$= 6x^3 - 5x^2 + 15x - 10 + \dfrac{6}{x + 3}$

13. $(x^4 - 6x^3 + x^2 - 9) \div (x + 1)$

$$\begin{array}{r|rrrrr} -1 & 1 & -6 & 1 & 0 & -9 \\ & & -1 & 7 & -8 & 8 \\ \hline & 1 & -7 & 8 & -8 & \underline{-1} \end{array}$$

$(x^4 - 6x^3 + x^2 - 9) \div (x + 1)$
$= x^3 - 7x^2 + 8x - 8 + \dfrac{-1}{x + 1}$

15. $(3x^5 + x - 1) \div (x + 1)$

$$\begin{array}{r|rrrrrr} -1 & 3 & 0 & 0 & 0 & 1 & -1 \\ & & -3 & 3 & -3 & 3 & -4 \\ \hline & 3 & -3 & 3 & -3 & 4 & \underline{-5} \end{array}$$

$(3x^5 + x - 1) \div (x + 1)$
$= 3x^4 - 3x^3 + 3x^2 - 3x + 4 + \dfrac{-5}{x + 1}$

17. $(2x^5 + x^4 - 11x^3 + 13x^2 - x - 8) \div (x - 1)$

$$\begin{array}{r|rrrrrr} 1 & 2 & 1 & -11 & 13 & -1 & -8 \\ & & 2 & 3 & -8 & 5 & 4 \\ \hline & 2 & 3 & -8 & 5 & 4 & \underline{-4} \end{array}$$

$(2x^5 + x^4 - 11x^3 + 13x^2 - x - 8) \div (x - 1)$
$= 2x^4 + 3x^3 - 8x^2 + 5x + 4 + \dfrac{-4}{x - 1}$

19. $(x^6 - 4) \div (x + 1)$

$$\begin{array}{r|rrrrrrr} -1 & 1 & 0 & 0 & 0 & 0 & 0 & -4 \\ & & -1 & 1 & -1 & 1 & -1 & 1 \\ \hline & 1 & -1 & 1 & -1 & 1 & -1 & \underline{-3} \end{array}$$

$(x^6 - 4) \div (x + 1)$

$= x^5 - x^4 + x^3 - x^2 + x - 1 + \dfrac{-3}{x + 1}$

21. $(x^3 + 2.5x^2 - 3.6x + 5.4) \div (x - 1.2)$

$$\begin{array}{r|rrrr} 1.2 & 1 & 2.5 & -3.6 & 5.4 \\ & & 1.2 & 4.44 & 1.008 \\ \hline & 1 & 3.7 & 0.84 & \underline{6.408} \end{array}$$

$(x^3 + 2.5x^2 - 3.6x + 5.4) \div (x - 1.2)$

$= x^2 + 3.7x + 0.84 + \dfrac{6.408}{x - 1.2}$

23. $(x^4 + 3x^3 - 2x^2 + bx + 5) \div (x + 3)$

$$\begin{array}{r|rrrrr} -3 & 1 & 3 & -2 & b & 5 \\ & & -3 & 0 & 6 & -3b-18 \\ \hline & 1 & 0 & -2 & b+6 & \underline{5-3b-18} \end{array}$$

Remainder $= 0 \Rightarrow 5 - 3b - 18 = 0$

$$b = -\frac{13}{3}$$

25. $(2x^3 - 3x^2 + 6x + 4) \div (2x + 1)$

$$\begin{array}{r|rrrr} -\dfrac{1}{2} & 2 & -3 & 6 & 4 \\ & & -1 & 2 & -4 \\ \hline & 2 & -4 & 8 & \underline{0} \end{array}$$

(continued)

25. (continued)

$(2x^3 - 3x^2 + 6x + 4) \div (2x + 1)$

$= \dfrac{2x^2 - 4x + 8}{2} = x^2 - 2x + 4$

27. We are using the basic property of fractions that for any nonzero polynomial a, b, and c, $\dfrac{ac}{bc} = \dfrac{a}{b}$.

Cumulative Review Problems

29. $2,000,000 \text{ gallon} \cdot \dfrac{0.134 \text{ cubic feet}}{\text{gallon}}$

$= 268,000 \text{ cubic feet}$

31. $p(x) = 2x^4 - 3x^2 + 6x - 1$

$p(-3) = 2(-3)^4 - 3(-3)^2 + 6(-3) - 1$

$p(-3) = 116$

5.4 Exercises

1. $-30 - 15y = -15(2 + y)$

3. $xy - 3x^2y = xy(1 - 3x)$

5. $3c^2x^3 - 9cx - 6c = 3c(cx^3 - 3x - 2)$

7. $2x^3 - 8x^2 + 12x = 2x(x^2 - 4x + 6)$

9. $9a^2b^2 - 36ab + 45ab^2$
   $= 9ab(ab - 4 + 5b)$

11. $12xy^3 - 24x^3y^2 + 36x^2y^4 - 60x^4y^3$
   $= 12xy^2(y - 2x^2 + 3xy^2 - 5x^3y)$

13. $3x(x + y) - 2(x + y)$
   $= (x + y)(3x - 2)$

15. $5b(a-3b)+8(-3b+a)$
$$=5b(a-3b)+8(a-3b)$$
$$=(a-3b)(5b+8)$$

17. $3x(a+5b)+(a+5b)=(a+5b)(3x+1)$

19. $2a^2(3x-y)-5b^3(3x-y)$
$$=(3x-y)(2a^2-5b^3)$$

21. $3x(5x+y)-8y(5x+y)-(5x+y)$
$$=(5x+y)(3x-8y-1)$$

23. $x^3+5x^2+3x+15$
$$=x^2(x+5)+3(x+5)$$
$$=(x+5)(x^2+3)$$

25. $8x+8-5xy-5y$
$$=8(x+1)-5y(x+1)$$
$$=(x+1)(8-5y)$$

27. $7ax-7ay+x-y$
$$=7a(x-y)+(x-y)$$
$$=(x-y)(7a+1)$$

29. $yz^2-15-3z^2+5y$
$$=yz^2-3z^2+5y-15$$
$$=z^2(y-3)+5(y-3)$$
$$=(y-3)(z^2+5)$$

31. $40x^2+18by^2-15bx-48xy^2$
$$=40x^2-15bx-48xy^2+18by^2$$
$$=5x(8x-3b)-6y^2(8x-3b)$$
$$=(3x-3b)(5x-6y^2)$$

33. $\dfrac{1}{3}x^3+\dfrac{1}{2}x^2+\dfrac{1}{6}x=x\left(\dfrac{1}{3}x^2+\dfrac{1}{2}x+\dfrac{1}{6}\right)$

Cumulative Review Problems

35. $6x-2y=-12$

x	y
-2	0
-1	3
0	6

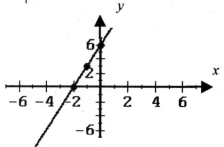

37. $m=\dfrac{y_2-y_1}{x_2-x_1}=\dfrac{-1-3}{6-2}=-1$

39. $x=$ number of multiple choice questions answered correctly
$$4x+5(22-4-x)=82$$
$$4x+110-20-5x=82$$
$$-x=-8$$
$$x=8$$
The student answered 8 multiple-choice questions correctly.

5.5 Exercises

1. $x^2+3x-28=(x-4)(x+7)$

3. $x^2+x-30=(x+6)(x-5)$

5. $x^2+10x+24=(x+6)(x+4)$

7. $a^2+4a-45=(a+9)(a-5)$

9. $x^2-9xy+20y^2=(x-4y)(x-5y)$

11. $x^2 - 15xy + 14y^2 = (x - 14y)(x - y)$

13. $x^4 - 3x^2 - 40 = (x^2 - 8)(x^2 + 5)$

15. $x^4 + 16x^2y^2 + 63y^4$
$= (x^2 + 7y^2)(x^2 + 9y^2)$

17. $2x^2 + 26x + 44 = 2(x^2 + 13x + 22)$
$= 2(x + 11)(x + 2)$

19. $x^3 + x^2 - 20x = x(x^2 + x - 20)$
$= x(x + 5)(x - 4)$

21. $30x^2 - x - 1 = (6x + 1)(5x - 1)$

23. $6x^2 - 7x - 5 = (3x - 5)(2x + 1)$

25. $3a^2 - 8a + 5 = (3a - 5)(a - 1)$

27. $8a^2 + 14a - 9 = (4a + 9)(2a - 1)$

29. $2x^2 + 13x + 15 = (2x + 3)(x + 5)$

31. $3x^4 - 8x^2 - 3 = (3x^2 + 1)(x^2 - 3)$

33. $6x^2 + 35xy + 11y^2 = (3x + y)(2x + 11y)$

35. $7x^2 + 11xy - 6y^2 = (7x - 3y)(x + 2y)$

37. $4x^3 + 4x^2 - 15x = x(4x^2 + 4x - 15)$
$= x(2x + 5)(2x - 3)$

39. $12x^3 + 66x^2 + 30x = 6x(2x^2 + 11x + 5)$
$= 6x(2x + 1)(x + 5)$

41. $x^2 - 2x - 63 = (x - 9)(x + 7)$

43. $6x^2 + x - 2 = (3x + 2)(2x - 1)$

45. $x^2 - 20x + 51 = (x - 17)(x - 3)$

47. $15x^2 + x - 2 = (5x + 2)(3x - 1)$

49. $2x^2 + 4x - 96 = 2(x^2 + 2x - 48)$
$= 2(x + 8)(x - 6)$

51. $18x^2 + 21x + 6 = 3(6x^2 + 7x + 2)$
$= 3(3x + 2)(2x + 1)$

53. $7x^3 - 28x^2 - 35x = 7x(x^2 - 4x - 5)$
$= 7x(x - 5)(x + 1)$

55. $6x^3 + 26x^2 - 20x = 2x(3x^2 + 13x - 10)$
$= 2x(3x - 2)(x + 5)$

57. $3x^4 - 2x^2 - 5 = (3x^2 - 5)(x^2 + 1)$

59. $7x^2 - 22xy + 3y^2 = (7x - y)(x - 3y)$

61. $x^6 - 10x^3 - 39 = (x^3 - 13)(x^3 + 3)$

63. $4x^3y + 2x^2y - 2xy = 2xy(2x^2 + x - 1)$
$= 2xy(2x - 1)(x + 1)$

65. $30x^2 + 19x - 5 = (6x + 5)(5x - 1)$
One possibility would be $6x + 5$ rows
with $5x - 1$ trees in each row. Another
possibility would be $5x - 1$ rows
with $6x + 5$ trees in each row.

Cumulative Review Problems

67. $A = \pi r^2 \approx 3.14(3)^2 = 28.26$

69. (a) $m = \dfrac{48}{156} = \dfrac{4}{13} \approx 30.8\%$

(b) Yes, the hill violates the city ordinance because the hill has a slope of 30.8% which is greater than 30%.

71. $x =$ number of racks
$y =$ number of helmets
$x + y = 120$
$60x + 70y = 7950$

$-60x - 60y = -7200$
$\underline{60x + 70y = 7950}$
$10y = 750$
$y = 75$

$x + 75 = 120$

$x = 45$

They should stock 45 bike racks and 75 helmets.

5.6 Exercises

1. $16x^2 - 81 = (4x + 9)(4x - 9)$

3. $64x^2 - 1 = (8x + 1)(8x - 1)$

5. $81x^4 - 1 = (9x^2 + 1)(9x^2 - 1)$
$\qquad = (9x^2 + 1)(3x + 1)(3x - 1)$

7. $49m^2 - 9n^2 = (7m + 3n)(7m - 3n)$

9. $1 - 81x^2 y^2 = (1 + 9xy)(1 - 9xy)$

11. $100y^2 - 81 = (10y + 9)(10y - 9)$

13. $32x^2 - 18 = 2(16x^2 - 9)$
$\qquad = 2(4x + 3)(4x - 3)$

15. $5x - 20x^3 = 5x(1 - 4x^2)$
$\qquad = 5x(1 + 2x)(1 - 2x)$

17. $49x^2 - 14x + 1 = (7x - 1)^2$

19. $16y^2 - 8y + 1 = (4y - 1)^2$

21. $81w^2 + 36wt + 4t^2 = (9w + 2t)^2$

23. $36x^2 - 60xy + 25y^2 = (6x - 5y)^2$

25. $8x^2 + 24x + 18 = 2(4x^2 + 12x + 9)$
$\qquad = 2(2x + 3)^2$

27. $3x^3 - 24x^2 + 48x = 3x(x^2 - 8x + 16)$
$\qquad = 3x(x - 4)^2$

29. $8x^3 + 27 = (2x + 3)(4x^2 - 6x + 9)$

31. $x^3 + 125 = (x + 5)(x^2 - 5x + 25)$

33. $64x^3 - 1 = (4x - 1)(16x^2 + 4x + 1)$

35. $125x^3 - 8 = (5x - 2)(25x^2 + 10x + 4)$

37. $1 - 27x^3 = (1 - 3x)(1 + 3x + 9x^2)$

39. $64x^3 + 125 = (4x + 5)(16x^2 - 20x + 25)$

41. $64s^6 + t^6 = (4s^2 + t^2)(16s^4 - 4s^2 t^2 + t^4)$

43. $6y^3 - 6 = 6(y^3 - 1)$
$\qquad = 6(y - 1)(y^2 + y + 1)$

45. $3x^3 - 24 = 3(x^3 - 8)$
$\qquad = 3(x - 2)(x^2 + 2x + 4)$

47. $x^5 - 8x^2 y^3 = x^2(x^3 - 2y^3)$
$$= x^2(x - 2y)(x^2 + 2xy + 4y^2)$$

49. $25w^6 - 1 = (5w^3 + 1)(5w^3 - 1)$

51. $8w^6 + 8w^3 + 2 = 2(4w^6 + 4w^3 + 1)$
$$= 2(2w^3 + 1)$$

53. $24a^3 - 3b^3 = 3(8a^3 - b^3)$
$$= 3(2a - b)(4a^2 + 2b + b^2)$$

55. $125m^3 + 8n^3$
$$= (5m + 2n)(25m^2 - 10mn + 4n^2)$$

57. $9x^2 - 100y^2 = (3x + 10y)(3x - 10y)$

59. $4w^2 - 20wz + 25z^2 = (2x - 5z)^2$

61. $36a^2 - 81b^2 = 9(4a^2 - 9b^2)$

63. $64x^5 + x^2 y^3 z^3$
$$= x^2(64x^3 + y^3 z^3)$$
$$= x^2(4x + yz)(16x^2 - 4xyz + y^2 z^2)$$

65. $81x^4 - 36x^2 + 4 = (9x^2 - 2)^2$

67. $16x^4 - 81y^4$
$$= (4x^2 + 9y^2)(4x - 9y^2)$$
$$= (4x^2 + 9y^2)(2x + 3y)(2x - 3y)$$

69. $25x^2 + 25x + 4 = (5x)^2 + 25x + 2^2$
$$25 \neq 2(5)(2) = 20$$
$$25x^2 + 25x + 4 = (5x + 4)(5x + 1)$$

71. $4x^2 - 15x + 9 = (2x)^2 - 15x + (3)^2$
$$2(2x)(3) = 12x \neq 15x$$
$$4x^2 - 15x + 9 = (4x - 3)(x - 3)$$

73. $A = (4x)(4x) - y^2$
$$A = 16x^2 - y^2$$
$$A = (4x - y)(4x + y)$$

Cumulative Review Problems

75. $3200x + 29,000$
$$= 1200x + 27,000 + 12,000$$
$$= 1200x + 39,000$$
$$3200x + 29,000 = 1200x + 39,000$$
$$2000x = 10,000$$

$x = 5$
In 5 years, the year 2001, the
bachelor's degree in math will be
offered \$12,000 per year more than a
bachelor's degree in marketing.

77. $x = $ length of second side
$$\frac{2}{3}x + x + x - 14 = 66$$
$$\frac{8}{3}x = 80$$
$$x = 30$$
$$\frac{2}{3}x = 20$$
$$x - 14 = 16$$
The first side is 20 cm, the second side
is 30 cm, and the third side is 16 cm.

5.7 Exercises

1. In any factoring problem the first step
is to factor out <u>a common factor if</u>
<u>possible.</u>

3. $3xy - 6yz = 3y(x - 2z)$

5. $3x^2 - 8x + 5 = (3x - 5)(x - 1)$

7. $8x^3 - 125y^3$
$= (2x - 5y)(4x^2 + 10xy + 25y^2)$

9. $x^2 + 2xy - xz = x(x + 2y - z)$

11. $3x^2 - x - 1$ is prime.

13. $6x^2 - 23x - 4 = (6x + 1)(x - 4)$

15. $3x^2 - 3x - xy + y = 3x(x - 1) - y(x - 1)$
$= (x - 1)(3x - y)$

17. $81a^4 - 1 = (9a^2 + 1)(9a^2 - 1)$
$= (9a^2 + 1)(3a + 1)(3a - 1)$

19. $2x^5 - 16x^3 - 18x$
$= 2x(x^4 - 8x^2 - 9)$
$= 2x(x^2 + 1)(x^2 - 9)$
$= 2x(x^2 + 1)(x + 3)(x - 3)$

21. $6x^3 - 9x^2 - 15x = 3x(2x^2 - 3x - 5)$
$= 3x(2x - 5)(x + 1)$

23. $4x^2 - 8x - 6 = 2(2x^2 - 4x - 3)$

25. $y^2 - 5y + 7$ is prime.

27. $6a^2 - 6a - 36 = 6(a^2 - a - 6)$
$= 6(a - 3)(a + 2)$

29. $64 + 49y^2$ is prime.

31. $2x^4 - 3x^2 - 5 = (2x^2 - 5)(x^2 + 1)$

33. $12x^2 + 11x - 2$ is prime.

35. $4x^4 + 20x^2y^4 + 25y^8 = (2x^2 + 5y^4)^2$

37. $S = 5x(x - 10) + 8y(x - 10)$
$S = (x - 10)(5x + 8y)$ square feet
$S = 5x^2 - 50x + 8xy - 80y$ square feet

Cumulative Review Problems

39. $3x - 2 \le -5 + 2(x - 3)$
$3x - 2 \le -5 + 2x - 6$
$x \le -9$

41. $\left| \dfrac{1}{3}(5 - 4x) \right| > 4$

$\dfrac{1}{3}(5 - 4x) < -4$ or $\dfrac{1}{3}(5 - 4x) > 4$

$5 - 4x < -12$ $5 - 4x > 12$

$-4x < -17$ $-4x > 7$

$x > \dfrac{17}{4}$ $x < -\dfrac{7}{4}$

43. $\dfrac{255 + 206 + 254 + 320}{4} = 258.75$

The average value of the net receipts for a 2-year period for the Republican Party was $258.75 million.

45. $189 + (0.436)(189) = 271.404$

The expected net receipts for the Democratic Party in the 2001-2002 period would be approximately $271.4 million.

5.8 Exercises

1. $x^2 - 13x = -36$

$x^2 - 13x + 26 = 0$

$(x-9)(x-4) = 0$

$x - 9 = 0$ or $x - 4 = 0$

$x = 9 \qquad\qquad x = 4$

Check:

$9^2 - 13(9) \overset{?}{=} -36$

$-36 = -36$

$4^2 - 13(4) \overset{?}{=} -36$

$-36 = -36$

3. $5x^2 - 6x = 0$

$x(5x - 6) = 0$

$x = 0$ or $5x - 6 = 0$

$$x = \frac{6}{5}$$

Check:

$5(0)^2 - 6(0) \overset{?}{=} 0$

$0 = 0$

$5\left(\frac{6}{5}\right)^2 - 6\left(\frac{6}{5}\right) \overset{?}{=} 0$

$0 = 0$

5. $3x^2 - x - 2 = 0$

$(3x + 2)(x - 1) = 0$

$3x + 2 = 0$ or $x - 1 = 0$

$$x = -\frac{2}{3} \qquad\qquad x = 1$$

(continued)

5. (continued)

Check:

$3\left(-\frac{2}{3}\right)^2 - \left(-\frac{2}{3}\right) - 2 \overset{?}{=} 0$

$0 = 0$

$3(1)^2 - 1 - 2 \overset{?}{=} 0$

$0 = 0$

7. $3x^2 - 2x - 8 = 0$

$(3x + 4)(x - 2) = 0$

$3x + 4 = 0$ or $x - 2 = 0$

$$x = -\frac{4}{3} \qquad\qquad x = 2$$

Check:

$3\left(-\frac{4}{3}\right)^2 - 2\left(-\frac{4}{3}\right) - 8 \overset{?}{=} 0$

$0 = 0$

$3(2)^2 - 2(2) - 8 \overset{?}{=} 0$

$0 = 0$

9. $8x^2 - 3 = 2x$

$8x^2 - 3 = 2x$

$8x^2 - 2x - 3 = 0$

$(4x - 3)(2x + 1) = 0$

$4x + 3 = 0$ or $2x + 1 = 0$

$$x = -\frac{3}{4} \qquad\qquad x = -\frac{1}{2}$$

Check:

$8\left(\frac{3}{4}\right)^2 - 3 \overset{?}{=} 2\left(\frac{3}{4}\right)$

$$\frac{3}{2} = \frac{3}{2}$$

(continued)

9. (continued)

$$8\left(-\frac{1}{2}\right)^2 - 3 \overset{?}{=} 2\left(-\frac{1}{2}\right)$$

$$-1 = -1$$

11. $8x^2 = 11x - 3$

$$8x^2 - 11x + 3 = 0$$

$$(8x - 3)(x - 1) = 0$$

$$8x - 3 = 0 \text{ or } x - 1 = 0$$

$$x = \frac{3}{8} \qquad\qquad x = 1$$

Check:

$$8\left(\frac{3}{8}\right)^2 \overset{?}{=} 11\left(\frac{3}{8}\right) - 3$$

$$\frac{9}{8} = \frac{9}{8}$$

$$8(1)^2 \overset{?}{=} 11(1) - 3$$

$$8 = 8$$

13. $x^2 + \frac{5}{3}x = \frac{2}{3}x$

$$x^2 + x = 0$$

$$x(x + 1) = 0$$

$$x = 0 \text{ or } x + 1 = 0$$

$$x = -1$$

Check:

$$0^2 + \frac{5}{3}(0) \overset{?}{=} \frac{2}{3}(0)$$

$$0 = 0$$

$$1^2 + \frac{5}{3}(-1) \overset{?}{=} \frac{2}{3}(-1)$$

$$-\frac{2}{3} = -\frac{2}{3}$$

15. $2x^2 + 3x = -14x^2 + 5x$

$$16x^2 - 2x = 0$$

$$2x(8x - 1) = 0$$

$$2x = 0 \text{ or } 8x - 1 = 0$$

$$x = 0 \qquad\qquad x = \frac{1}{8}$$

Check:

$$2(0)^2 + 3(0) \overset{?}{=} -14(0)^2 + 5(0)$$

$$0 = 0$$

$$2\left(\frac{1}{8}\right)^2 + 3\left(\frac{1}{8}\right) \overset{?}{=} -14\left(\frac{1}{8}\right)^2 + 5\left(\frac{1}{8}\right)$$

$$\frac{13}{32} = \frac{13}{32}$$

17. $x^3 + 5x^2 + 6x = 0$

$$x(x^2 + 5x + 6) = 0$$

$$x(x + 3)(x + 2)$$

$$x = 0 \text{ or } x = -3 \text{ or } x = -2$$

Check:

$$0^3 + 5(0)^2 + 6(0) \overset{?}{=} 0$$

$$0 = 0$$

$$(-3)^3 + 5(-3)^2 + 6(-3) \overset{?}{=} 0$$

$$0 = 0$$

$$(-2)^3 + 5(-2)^2 + 6(-2) \overset{?}{=} 0$$

$$0 = 0$$

19. $x^3 = x^2 + 20x$

$$x^3 - x^2 - 20x = 0$$

$$x(x^2 - x - 20) = 0$$

$$x(x - 5)(x + 4) = 0$$

$$x = 0 \text{ or } x - 5 = 0 \text{ or } x + 4 = 0$$

$$x = 0 \qquad x = 5 \qquad x = -4$$

(continued)

19. (continued)
Check:

$$0^3 \overset{?}{=} 0^2 + 20(0)$$

$$0 = 0$$

$$5^3 \overset{?}{=} 5^2 + 20(5)$$

$$125 = 125$$

$$(-4)^3 \overset{?}{=} (-4)^2 + 20(-4)$$

$$-64 = -64$$

21. $3x^3 - 10x = 17x$

$$3x^3 - 27x = 0$$

$$3x(x^2 - 9) = 3x(x+3)(x-3)$$

$$3x = 0 \text{ or } x+3 = 0 \text{ or } x-3 = 0$$

$$x = 0 \qquad x = -3 \qquad x = 3$$

Check:

$$3(0)^3 - 10(0) \overset{?}{=} 17(0)$$

$$0 = 0$$

$$3(-3)^3 - 10(-3) \overset{?}{=} 17(-3)$$

$$-51 = -51$$

$$3(3)^3 - 10(3) \overset{?}{=} 17(3)$$

$$51 = 51$$

23. $3x^3 + 15x^2 = 42x$

$$3x^3 + 15x^2 - 42x = 0$$

$$3x(x^2 + 5x - 14) = 0$$

$$3x(x+7)(x-2) = 0$$

$$3x = 0 \text{ or } x+7 = 0 \text{ or } x-2 = 0$$

$$x = 0 \qquad x = -7 \qquad x = 2$$

Check:

$$3(0)^3 + 15(0)^2 \overset{?}{=} 42(0)$$

$$0 = 0$$

(continued)

23. (continued)

$$3(-7)^3 + 15(-7)^2 \overset{?}{=} 42(-7)$$

$$-294 = -294$$

$$3(2)^3 + 15(2)^2 \overset{?}{=} 42(2)$$

$$84 = 84$$

25. $\dfrac{7x^2 - 3}{2} = 2x$

$$7x^2 - 3 = 4x$$

$$7x^2 - 4x - 3 = 0$$

$$(7x+3)(x-1) = 0$$

$$7x + 3 = 0 \text{ or } x - 1 = 0$$

$$x = -\frac{3}{7} \qquad\qquad x = 1$$

Check:

$$\dfrac{7\left(-\dfrac{3}{7}\right)^2 - 3}{2} \overset{?}{=} 2\left(-\dfrac{3}{7}\right)$$

$$-\frac{6}{7} = -\frac{6}{7}$$

$$\dfrac{7(1)^2 - 3}{2} \overset{?}{=} 2(1)$$

$$2 = 2$$

27. $2(x+3) = -3x + 2(x^2 - 3)$

$$2x + 6 = -3x + 2(x^2 - 3)$$

$$2x^2 - 5x - 12 = 0$$

$$(2x+3)(x-4) = 0$$

$$2x + 3 = 0 \text{ or } x - 4 = 0$$

$$x = -\frac{3}{2} \qquad\qquad x = 4$$

(continued)

27. (continued)
Check:

$$2\left(-\frac{3}{2}+3\right)\overset{?}{=}-3\left(-\frac{3}{2}\right)+2\left(\left(-\frac{3}{2}\right)^2-3\right)$$

$$3 = 3$$

$$2(4+3)\overset{?}{=}-3(4)+2(4^2-3)$$

$$14 = 14$$

29. $7x^2+6=2x^2+2(4x+3)$

$$5x^2+6=8x+6$$

$$5x^2-8x=0$$

$$x(5x-8)=0$$

$$x=0 \ \text{ or } \ 5x-8=0$$

$$x=\frac{8}{5}$$

Check:

$$7(0)^2+6\overset{?}{=}2(0)^2+2(4(0)+3)$$

$$6 = 6$$

$$7\left(\frac{8}{5}\right)^2+6\overset{?}{=}2\left(\frac{8}{5}\right)^2+2\left(4\left(\frac{8}{5}\right)+3\right)$$

$$\frac{598}{25}=\frac{598}{25}$$

31. $x^2+bx-12=0$

$$x=-4 \Rightarrow x+4 \text{ is a factor.}$$

$$(x+4)(x+k)=0$$

$$x^2+(4+k)x+4k=0 \text{ from which}$$

$$4k=-12$$

$$k=-3$$

$$b=4+k=4-3$$

$$b=1$$

$$x+k=0, \ x-3=0$$

$$x=3 \text{ is the other solution.}$$

33. $A=\frac{1}{2}bh=87$

$$\frac{1}{2}b(4b+5)=87$$

$$4b^2+5b-174=0$$

$$(b-6)(4b+29)=0$$

$$b=6 \ \text{ or } \ 4b+29=0$$

$$b=-\frac{29}{4}, \text{ reject, } b>0$$

$$h=4b+5=4(6)+5=29$$

The base of the triangular logo is
6 mm and the altitude is 29 mm.

35. $A=\frac{1}{2}bh=119$

$$\frac{1}{2}(4h+6)h=119$$

$$4h^2+6h=238$$

$$2h^2+3h-119=0$$

$$(h-7)(2h+17)=0$$

$$h-7=0 \ \text{ or } \ 2h+17=0$$

$$h=7 \qquad\qquad h=-\frac{17}{2}$$

$$\text{reject, } h>0$$

$$b=4h+6=4(7)+6=34$$

The altitude is 7 yards and the base is
34 yards.

37. $A=LW=896$
$$W(W+4)=896$$

$$W^2+4W-896=0$$

$$(W-28)(W+32)=0$$

$$W-28=0 \ \text{ or } \ W+32=0$$

$$W=28 \qquad\qquad W=-32$$

$$\text{reject, } W>0$$

$$L=W+4=32 \text{ (continued)}$$

37. (continued)
  (a) The width is 28 cm and the length is 32 cm.
  (b) The width is 280 mm and the length is 320 mm.

39. $s^2 = 4s + 165$

$s^2 - 4s - 165 = 0$

$(s - 15)(s + 11) = 0$

$s - 15 = 0$ or $s + 15 = 0$

$s = 15 \qquad\qquad s = -15$

$\qquad\qquad\qquad$ reject, $s > 0$

Each side of the room is 15 feet.

41. $V = LWH = 198$
$\quad L \cdot 2(L + 2) = 198$

$L^2 + 2L - 99 = 0$

$(L + 11)(L - 9) = 0$

$L + 11 = 0$ or $L - 9 = 0$

$L = -11 \qquad\qquad L = 9$

reject, $L > 0$

$H = L + 2 = 11$
The length is 9 inches and the height is 11 inches.

43. $A = LW = 85$
$\quad (2W + 7)W = 85$

$2W^2 + 7W - 85 = 0$

$(2W + 17)(W - 5) = 0$

$2W + 17 = 0$ or $W - 5 = 0$

$W = -\dfrac{17}{5} \qquad\qquad W = 5$

reject, $W > 0$

$L = 2W + 7 = 17$
The length of the fishing area is 17 miles and the width of the fishing area is 5 miles.

45. $x = $ length of old side
$2x + 1 = $ length on new side

$x^2 + 176 = (2x + 1)^2$

$3x^2 + 4x - 175 = 0$

$(3x + 25)(x - 7) = 0$

$3x + 25 = 0$ or $x - 7 = 0$

$x = -\dfrac{25}{7} \qquad\qquad x = 7$

reject, $x > 0$

The old target is a square 7 cm on a side and the new target is a square 15 cm on a side.

47. $P = 2n^2 - 19n - 10$

$410 = 2n^2 - 19n - 10$

$2n^2 - 19n - 420 = 0$

$(2n + 21)(n - 20) = 0$

$2n + 21 = 0$ or $n - 20 = 0$

$n = -\dfrac{21}{2} \qquad\qquad n = 20$

reject, $n > 0$
20 units are produced when the profit is \$410.

49. $P = 2n^2 - 19n - 10$

$-52 = 2n^2 - 19n + 42$

$2n^2 - 19n + 42 = 0$

$(2n - 7)$ or $(n - 6) = 0$

$n = \dfrac{7}{2} \qquad\qquad n = 6$

reject, n is a positive integer.
Producing 6 units will result in a loss of \$52.

51. $N = 28x^2 + 80x + 560$

$N = 28(20)^2 + 80(20) + 560$

$N = 13,360$

There were 13,360 mutual funds in the year 2000.

53. $N = 28x^2 + 80x + 560$

$668 = 28x^2 + 80x + 560$

$7x^2 + 20x - 27 = 0$

$(7x + 27)(x - 1) = 0$

$7x + 27 = 0$ or $x - 1 = 0$

$x = -\dfrac{27}{7}$ $\qquad$ $x = 1$

reject, $x > 0$

There were 668 mutual funds in 1981.

Cumulative Review Problems

55. $(2x^3 y^2)^3 (5xy^2)^2$

$= 2^3 (x^3)^3 (y^2)^3 5^2 x^2 (y^2)^2$

$= 8x^{3(3)} y^{2(3)} 25x^2 y^{2(2)}$

$= 200x^{9+2} y^{6+4}$

$= 200x^{11} y^{10}$

57. $(-3x^{-2} y^4 z)^{-2}$

$= (-3)^{-2} (x^{-2})^{-2} (y^4)^{-2} z^{-2}$

$= \dfrac{1}{(-3)^2} x^{-2(2)} y^{4(-2)} \dfrac{1}{z^2}$

$= \dfrac{1}{9z^2} x^4 y^{-8}$

$= \dfrac{x^4}{9y^8 z^2}$

Putting Your Skills To Work

1. $y = -x^3 + 23x^2 - 130x$

$y = -x(x - 10)(x - 13) = 0$ for

$x = 0, x = 10, x = 13$

x	0	10	13
y	0	0	0

2. Substitute the values

$x = 2, 4, 6, 8, 11, 12, 14$ into

$y = -x^3 + 23x^2 - 130x$ to obtain

x	2	4	6	8	11	12	14
y	−176	−216	−168	−80	22	24	−56

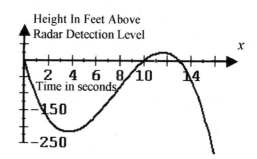

3. Substitute the values

$x = -3, -2, -2$ into

$y = -x^3 + 23x^2 - 130x$ to obtain

x	−3	−2	−1
y	624	360	154

The aircraft is above the minimum radar distance for time prior to first dropping off the radar screen.

4. Substituting the values

$x = 18, 19, 20$ gives

x	18	19	20
y	−720	−1026	−1400

A time between 18 and 19 seconds.

Chapter 5 Review Problems

1. $(x^2 - 3x + 5) + (-2x^2 - 7x + 8)$
$= x^2 - 3x + 5 - 2x^2 - 7x + 8$
$= -x^2 - 10x + 13$

2. $(-4x^2 - 7xy + y) + (5x^2y + 2xy - 9y)$
$= -4x^2 - 7xy + y + 5x^2y + 2xy - 9y$
$= x^2y - 5xy - 8y$

3. $(-6x^2 + 7xy - 3y^2) - (5x^2 - 3xy - 9y^2)$
$= -6x^2 + 7xy - 3y^2 - 5x^2 + 3xy + 9y^2$
$= -11x^2 + 10xy + 6y^2$

4. $(-13x^2 + 9x - 14) - (-2x^2 - 6x + 1)$
$= -13x^2 + 9x - 14 + 2x^2 + 6x - 1)$
$-11x^2 + 15x - 15$

5. $(7x - 2) + (5 - 3x) + (2 - 2x)$
$= 7x - 2 + 5 - 3x + 2 - 2x$
$= 2x + 5$

6. $(5x - 2x^2 - x^3) - (2x - 3 + 5x^2)$
$= 5x - 2x^2 - x^3 - 2x + 3 - 5x^2$
$= -x^3 - 7x^2 + 3x + 3$

7. $p(x) = 3x^3 - 2x^2 - 6x + 1$
$p(-4) = 3(-4)^3 - 2(-4)^2 - 6(-4) + 1$
$p(-4) = -199$

8. $p(x) = 3x^3 - 2x^2 - 6x + 1$
$p(3) = 3(3)^3 - 2(3)^2 - 6(3) + 1$
$p(3) = 46$

9. $g(x) = -2x^4 + x^3 - 5x - 2$
$g(2) = -2(2)^4 + (2)^3 - 5(2) - 2 = 36$

10. $g(x) = -2x^4 + x^3 - 5x - 2$
$g(-3) = -2(-3)^4 + (-3)^3 - 5(-3) - 2$
$g(-3) = -176$

11. $h(x) = -x^3 - 6x^2 + 12x - 4$
$h(3) = -(3)^3 - 6(3)^2 + 12(3) - 4$
$h(3) = -49$

12. $h(x) = -x^3 - 6x^2 + 12x - 4$
$h(-2) = -(-2)^3 - 6(-2)^2 + 12(-2) - 4$
$h(-2) = -44$

13. $3xy(x^2 - xy + y^2)$
$= 3x^3y - 3x^2y^2 + 3xy^3$

14. $(3x^2 + 1)(2x - 1) = 6x^3 - 3x^2 + 2x - 1$

15. $(5x^2 + 3)^2 = 25x^4 + 30x^2 + 9$

16. $(x - 3)(2x - 5)(x + 2)$
$= (2x^2 - 11x + 15)(x + 2)$
$= 2x^3_* + 4x^2 - 11x^2 - 22x + 15x + 30$
$= 2x^3 - 7x^2 - 7x + 30$

17. $(x^2 - 3x + 1)(-2x^2 + x - 2)$
$= -2x^4 + x^3 - 2x^2$
$\qquad + 6x^3 - 3x^2 + 6x$
$\qquad\qquad - 2x^2 + x - 2$
$= -2x^4 + 7x^3 - 7x^2 + 7x - 2$

18. $(3x - 5)(3x^2 + 2x - 4)$
$= 9x^3 + 6x^2 - 12x - 15x^2 - 10x + 20$
$= 9x^3 - 9x^2 - 22x + 20$

19. $(25x^3y - 15x^2y - 100xy) \div (-5xy)$

$$\frac{25x^3y - 15x^2y - 100xy}{-5xy}$$

$$= \frac{25x^3y}{-5xy} - \frac{15x^2y}{-5xy} - \frac{100xy}{-5xy}$$

$$= -5x^2 + 3x + 20$$

20. $(12x^2 - 16x - 4) \div (3x + 2)$

$$
\begin{array}{r}
4x - 8 \\
3x+2{\overline{\smash{\big)}\,12x^2 - 16x - 4}} \\
\underline{12x^2 + 8x} \\
-24x - 4 \\
\underline{-24x - 16} \\
12
\end{array}
$$

$(12x^2 - 16x - 4) \div (3x + 2)$

$$= 4x - 8 + \frac{12}{3x + 2}$$

21. $(2x^3 - 7x^2 + 2x + 8) \div (x - 2)$

$$
\begin{array}{r}
2x^2 - 3x - 4 \\
x-2{\overline{\smash{\big)}\,2x^3 - 7x^2 + 2x + 8}} \\
\underline{2x^3 - 4x^2} \\
-3x^2 + 2x \\
\underline{-3x^2 + 6x} \\
-4x + 8 \\
\underline{-4x + 8}
\end{array}
$$

$(2x^3 - 7x^2 + 2x + 8) \div (x - 2)$

$$= 2x^2 - 3x - 4$$

22. $(3y^3 - 2y + 5) \div (y - 3)$

$$
\begin{array}{r}
3y^2 + 9y + 25 \\
y-3{\overline{\smash{\big)}\,3y^3 + 0y^2 - 2y + 5}} \\
\underline{3y^3 - 9y^2} \\
9y^2 - 2y \\
\underline{9y^2 - 27y} \\
25y + 5 \\
\underline{25y - 75} \\
80
\end{array}
$$

$(3y^3 - 2y + 5) \div (y - 3)$

$$= 3y^2 + 9y + 25 + \frac{80}{y - 3}$$

23. $(15a^4 - 3a^3 + 4a^2 + 4) \div (3a^2 - 1)$

$$
\begin{array}{r}
5a^2 - a + 3 \\
3a^2-1{\overline{\smash{\big)}\,15a^4 - 3a^3 + 4a^2 + 0a + 4}} \\
\underline{15a^4 \qquad\quad - 5a^2} \\
-3a^3 + 9a^2 \\
\underline{-3a^3 \qquad\quad + a} \\
9a^2 \quad - a \\
\underline{9a^2 \qquad\quad - 3} \\
-a + 7
\end{array}
$$

$(15a^4 - 3a^3 + 4a^2 + 0a + 4) \div (3a^2 - 1)$

$$= 5a^2 - a + 3 + \frac{-a + 7}{3a^2 - 1}$$

24. $(x^4 - x^3 - 7x^2 - 7x - 2) \div (x^3 - 3x - 2)$

$$
\begin{array}{r}
x^2 + 2x + 1 \\
x^3 - 3x - 2 \overline{\smash{)}\ x^4 - x^3 - 7x^2 - 7x - 2} \\
\underline{x^4 - 3x^3 - 2x^2} \\
2x^3 - 5x^2 - 7x \\
\underline{2x^3 - 6x^2 - 4x} \\
x^2 - 3x - 2 \\
\underline{x^2 - 3x - 2}
\end{array}
$$

$(x^4 - x^3 - 7x^2 - 7x - 2) \div (x^3 - 3x - 2)$
$= 3x^3 - x^2 + x - 1$

25. $(2x^4 - 13x^3 + 16x^2 - 9x + 20) \div (x - 5)$

$$
\begin{array}{r}
2x^3 - 3x^2 + x - 4 \\
x - 5 \overline{\smash{)}\ 2x^4 - 13x^3 + 16x^2 - 9x + 2} \\
\underline{2x^4 - 10x^3} \\
-3x^3 + 16x^2 \\
\underline{-3x^3 + 15x^2} \\
x^2 - 9x \\
\underline{x^2 - 5x} \\
-4x + 20 \\
\underline{-4x + 20}
\end{array}
$$

$(2x^4 - 13x^3 + 16x^2 - 9x + 2) \div (x - 5)$
$= 2x^3 - 3x^2 + x - 4$

26. $(3x^4 + 5x^3 - x^2 + x - 2) \div (x + 2)$

$$
\begin{array}{r}
3x^3 - x^2 + x - 1 \\
x + 2 \overline{\smash{)}\ 3x^4 + 5x^3 - x^2 + x - 2} \\
\underline{3x^4 + 6x^3} \\
-x^3 - x^2 \\
\underline{-x^3 - 2x^2} \\
x^2 + x \\
\underline{x^2 + 2x} \\
-x - 2 \\
\underline{-x - 2}
\end{array}
$$

$(3x^4 + 5x^3 - x^2 + x - 2) \div (x + 2)$
$= 3x^3 - x^2 + x - 1$

27. $5x^2 - 11x + 2 = (5x - 1)(x - 2)$

28. $9x^2 - 121 = (3x + 11)(3x - 11)$

29. $36x^2 + 25$ is prime.

30. $x^2 - 8wy + 4wx - 2xy$
$= x^2 + 4wx - 8wy - 2xy$
$= x(x + 4w) - 2y(4w + x)$
$= x(x + 4w) - 2y(x + 4w)$
$= (x + 4w)(x - 2y)$

31. $x^3 + 8x^2 + 12x = x(x^2 + 8x + 12)$
$\qquad\qquad\qquad = x(x + 6)(x + 2)$

32. $2x^2 - 7x - 3$ is prime.

33. $x^2 + 6xy - 27y^2$
$= (x + 9y)(x - 3y)$

34. $27x^4 - x$

$= x(27x^3 - 1)$

$= x(3x - 1)(9x^2 + 3x + 1)$

35. $21a^2 + 20ab + 4b^2$

$= (7a + 2b)(3a + 2b)$

36. $-3a^3b^3 + 2a^2b^4 - a^2b^3$

$= -a^2b^3(3a - 2b + 1)$

37. $a^4b^4 + a^3b^4 - 6a^2b^4$

$= a^2b^4(a^2 + a - 6)$

$= a^2b^4(a + 3)(a - 2)$

38. $3x^4 - 5x^2 - 2$

$= (3x^2 + 1)(x^2 - 2)$

39. $9a^2b + 15ab - 14b = b(9a^2 + 15a - 14)$

$= b(3a + 7)(3a - 2)$

40. $2x^4 + 7x^2 - 6$ is prime.

41. $12x^2 + 12x + 3 = 3(4x^2 + 4x + 1)$

$= 3(2x + 1)^2$

42. $4y^4 - 13y^3 + 9y^2 = y^2(4y^2 - 13y + 9)$

$= y^2(4y - 9)(y - 1)$

43. $y^4 + 2y^3 - 35y^2 = y^2(y^2 + 2y - 35)$

$= y^2(y + 7)(y - 5)$

44. $4x^2y^2 - 12x^2y - 8x^2$

$= 4x^2(y^2 - 3y - 2)$

45. $3x^4 - 7x^2 - 6 = (3x^2 + 2)(x^2 - 3)$

46. $a^2 + 5ab^3 + 4b^6 = (a + b^2)(a + 4b^3)$

47. $3x^2 - 12 - 8x + 2x^3$

$= 2x^3 + 3x^2 - 8x - 12$

$= x^2(2x + 3) - 4(2x + 3)$

$= (x^2 - 4)(2x + 3)$

$= (x + 2)(x - 2)(2x + 3)$

48. $2x^4 - 12x^2 - 54$

$= 2(x^4 - 6x^2 - 27)$

$= 2(x^2 - 9)(x^2 + 3)$

$= 2(x + 3)(x - 3)(x^2 + 3)$

49. $8a + 8b - 4bx - 4ax$

$= 4(2a + 2b - ax - bx)$

$= 4(2(a + b) - x(a + b))$

$= 4(a + b)(2 - x)$

50. $8x^4 + 34x^2y^2 + 21y^4$

$= (4x^2 + 3y^2)(2x^2 + 7y^2)$

51. $4x^3 + 10x^2 - 6x = 2x(2x^2 + 5x - 3)$

$= 2x(2x - 1)(x + 3)$

52. $2a^2x - 15ax + 7x = x(2a^2 - 15x + 7)$

$= x(2a - 1)(x - 7)$

53. $16x^4y^2 - 56x^2y + 49 = (4x^2y - 7)^2$

54. $128x^3y - 2xy$

$= 2xy(64x^2 - 1)$

$= 2xy(8x - 1)(8x + 1)$

55. $5xb - 28y + 4by - 35x$

$= 5xb - 35x + 4by - 28y$

$= 5x(b - 7) + 4y(b - 7)$

$= (b - 7)(5x + 4y)$

56. $27abc^2 - 12ab$
$= 3ab(9c^2 - 4)$
$= 3ab(3c + 2)(3c - 2)$

57. $5x^2 - 9x - 2 = 0$
$(5x + 1)(x - 2) = 0$
$5x + 1 = 0$ or $x - 2 = 0$
$x = -\dfrac{1}{5}$ $x = 2$

58. $2x^2 - 11x + 12 = 0$
$(2x - 3)(x - 4) = 0$
$2x - 3 = 0$ or $x - 4 = 0$
$x = \dfrac{3}{2}$ $x = 4$

59. $(2x - 1)(3x - 5) = 20$
$6x^2 - 13x + 5 = 20$
$6x^2 - 13x - 15 = 0$
$(6x + 5)(x - 3) = 0$
$6x + 5 = 0$ or $x - 3 = 0$
$x = -\dfrac{5}{6}$ $x = 3$

60. $7x^2 = 21x$
$7x^2 - 21x = 0$
$7x(x - 21) = 0$
$7x = 0$ or $x - 21 = 0$
$x = 0$ $x = 21$

61. $x^3 + 7x^2 = -12x$
$x^3 + 7x^2 + 12x = 0$
$x(x^2 + 7x + 12) = 0$
$x(x + 4)(x + 3) = 0$
$x = 0$ or $x + 4 = 0$ or $x + 3 = 0$
$x = 0$ $x = -4$ $x = -3$

62. $3x^2 + 14x + 3 = -1 + 4(x + 1)$
$3x^2 + 14x + 3 = -1 + 4x + 4$
$3x^2 + 10x = 0$
$x(3x + 10) = 0$
$x = 0$ or $3x + 10 = 0$
$x = 0$ $x = -\dfrac{10}{3}$

63. $A = \dfrac{1}{2}bh = 77$
$b(b + 3) = 154$
$b^2 + 3b - 154 = 0$
$(b + 14)(b - 11) = 0$
$b + 14 = 0$ or $b - 11 = 0$
$b = -14$ $b = 11$
reject, $b > 0$ $h = b + 3 = 14$

The base of the triangle is 11 meters and the altitude is 14 meters.

64. $A = LW = 40$
$(3W - 2)W = 40$
$3W^2 - 2W - 40 = 0$
$(3W + 10)(W - 4) = 0$
$3W + 10 = 0$ or $W - 4 = 0$
$W = -\dfrac{10}{3}$ $W = 4$
reject, $W > 0$ $L = 3W - 2 = 10$

The length is 10 miles and the width is 4 miles.

65. $x =$ length of old side

$2x + 3 =$ length of new side

$x^2 + 24 = (2x + 3)^2$

$x^2 + 24 = 4x^2 + 12x + 9$

$3x^2 + 12x - 15 = 0$

$x^2 + 4x - 5 = 0$

$(x + 5)(x - 1) = 0$

$x + 5 = 0 \quad$ or $\quad x - 1 = 0$

$x = -5 \qquad\qquad x = 1$

reject, $x > 0, \ 2x + 3 = 5$

The old side is 1 yard and the new side
is 5 yards.

66. $P = 3x^2 - 7x - 10 = 30$

$3x^2 - 7x - 40 = 0$

$(3x + 8)(x - 5) = 0$

$3x + 8 = 0 \ $ or $\ x - 5 = 0$

$x = -\dfrac{8}{3} \qquad\qquad x = 5$

reject, $x > 0$

Five calculators should be made.

Chapter 5 Test

1. $(3x^2y - 2xy^2 - 6) + (5 + 2xy^2 - 7x^2y)$

$= 3x^2y - 2xy^2 - 6 + 5 + 2xy^2 - 7x^2y$

$= -4x^2y - 1$

2. $(5a^2 - 3) - (2 + 5a) - (4a - 3)$

$= 5a^2 - 3 - 2 - 5a - 4a + 3$

$= 5a^2 - 9a - 2$

3. $-2x(x + 3y - 4)$

$= -2x^2 - 6xy + 8x$

4. $(2x - 3y^2)^2 = 4x^2 - 6xy^2 + 9y^4$

5. $(x^2 + 6x - 2)(x^2 - 3x - 4)$

$= x^4 - 3x^3 - 4x^2$

$\qquad + 6x^3 - 18x^2 - 24x$

$\qquad\qquad - 2x^2 + 6x + 8$

$= x^4 + 3x^3 - 24x^2 - 18x + 8$

6. $(-15x^3 - 12x^2 + 21x) \div (-3x)$

$= \dfrac{-15x^3}{-3x} - \dfrac{12x^2}{-3x} + \dfrac{21x}{-3x}$

$= 5x^2 + 4x - 7$

7. $(2x^4 - 7x^3 + 7x^2 - 9x + 10) \div (2x - 5)$

$$
\require{enclose}
\begin{array}{r}
x^3 - x^2 + x - 2 \\
2x - 5 \enclose{longdiv}{2x^4 - 7x^3 + 7x^2 - 9x + 10} \\
\underline{2x^4 - 5x^3} \phantom{{}+ 7x^2 - 9x + 10} \\
-2x^3 + 7x^2 \phantom{{}- 9x + 10} \\
\underline{-2x^3 + 5x^2} \phantom{{}- 9x + 10} \\
2x^2 - 9x \phantom{{}+ 10} \\
\underline{2x^2 - 5x} \phantom{{}+ 10} \\
-4x + 10 \\
\underline{-4x + 10}
\end{array}
$$

$(2x^4 - 7x^3 + 7x^2 - 9x + 10) \div (2x - 5)$

$= x^3 - x^2 + x - 2$

8. $(x^3 - x^2 - 5x + 2) \div (x + 2)$

$$
\begin{array}{r}
x^2 - 3x + 1 \\
x+2\overline{\smash{)}x^3 -\ x^2 - 5x + 2} \\
\underline{x^3 + 2x^2} \\
-3x^2 - 5x \\
\underline{-3x^2 - 6x} \\
x + 2 \\
\underline{x + 2}
\end{array}
$$

$(x^3 - x^2 - 5x + 2) \div (x + 2)$
$= x^2 - 3x + 1$

9. $(x^4 + x^3 - x - 3) \div (x + 1)$

$$
\begin{array}{r}
x^3 - 1 \\
x+1\overline{\smash{)}x^4 + x^3 - x - 3} \\
\underline{x^4 + x^3} \\
-x - 3 \\
\underline{-x - 1} \\
-2
\end{array}
$$

$(x^4 + x^3 - x - 3) \div (x + 1)$
$= x^3 - 1 + \dfrac{-2}{x+1}$

10. $(2x^5 - 7x^4 - 15x^2 - x + 5) \div (x - 4)$

$$
\begin{array}{r}
2x^4 + x^3 + 4x^2 + x + 3 \\
x-4\overline{\smash{)}2x^5 - 7x^4 + 0x^3 - 15x^2 - x + 5} \\
\underline{2x^4 - 8x^4} \\
x^4 + 0x^3 \\
\underline{x^4 - 4x^3} \\
4x^3 - 15x^2 \\
\underline{4x^3 - 16x^2} \\
x^2 - x \\
\underline{x^2 - 4x} \\
3x + 5 \\
\underline{3x - 12} \\
17
\end{array}
$$

$(2x^5 - 7x^4 - 15x^2 - x + 5) \div (x - 4)$
$= 2x^4 + x^3 + 4x^2 + x + 3 + \dfrac{17}{x-4}$

11. $121x^2 - 25y^2$
$= (11x + 5y)(11x - 5y)$

12. $9x^2 + 30xy + 25y^2 = (3x + 5y)^3$

13. $x^3 - 26x^2 + 48x$
$= x(x^2 - 26x + 48)$
$= x(x - 2)(x - 24)$

14. $24x^2 + 10x - 4$
$= 2(12x^2 + 5x - 2)$
$= 2(4x - 1)(3x + 2)$

15. $4x^3y + 8x^2y^2 + 4x^2y$
$= 4x^2y(x + 2y + 1)$

16. $x^2 - 6wy + 3xy - 2wx$
$$= x^2 + 3xy - 2wx - 6wy$$
$$= x(x + 3y) - 2w(x + 3y)$$
$$= (x + 3y)(x - 2)$$

17. $2x^2 - 3x + 2$ is prime.

18. $3x^4 + 36x^3 + 60x^2 = 3x^2(x^2 + 12x + 20)$
$$= 3x^2(x + 10)(x + 2)$$

19. $18x^2 + 3x - 15 = 3(6x^2 + x - 5)$
$$= 3(6x - 5)(x + 1)$$

20. $25x^2 y^4 - 16y^4 = y^4(25x^2 - 16)$
$$= y^4(5x^2 - 4)(5x^2 + 4)$$

21. $54a^4 - 16a = 2a(27a^3 - 8)$
$$= 2a(3a - 2)(9a^2 + 6a + 4)$$

22. $9x^5 - 6x^3 y + xy^2 = x(9x^4 - 6x^2 y + y^2)$
$$= x(3x^2 - y)^2$$

23. $3x^4 + 17x^2 + 10 = (3x^2 + 2)(x^2 + 5)$

24. $x^2 - 8xy + 12y^2 = (x - 6y)(x - 2y)$

25. $3x - 10ay + 6y - 5ax$
$$= 3x - 5ax + 6y - 10ay$$
$$= x(3 - 5a) + 2y(3 - 5a)$$
$$= (3 - 5a)(x + 2y)$$

26. $16x^4 - 1 = (4x^2 - 1)(4x^2 + 1)$
$$= (2x + 1)(2x - 1)(4x^2 + 1)$$

27. $p(x) = -2x^3 - x^2 + 6x - 10$
$$p(2) = -2(2)^3 - (2)^2 + 6(2) - 10 = -18$$

28. $p(x) = -2x^3 - x^2 + 6x - 10$
$$p(-3) = -2(-3)^3 - (-3)^2 + 6(-3) - 10$$
$$p(-3) = 17$$

29. $x^2 = 5x + 14$
$$x^2 - 5x - 14 = 0$$
$$(x - 7)(x + 2) = 0$$
$$x - 7 = 0 \ \text{ or } \ x + 2 = 0$$
$$x = 7 \qquad\qquad x = -2$$

30. $3x^2 - 11x - 4 = 0$
$$(3x + 1)(x - 4) = 0$$
$$3x + 1 = 0 \ \text{ or } \ x - 4 = 0$$
$$x = -\frac{1}{3} \qquad\qquad x = 4$$

31. $7x^2 + 6x = 8x$
$$7x^2 - 2x = 0$$
$$x(7x - 2) = 0$$
$$x = 0 \ \text{ or } \ 7x - 2 = 0$$
$$x = 0 \qquad\qquad x = \frac{2}{7}$$

32. $A = \frac{1}{2}bh = \frac{1}{2}b(b - 4) = 70$
$$b^2 - 4b = 140$$
$$b^2 - 4b - 140 = 0$$
$$(b + 10)(b - 14) = 0$$
$$b + 10 = 0 \ \text{ or } \ b - 14 = 0$$
$$b = -10 \quad\text{ or }\quad b = 14$$
reject, $b > 0$ $\qquad$ $h = b - 4 = 10$

The base of the triangle is 14 inches and the altitude of the triangle is 10 inches.

Cumulative Test for Chapters 1-5

1. $3(5 \cdot 2) = (3 \cdot 5)2$ illustrates the associative property of multiplication.

2. $\dfrac{2+6(-2)}{(2-4)^3+3} = \dfrac{2-12}{(-2)^3+3}$

   $\qquad = \dfrac{-10}{-8+3} = \dfrac{-10}{-5} = 2$

3. $2\sqrt{16} + 3\sqrt{49} = 2(4) + 3(7)$

   $\qquad\qquad\qquad = 8 + 21 = 29$

4. $5x + 7y = 2$

   $5x = 2 - 7y$

   $x = \dfrac{2-7y}{5}$

5. $2(3x-1) - 4 = 2x - (6-x)$

   $6x - 2 - 4 = 2x - 6 + x$

   $3x = 0$

   $x = 0$

6. $m = \dfrac{y_2 - y_1}{x_2 - x_1} = \dfrac{-3-5}{-2-1} = \dfrac{-8}{-3} = \dfrac{8}{3}$

7. $y = -\dfrac{2}{3}x + 4$

x	y
-3	6
0	4
3	2

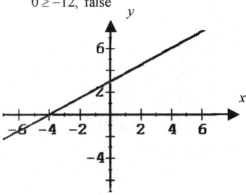

8. $3x - 4y \geq -12$

   Test point: $(0,0)$

   $3(0) + 4(0) \geq -12$

   $0 \geq -12$, false

9. $-3(x+2) < 5x - 2(4+x)$

   $-3x - 6 < 5x - 8 - 2x$

   $-6x < -2$

   $x > \dfrac{1}{3}$

10. $\quad A = 2L + 2W = 46$

    $2(2W + 5) + 2W = 46$

    $4W + 10 + 2W = 46$

    $6W = 36$

    $W = 6$

    $L = 2W + 5 = 17$

    The length of the rectangle is 17 meters and the width of the rectangle is 6 meters.

11. $(2a^2 - 3ab + 4b^2) - (-3a^2 + 6ab - 8b^2)$

    $= 2a^2 - 3ab + 4b^2 + 3a^2 - 6ab + 8b^2$

    $= 5a^2 - 9ab + 12b^2$

12. $-3xy^2(2x + 3y - 5wy)$

    $= -6x^2y^2 - 9xy^3 + 14wxy^3$

13. $(5x-2)(2x^2-3x-4)$

$\quad = 10x^3 - 15x^2 - 20x - 4x^2 + 6x + 8$

$\quad = 10x^3 - 19x^2 - 14x + 8$

14. $(-21x^3 + 14x^2 - 28x) \div (7x)$

$\quad \dfrac{-21x^3}{7x} + \dfrac{14x^2}{7x} - \dfrac{-28x}{7x} = -3x^2 + 2x + 4$

15. $(2x^3 - 3x^2 + 3x - 4) \div (x - 2)$

$$\begin{array}{r} 2x^2 + x + 5 \\ x-2\overline{\smash)2x^3 - 3x^2 + 3x - 4} \\ \underline{2x^3 - 4x^2} \\ x^2 + 3x \\ \underline{x^2 - 2x} \\ 5x - 4 \\ \underline{5x - 10} \\ 6 \end{array}$$

$\quad (2x^3 - 3x^2 + 3x - 4) \div (x - 2)$

$\quad = 2x^2 + x + 5 + \dfrac{6}{x-2}$

16. $2x^3 - 10x^2 = 2x^2(x - 5)$

17. $64x^2 - 49 = (8x + 7)(8x - 7)$

18. $9x^3 - 24x^2 + 16x$

$\quad = x(9x^2 - 24x + 16)$

$\quad = x(3x - 4)^2$

19. $25x^2 + 60x + 36 = (5x + 6)^2$

20. $3x^2 - 15x - 42 = 3(x^2 - 5x - 14)$

$\quad\quad\quad\quad\quad\quad = 3(x - 7)(x + 2)$

21. $2x^2 + 24x + 40$

$\quad = 2(x^2 + 12x + 20)$

$\quad = 2(x + 10)(x + 2)$

22. $16x^2 + 9$ is prime.

23. $6x^3 + 11x^2 + 3x$

$\quad = x(x^2 + 11x + 3)$

$\quad = x(3x + 1)(2x + 3)$

24. $27x^4 + 64x$

$\quad = x(27x^3 + 64)$

$\quad = x(3x + 4)(9x^2 - 12x + 16)$

25. $14mn + 7n + 10mp + 5p$

$\quad = 7n(2m + 1) + 5p(2m + 1)$

$\quad = (2m + 1)(7n + 5p)$

26. $3x^2 - 4x - 4 = 0$

$\quad (3x + 2)(x - 2) = 0$

$\quad 3x + 2 = 0 \ \text{ or } \ x - 2 = 0$

$\quad x = -\dfrac{2}{3} \quad\quad\quad\quad x = 2$

27. $x^2 + 11x = 26$

$\quad x^2 + 11x - 26 = 0$

$\quad (x + 13)(x - 2) = 0$

$\quad x + 13 = 0 \ \text{ or } \ x - 2 = 0$

$\quad x = -13 \quad\quad\quad\quad x = 2$

28. $A = \dfrac{1}{2}bh = \dfrac{1}{2}b(2b+1) = 68$

$2b^2 + b - 136 = 0$

$(2b+17)(b-8) = 0$

$2b+17 = 0 \quad \text{or} \quad b-8 = 0$

$b = -\dfrac{17}{2} \qquad\qquad b = 8$

reject, $b > 0$ $\qquad h = 2b+1 = 17$

The base of the triangle is **8** meters and the altitude of the triangle is 17 meters.

Chapter 6

Pretest Chapter 6

1. $\dfrac{49x^2-9y^2}{7x^2+4xy-3y^2} = \dfrac{(7x-3y)(7x+3y)}{(x+y)(7x-3y)}$

   $= \dfrac{7x+3y}{x=y}$

2. $\dfrac{2x^3-5x^2-3x}{x^3-8x^2+15x} = \dfrac{x(2x^2-5x-3)}{x(x^2-8+15)}$

   $= \dfrac{(2x+1)(x-3)}{(x-5)(x-3)}$

   $= \dfrac{2x+1}{x-5}$

3. $\dfrac{2a^2+5a+3}{a^2+a+1} \cdot \dfrac{a^3-1}{2a^2+a-3} \cdot \dfrac{6a-30}{3a+3}$

   $= \dfrac{(2a+3)(a+1)}{a^2+a+1} \cdot \dfrac{(a-1)(a^2+a+1)}{(2a+3)(a-1)} \cdot \dfrac{6(a-5)}{3(a+1)}$

   $= 2(a-5)$

4. $\dfrac{5x^3y^2}{x^2y+10xy^2+25y^3} \div \dfrac{2x^4y^5}{3x^3-75xy^2}$

   $= \dfrac{5x^3y^2}{y(x^2+10xy+5y^2)} \cdot \dfrac{3x(x^2-25y^2)}{2x^4y^5}$

   $= \dfrac{15}{2y^4} \cdot \dfrac{(x-5y)(x+5y)}{(x+5y)(x+5y)}$

   $= \dfrac{15}{2y^4} \cdot \dfrac{(x-5y)}{(x+5y)}$

5. $\dfrac{x}{3x-6} - \dfrac{4}{3x} = \dfrac{x^2}{3x(x-2)} - \dfrac{4(x-2)}{3x(x-2)}$

   $= \dfrac{x^2-4x+8}{3x(x-2)}$

6. $\dfrac{2}{x+5} + \dfrac{3}{x-5} + \dfrac{7x}{x^2-25}$

   $= \dfrac{2(x-5)}{(x+5)(x-5)} + \dfrac{3(x+5)}{(x-5)(x+5)}$

   $\qquad + \dfrac{7x}{(x-5)(x+5)}$

   $= \dfrac{2x-10+3x+15+7x}{(x-5)(x+5)}$

   $= \dfrac{12x+5}{(x-5)(x+5)}$

7. $\dfrac{y+1}{y^2+y-12} - \dfrac{y-3}{y^2+7y+12}$

   $= \dfrac{(y+1)(y+3)}{(y+4)(y-3)(y+3)}$

   $\qquad - \dfrac{(y-3)(y-3)}{(y+4)(y+3)(y-3)}$

   $= \dfrac{y^2+4y+3-y^2+6y-9}{(y+4)(y+3)(y-3)}$

   $= \dfrac{10y-6}{(y+4)(y+3)(y-3)}$

8. $\dfrac{\dfrac{1}{12x}+\dfrac{5}{3x}}{\dfrac{2}{3x^2}}$

   $= \dfrac{x+20x}{8}$

   $= \dfrac{21x}{8}$

9. $\dfrac{\dfrac{x}{4x^2-1}}{3-\dfrac{2}{2x+1}}$

$=\dfrac{\dfrac{x}{(2x+1)(2x-1)}}{3-\dfrac{2}{2x+1}}\cdot\dfrac{(2x+1)(2x-1)}{(2x+1)(2x-1)}$

$=\dfrac{x}{3(2x+1)(2x-1)-2(2x-1)}$

$=\dfrac{x}{12x^2-1-4x+2}=\dfrac{x}{12x^2-4x+1}$

$=\dfrac{x}{(2x-1)(6x+1)}$

10. $\dfrac{3}{y+5}-\dfrac{1}{y-5}=\dfrac{5}{y^2-25}$

$\dfrac{3}{y+5}-\dfrac{1}{y-5}=\dfrac{5}{(y-5)(y+5)}$

$3(y-5)-(y+5)=5$

$3y-15-y-5=5$

$2y=25$

$y=\dfrac{25}{2}$

11. $\dfrac{1}{6y}-\dfrac{4}{9}=\dfrac{4}{9y}-\dfrac{1}{2}$

$3-8y=8-9y$

$y=5$

12. $\dfrac{d_1}{d_2}=\dfrac{w_1}{w_2}$

$d_2w_1=d_1w_2$

$d_2=\dfrac{d_1w_2}{w_1}$

13. $\qquad I=\dfrac{nE}{nr+R}$

$Inr+IR=nE$

$nE-Inr=IR$

$n(E-Ir)=IR$

$\qquad n=\dfrac{IR}{E-Ir}$

14. $\dfrac{\text{height}}{\text{shadow}}:\dfrac{x}{9}=\dfrac{49}{14}$

$x=\dfrac{49(9)}{14}=31.5$

The flagpole is 31.5 ft tall.

15. $\dfrac{3}{4}$ in $(x)=2$ ft $=24$ in

$x=32$

$32(3\text{ in})=96\text{ in}\cdot\dfrac{\text{ft}}{12\text{ in}}=8$ ft

$32(5\text{ in})=160\text{ in}\cdot\dfrac{\text{ft}}{12\text{ in}}=13\dfrac{1}{3}$ ft

The card is 8 ft by $13\dfrac{1}{3}$ ft.

6.1 Exercises

1. $2x-6\neq0$

$2x\neq6$

$x\neq3$

All real numbers except 3

3. $x^2-5x-36\neq0$

$(x-9)(x+4)\neq0$

$x\neq-4,9$

All real numbers except -4 and 9

5. $\dfrac{-18x^4y}{12x^2y^6}=-\dfrac{6x^2y\cdot3x^2}{6x^2y\cdot2y^5}=-\dfrac{3x^2}{2y^5}$

7. $\dfrac{3x^2-24x}{3x^2+12x}=\dfrac{3x(x-8)}{3x(x+4)}=\dfrac{x-8}{x-4}$

9. $\dfrac{9x^2}{12x^2-15x}=\dfrac{3x\cdot 3x}{3x(x-5)}=\dfrac{3x}{x-5}$

11. $\dfrac{2x^3y-10x^2y}{2x^3y+6x^2y}=\dfrac{2x^2y(x-5)}{2x^2y(x+3)}=\dfrac{x-5}{x+3}$

13. $\dfrac{2x+10}{2x^2-50}=\dfrac{2(x+5)}{2(x+5)(x-5)}=\dfrac{1}{x-5}$

15. $\dfrac{2y^2-8}{2y+4}=\dfrac{2(y+2)(y-2)}{2(y+2)}=y-2$

17. $\dfrac{30x-x^2-x^3}{x^3-x^2-20x}=\dfrac{-x(x^2+x-30)}{x(x^2-x-20)}$

$=-\dfrac{(x+6)(x-5)}{(x+4)(x-5)}$

$=-\dfrac{x+6}{x+4}$

19. $\dfrac{36-b^2}{3b^2-16b-12}=\dfrac{(6+b)(6-b)}{(3b+2)(b-6)}$

$=-\dfrac{6+b}{3b+2}$

21. $\dfrac{-8mn^5}{3m^4n^3}\cdot\dfrac{9m^3n^3}{6mn}=\dfrac{-2\cdot 4\cdot 3\cdot 3m^4n^4n^4}{2\cdot 3\cdot 3m^4n^4m}$

$=\dfrac{-4n^4}{m}$

23. $\dfrac{3a^2}{a+2}\cdot\dfrac{a^2-4}{3a}=\dfrac{3a\cdot a(a+2)(a-2)}{3a\cdot(a+2)}$

$=a(a-2)$

25. $\dfrac{x^2+5x+7}{x^2-5x+6}\cdot\dfrac{3x-6}{x^2+5x+7}$

$=\dfrac{3(x-2)}{(x-2)(x-3)}$

$=\dfrac{3}{x-3}$

27. $\dfrac{x^2-5xy-24y^2}{x-y}\cdot\dfrac{x^2+6xy-7y^2}{x+3y}$

$=\dfrac{(x-8y)(x+3y)}{(x-y)}\cdot\dfrac{(x+7y)(x-y)}{(x+3y)}$

$=(x-8y)(x+7y)$

29. $\dfrac{2y^2-5y-12}{4y^2+8y+3}\cdot\dfrac{2y^2+7y+3}{y^2-16}$

$=\dfrac{(2y+3)(y-4)}{(2y+3)(2y+1)}\cdot\dfrac{(2y+1)(y+3)}{(y-4)(y+4)}$

$=\dfrac{y+3}{y+4}$

31. $\dfrac{x^3-125}{x^5y}\cdot\dfrac{x^3y^2}{x^2+5x+25}$

$=\dfrac{(x-5)(x^2+5x+25)}{x^3y\cdot x^2}\cdot\dfrac{x^3y\cdot y}{(x^2+5x+25)}$

$=\dfrac{y(x-5)}{x^2}$

33. $\dfrac{2mn-m}{15m^3}\div\dfrac{2n-1}{3m^2}$

$=\dfrac{m(2n-1)}{5m\cdot 3m^2}\cdot\dfrac{3m^2}{(2n-1)}$

$=\dfrac{1}{5}$

35. $\dfrac{(b-3)^2}{b^2-b-6} \div \dfrac{3b-9}{3b+4}$

$= \dfrac{(b-3)^2}{(b-3)(b+2)} \cdot \dfrac{(3b+4)}{3(b-3)}$

$= \dfrac{3b+4}{3(b+2)}$

37. $\dfrac{x^2-xy-6y^2}{x^2+2} \div (x^2+2xy)$

$= \dfrac{(x+2y)(x-3y)}{(x^2+2)} \cdot \dfrac{1}{x(x+2y)}$

$= \dfrac{x-3y}{x(x^2+2)}$

39. $\dfrac{7x}{y^2} \div 21x^3 = \dfrac{7x}{y^2} \cdot \dfrac{1}{7x \cdot 3x^2}$

$= \dfrac{1}{3x^2 y^2}$

41. $\dfrac{3x^2-2x}{6x-4} = \dfrac{x(3x-2)}{2(3x-2)} = \dfrac{x}{2}$

43. $\dfrac{x^2y-49y}{x^2y^3} \cdot \dfrac{3x^2y-21xy}{x^2-14x+49}$

$= \dfrac{y(x+7)(x-7)}{x^2y^3} \cdot \dfrac{3xy(x-7)}{(x-7)(x-7)}$

$= \dfrac{3(x+7)}{xy}$

45. $\dfrac{-8+6x-x^2}{2x^3-8x} = -\dfrac{(x-4)(x-2)}{2x(x^2-4)}$

$= -\dfrac{(x-4)(x-2)}{2x(x+2)(x-2)}$

$= -\dfrac{x-4}{2x(x+2)}$

47. $\dfrac{a^2-a-12}{2a^2+5a-12} = \dfrac{(a-4)(a+3)}{(2a-3)(a+4)}$

49. Domain of $f(x) = \dfrac{2x+5}{3.6x^2+1.8x-4.3}$

is all real number except $x \approx -1.4$
and $x \approx 0.9$.

51. Domain of $D = \dfrac{10,000}{x^2+3x} = \dfrac{10,000}{x(x+3)}$ is

all integers except -3 and 0.

53. $D = \dfrac{10,000}{10^2+3(10)} = 76.9230769231\ldots$

If 10 unit are produced daily, the
demand is 77 toys per day.

55. $D = \dfrac{10,000}{30^2+3(0)} = 10.\overline{10}$

If 30 unit are produced daily, the
demand is 10 toys per day.

57. $D = \dfrac{10,000}{21^2+3(21)} = 19.84126984\ldots$

If 21 unit are produced daily, the
demand is 20 toys per day.

Cumulative Review Problems

59.

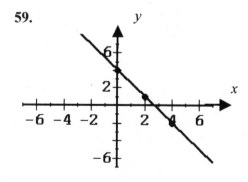

61. $m = \dfrac{y_2 - y_2}{x_2 - x_1} = \dfrac{5 - (-3)}{0 - (-2)} = 4$

$y - y_1 = m(x - x_1)$

$y - 5 = 4(x - 0)$

$4x - y = -5$

63. $\dfrac{2 \text{ inquiries}}{\text{minute}} \cdot \dfrac{60 \text{ minutes}}{\text{hour}} \cdot \dfrac{33 \text{ hours}}{\text{week}}$

$= 3960 \dfrac{\text{inquiries}}{\text{week}}$

65. $69,399(0.37) = $ deaths in spring 2000

$\dfrac{69,399}{1.07}(0.37) = $ deaths in spring 1999

$\dfrac{69,399}{1.07^2}(0.37) = $ deaths in spring 1998

72,103 people have died

6.2 Exercises

1. The factors are 5, x, and y. The factor y is repeated in one fraction three times. Since the highest power of y is 3, the LCD is $5xy^3$.

3. $x^2 + 7xy + 12y^2 = (x + 4y)(x + 3y)$

$x + 4y = x + 4y$

$LCD = (x + 3y)(x + 4y)$

5. $LCD = (x + 2)(3x + 4)^3$

7. $3x^2 + 2x = x(3x + 2)$

$18x^2 + 9x - 2 = (6x - 1)(3x + 2)$

$LCD = x(6x - 1)(3x + 2)$

9. $9y^2 - 49 = (3y + 7)(3y - 7)$

$9y^2 - 42y + 49 = (3y - 7)^2$

$LCD = (3y + 7)(3y - 7)^2$

11. $\dfrac{2}{(x + 2)} \cdot \dfrac{2x}{2x} - \dfrac{3}{2x} \cdot \dfrac{(x + 2)}{(x + 2)}$

$= \dfrac{4x - 3x - 6}{2x(x + 2)} = \dfrac{x - 6}{2x(x + 2)}$

13. $\dfrac{12}{5x^2} \cdot \dfrac{y}{y} + \dfrac{2}{5xy} \cdot \dfrac{x}{x} = \dfrac{12y + 2x}{5x^2 y}$

15. $\dfrac{3}{x^2 - 7x + 12} + \dfrac{5}{x^2 - 4x}$

$= \dfrac{3x}{(x - 4)(x - 3)x} + \dfrac{5(x - 3)}{x(x - 4)(x - 3)}$

$= \dfrac{3x + 5x - 15}{x(x - 4)(x - 3)}$

$= \dfrac{8x - 15}{x(x - 4)(x - 3)}$

17. $\dfrac{6x}{2x - 5} + 4 = \dfrac{6x + 4(2x - 5)}{2x - 5}$

$= \dfrac{6x + 8x - 20}{2x - 5}$

$= \dfrac{14x - 2}{2x - 5}$

19. $\dfrac{-5y}{y^2 - 1} + \dfrac{6}{y^2 - 2y + 1}$

$= \dfrac{-5y(y - 1)}{(y - 1)^2(y + 1)} + \dfrac{6(y + 1)}{(y - 1)^2(y + 1)}$

$= \dfrac{-5y^2 + 5y + 6y + 6}{(y - 1)^2(y + 1)}$

$= \dfrac{-5y^2 + 11y + 6}{(y - 1)^2(y + 1)}$

21. $\dfrac{a+5}{a^2-4}+\dfrac{a-3}{2a-4}$

$=\dfrac{2(a+5)}{2(a+2)(a-2)}+\dfrac{(a-3)(a+2)}{2(a-2)(a+2)}$

$=\dfrac{2a+10+a^2-a-6}{2(a+2)(a-2)}$

$=\dfrac{a^2+a+4}{2(a+2)(a-2)}$

23. $\dfrac{2x-1}{x-6}-1=\dfrac{2x-1-(x-6)}{x-6}$

$=\dfrac{2x-1-x+6}{x-6}$

$=\dfrac{x+5}{x-6}$

25. $\dfrac{3x}{x^2+3x-10}-\dfrac{2x}{x^2+x-6}$

$=\dfrac{3x(x+3)}{(x+5)(x-2)(x+3)}-\dfrac{2x(x+5)}{(x+3)(x-2)(x+5)}$

$=\dfrac{3x^2+9x-2x^2-10x}{(x+5)(x-2)(x+3)}$

$=\dfrac{x^2-x}{(x+5)(x-2)(x+3)}$

27. $\dfrac{3y^2}{y^2-1}-\dfrac{y+2}{y+1}$

$=\dfrac{3y^2}{(y+1)(y-1)}-\dfrac{(y+2)(y-1)}{(y+1)(y-1)}$

$=\dfrac{3y^2-y^2-y+2}{(y+1)(y-1)}$

$=\dfrac{2y^2-y+2}{(y+1)(y-1)}$

29. $a+3+\dfrac{2}{3a-5}$

$=\dfrac{(a+3)(3a-5)+2}{3a-5}$

$=\dfrac{3a^2+4a-15+2}{3a-5}$

$=\dfrac{3a^2+4a-13}{3a-5}$

31. $P(x)=R(x)-C(x)$

$P(x)=\dfrac{80-24x}{2-x}-\dfrac{60-12x}{3-x}$

$=\dfrac{8(10-3x)(3-x)-12(5-x)(2-x)}{(2-x)(3-x)}$

$=\dfrac{8(30-19x+3x^2)-12(10-7x+x^2)}{x^2-5x+6}$

$=\dfrac{240-152x+24x^2-120+84x-12x^2}{x^2-5x+6}$

$=\dfrac{12x^2-68x+120}{x^2-5x+6}$

33. $P(x)=R(x)-C(x)$

$P(x)=\dfrac{12x^2-68x+120}{x^2-5x+6}$

$P(10)=\dfrac{12(10)^2-68(10)+120}{(10)^2-5(10)+6}$

$P(10)=11.4285714286...$

If ten machines per day are manufactured, the daily profit will be $11,429.

35. $3x - 2 + \dfrac{5x}{3x-2} + \dfrac{2x^2}{(3x-2)^2}$

$= \dfrac{(3x-2)^3 + 5x(3x-2) + 2x^2}{(3x-2)^2}$

$= \dfrac{27x^3 - 54x^2 + 36x - 8 + 15x^2 - 10x + 2x^2}{(3x-2)^2}$

$= \dfrac{27x^3 - 37x^2 + 26x - 8}{(3x-2)^2}$

Cumulative Review Problems

37. Assume people reading the letters worked 24 hours per day.

$\dfrac{10 \text{ letters}}{\text{person} \cdot \text{hour}} \cdot 5 \text{ people} = \dfrac{50 \text{ letters}}{\text{hour}}$

$\dfrac{50 \text{ letters}}{\text{hour}} \cdot x = 73{,}000 \text{ letters}$

$x = 1460 \text{ hours} \cdot \dfrac{\text{day}}{24 \text{ hours}}$

$x = 60.8\overline{3} \text{ days}$

It would take 1460 hours or about 60.8 days to read all the letters.

39. $x = $ cost of Tony's car
$x + 1500 = $ cost of Alreda's car
$2x + 1000 = $ cost of Melissa's car
$x + x + 1500 + 2x + 1000 = 26{,}500$

$4x = 24{,}000$

$x = 6000$

$x + 1500 = 7500$

$2x + 1000 = 13{,}000$

Tony's car cost $6000, Melissa's car cost $13,000, and Alreda's car cost $7500.

41. $x = $ speed of boat in still water
$y = $ speed of current
$75 = (x - y)5 \Rightarrow x - y = 15$
$75 = (x + y)3 \Rightarrow \underline{x + y = 25}$
$\qquad\qquad\qquad 2x = 40$
$\qquad\qquad\qquad x = 20$
$\qquad\qquad 20 + y = 25$
$\qquad\qquad\qquad y = 5$

The speed of the boat in still water is 20 km/hr and the speed of the current is 5 km/hr.

6.3 Exercises

1. $\dfrac{1 - \dfrac{6}{5y}}{\dfrac{3}{y} + 1} \cdot \dfrac{5y}{5y} = \dfrac{5y - 6}{15 + 5y} = \dfrac{5y - 6}{5(y + 3)}$

3. $\dfrac{\dfrac{y}{6} - \dfrac{1}{2y}}{\dfrac{3}{2y} - \dfrac{1}{y}} \cdot \dfrac{6y}{6y} = \dfrac{y^2 - 3}{9 - 6} = \dfrac{y^2 - 3}{3}$

5. $\dfrac{\dfrac{2}{y^2 - 9}}{\dfrac{3}{y + 3} + 1} \cdot \dfrac{(y + 3)(y - 3)}{(y + 3)(y - 3)}$

$= \dfrac{2}{3(y - 3) + y^2 - 9}$

$= \dfrac{2}{3y - 9 + y^2 - 9}$

$= \dfrac{2}{y^2 + 3y - 18}$

$= \dfrac{2}{(y + 6)(y - 3)}$

7. $\dfrac{3-\dfrac{5}{x}}{\dfrac{x+2}{4}} \cdot \dfrac{4x}{4x} = \dfrac{12x-20}{x^2+2x} = \dfrac{4(3x-5)}{x(x+2)}$

9. $\dfrac{6}{2x-\dfrac{10}{x-4}} \cdot \dfrac{(x-4)}{(x-4)}$

$= \dfrac{6(x-4)}{2x^2-8x-10}$

$= \dfrac{6(x-4)}{2(x^2-4x-5)}$

$= \dfrac{3(x-4)}{(x^2-4x-5)}$

11. $\dfrac{\dfrac{1}{2x+3}+\dfrac{2}{4x^2+12x+9}}{\dfrac{5}{2x^2+3x}}$

$= \dfrac{\dfrac{1}{2x+3}+\dfrac{2}{(2x+3)(2x+3)}}{\dfrac{5}{x(2x+3)}} \cdot \dfrac{x(2x+3)^2}{x(2x+3)^2}$

$= \dfrac{x(2x+3)+2x}{5(2x+3)}$

$= \dfrac{2x^2+3x+2x}{5(2x+3)}$

$= \dfrac{2x^2+5x}{5(2x+3)}$

$= \dfrac{x(2x+5)}{5(2x+3)}$

13. $\dfrac{\dfrac{2}{a}+\dfrac{3}{b}}{\dfrac{4}{a+b}-\dfrac{1}{a}} \cdot \dfrac{ab(a+b)}{ab(a+b)}$

13. (continued)

$= \dfrac{2b(a+b)+3a(a+b)}{4ab-b(a+b)}$

$= \dfrac{2ab+2b^2+3a^2+3ab}{4ab-ab-b^2}$

$= \dfrac{3a^2+5ab+2b^2}{3ab-b^2}$

$= \dfrac{(3a+2b)(a+b)}{b(3a-b)}$

15. $\dfrac{\dfrac{1}{x-a}-\dfrac{1}{x}}{a} \cdot \dfrac{x(x-a)}{x(x-a)} = \dfrac{x-(x-a)}{ax(x-a)}$

$= \dfrac{x-x+a}{ax(x-a)}$

$= \dfrac{a}{ax(x-a)}$

$= \dfrac{1}{x(x-a)}$

17. $1-\dfrac{1}{1-\dfrac{1}{y-2}} \cdot \dfrac{y-2}{y-2} = 1-\dfrac{y-2}{y-2-1}$

$= 1-\dfrac{y-2}{y-3}$

$= \dfrac{y-3-y+2}{y-3}$

$= -\dfrac{1}{y-3}$

19. $\dfrac{\dfrac{x^2-1}{6x^2+3x}}{\dfrac{x-1}{2x^2}} = \dfrac{(x+1)(x-1)}{3x(2x+1)} \cdot \dfrac{2x^2}{(x-1)}$

$= \dfrac{2x(x+1)}{3(2x+1)}$

(continued)

Cumulative Review Problems

21. $|2 - 3x| = 4$

$2 - 3x = 4$ or $2 - 3x = -4$

$3x = -2 \qquad -3x = -6$

$x = -\dfrac{2}{3} \qquad\quad x = 2$

23. $|7x - 3 - 2x| < 6$

$|5x - 3| < 6$

$-6 < 5x - 3 < 6$

$-3 < 5x < 9$

$-\dfrac{3}{5} < x < \dfrac{9}{5}$

25. (a) $\dfrac{\$4000}{\text{inch}} \cdot \dfrac{12 \text{ inches}}{\text{foot}} \cdot \dfrac{5280 \text{ feet}}{\text{mile}}$

$= \dfrac{\$253,440,000}{\text{mile}}$

$\$253,440,000$ was spent per mile in the 1970's.

(b) $\$4,860,000,000 - \$570,000,000$

$= \$4,290,000,000$

$\dfrac{\$660,000,000}{\text{mile}} \cdot x = \$4,290,000,000$

$x = 6.5$ miles

A total of 6.5 miles can be built with the budget limit.

6.4 Exercises

1. $\dfrac{2}{x} + \dfrac{3}{2x} = \dfrac{7}{6}$

$12 + 9 = 7x$

$7x = 21$

$x = 3$

(continued)

1. (continued)

check:

$\dfrac{2}{3} + \dfrac{3}{2(3)} \overset{?}{=} \dfrac{7}{6}$

$\dfrac{7}{6} = \dfrac{7}{6}$

3. $3 - \dfrac{2}{x} = \dfrac{1}{4x}$

$12x - 8 = 1$

$12x = 9$

$x = \dfrac{3}{4}$

check:

$3 - \dfrac{2}{\frac{3}{4}} \overset{?}{=} \dfrac{1}{4 \cdot \frac{3}{4}}$

$\dfrac{1}{3} = \dfrac{1}{3}$

5. $\dfrac{7}{2x} - 1 = \dfrac{3}{x}$

$7 - 2x = 6$

$2x = 1$

$x = \dfrac{1}{2}$

check:

$\dfrac{7}{2 \cdot \frac{1}{2}} - 1 \overset{?}{=} \dfrac{3}{\frac{1}{2}}$

$6 = 6$

7. $\dfrac{2}{y} = \dfrac{5}{y - 3}$

$2y - 6 = 5y$

$3y = -6$

$y = -2$

(continued)

7. (continued)
check:

$$\frac{2}{-2} \overset{?}{=} \frac{5}{-2-3}$$

$$-1 = -1$$

9. $\quad \dfrac{y+6}{y+3} - 2 = \dfrac{3}{y+3}$

$$y + 6 - 2y - 6 = 3$$

$$-y = 3$$

$$y = -3$$

No solution.

11. $\quad \dfrac{3}{x} + \dfrac{4}{2x} = \dfrac{4}{x-1}$

$$6(x-1) + 4(x-1) = 8x$$

$$6x - 6 + 4x - 4 = 8x$$

$$2x = 10$$

$$x = 5$$

check:

$$\frac{3}{5} + \frac{4}{2 \cdot 5} \overset{?}{=} \frac{4}{5-1}$$

$$1 = 1$$

13. $\quad \dfrac{2x+3}{x+3} = \dfrac{2x}{x+1}$

$$(2x+3)(x+1) = 2x(x+3)$$

$$2x^2 + 5x + 3 = 2x^2 + 6x$$

$$x = 3$$

check:

$$\frac{2 \cdot 3 + 3}{3 + 3} \overset{?}{=} \frac{2 \cdot 3}{3 + 1}$$

$$\frac{3}{2} = \frac{3}{2}$$

15. $\quad \dfrac{3y}{y+1} + \dfrac{4}{y-2} = 3$

(continued)

15. (continued)

$$3y(y-2) + 4(y+1) = 3(y-2)(y+1)$$

$$3y^2 - 6y + 4y + 4 = 3y^2 - 3y - 6$$

$$y = -10$$

check:

$$\frac{3(-10)}{-10+1} + \frac{4}{-10-2} \overset{?}{=} 3$$

$$3 = 3$$

17. $\quad \dfrac{3}{2x-1} + \dfrac{3}{2x+1} = \dfrac{8x}{4x^2-1}$

$$\frac{3}{2x-1} + \frac{3}{2x+1} = \frac{8x}{(2x+1)(2x-1)}$$

$$3(2x+1) + 3(2x-1) = 8x$$

$$6x + 3 + 6x - 3 = 8x$$

$$4x = 0$$

$$x = 0$$

check:

$$\frac{3}{2(0)-1} + \frac{3}{2(0)+1} \overset{?}{=} \frac{8(0)}{4(0)^2-1}$$

$$0 = 0$$

19. $\quad \dfrac{5}{y-3} + 2 = \dfrac{3}{3y-9} = \dfrac{3}{3(y-3)}$

$$15 + 6(y-3) = 3$$

$$12 + 6y - 18 = 0$$

$$6y = 6$$

$$y = 1$$

check:

$$\frac{5}{1-3} + 2 \overset{?}{=} \frac{3}{3(1)-9}$$

$$-\frac{1}{2} = -\frac{1}{2}$$

21. $1 - \dfrac{10}{z-3} = \dfrac{-5}{3z-9} = \dfrac{-5}{3(z-3)}$

$3(z-3) - 30 = -5$

$3z - 9 = 25$

$3z = 34$

$z = \dfrac{34}{3}$

check:

$1 - \dfrac{10}{\dfrac{34}{3} - 3} \overset{?}{=} \dfrac{-5}{3 \cdot \dfrac{34}{3} - 9}$

$-\dfrac{1}{5} = -\dfrac{1}{5}$

23. $\dfrac{4}{y+5} - \dfrac{32}{y^2 - 25} = \dfrac{-2}{y-5}$

$\dfrac{4}{y+5} - \dfrac{32}{(y-5)(y+5)} = \dfrac{-2}{y-5}$

$4(y-5) - 32 = -2(y+5)$

$4y - 20 - 32 = -2y - 10$

$6y = 42$

$y = 7$

check:

$\dfrac{4}{7+5} - \dfrac{32}{7^2 - 25} \overset{?}{=} \dfrac{-2}{7-5}$

$-1 = -1$

25. $\dfrac{4}{z^2 - 9} = \dfrac{2}{z^2 - 3z}$

$\dfrac{4}{(z-3)(z+3)} = \dfrac{2}{z(z-3)}$

$4z = 2(z+3)$

$4z = 2z + 6$

$2z = 6$

$z = 3$

27. $\dfrac{2x+3}{2} + \dfrac{1}{x+1} = x$

$(2x+3)(x+1) + 2 = 2x(x+1)$

$2x^2 + 5x + 3 + 2 = 2x^2 + 2x$

$3x = -5$

$x = -\dfrac{5}{3}$

check:

$\dfrac{2\left(-\dfrac{5}{3}\right) + 3}{2} + \dfrac{1}{-\dfrac{5}{3} + 1} \overset{?}{=} -\dfrac{5}{3}$

$-\dfrac{5}{3} = -\dfrac{5}{3}$

29. When the solved-for value of the variable causes the denominator of any fraction to equal to 0, or when the variable drops out and leaves a false statement.

31. $\dfrac{5}{x+3.6} - \dfrac{4.2}{x-7.6} = \dfrac{3.3}{x^2 - 4x - 27.36}$

$\dfrac{5}{x+3.6} - \dfrac{4.2}{x-7.6} = \dfrac{3.3}{(x+3.6)(x-7.6)}$

$5(x-7.6) - 4.2(x+3.6) = 3.3$

$5x - 38 - 4.2x - 15.12 = 3.3$

$0.8x = 56.42$

$x = 70.525$

Cumulative Review Problems

33. $7x^2 - 63 = 7(x^2 - 9) = 7(x+3)(x-3)$

35. $64x^3 - 27y^3$

$= (4x - 3y)(16x^2 + 12xy + 9y^2)$

37. 25% of the 160,000 couples,
= (0.25)(160,000), or 80,000 couples
will attend counseling for more than
one year but less than two years. 25%
of these 80,000 or (0.25)(80,000) =
20,000 will remain married.

6.5 Exercises

1. $\quad x = \dfrac{y-b}{m}$

$mx = y - b$

$m = \dfrac{y-b}{x}$

3. $\quad \dfrac{1}{f} = \dfrac{1}{a} + \dfrac{1}{b}$

$ab = bf + af$

$ab - bf = af$

$b(a - f) = af$

$b = \dfrac{af}{a-f}$

5. $\quad R = \dfrac{ab}{3x}$

$Rx = \dfrac{ab}{3}$

$x = \dfrac{ab}{3R}$

7. $\quad F = \dfrac{xy + xz}{2}$

$xy + xz = 2F$

$x(y + z) = 2F$

$x = \dfrac{2F}{y+z}$

9. $\quad \dfrac{3V}{4\pi} = r^3$

$3V = 4\pi r^3$

$V = \dfrac{4}{3}\pi r^3$

11. $\quad \dfrac{E}{e} = \dfrac{R+r}{r}$

$e(R + r) = Er$

$e = \dfrac{Er}{R+r}$

13. $\quad \dfrac{P_1 V_1}{T_1} = \dfrac{P_2 V_2}{T_2}$

$P_2 V_2 T_1 = P_1 V_1 T_2$

$T_1 = \dfrac{P_1 V_1 T_2}{P_2 V_2}$

15. $\quad F = \dfrac{Gm_1 m_2}{d^2}$

$Fd^2 = Gm_1 m_2$

$d^2 = \dfrac{Gm_1 m_2}{F}$

17. $\quad E = T_1 - \dfrac{T_1}{T_2}$

$ET_2 = T_1 T_2 - T_1$

$T_1(T_2 - 1) = ET_2$

$T_1 = \dfrac{ET_2}{T_2 - 1}$

19. $\quad m = \dfrac{y_2 - y_1}{x_2 - x_1}$

$mx_2 - mx_1 = y_2 - y_1$

$mx_1 = mx_2 + y_1 - y_2$

$x_1 = \dfrac{mx_2 + y_1 - y_2}{m}$

21. $\dfrac{2D - at^2}{2t} = V$

$2D - at^2 = 2Vt$

$2D = 2Vt + at^2$

$D = \dfrac{2Vt + at^2}{2}$

23. $Q = \dfrac{kA(t_1 - t_2)}{L}$

$QL = kAt_1 - kAt_2$

$kAt_2 = kAt_1 - QL$

$t_2 = \dfrac{kAt_1 - QL}{kA}$

25. $\dfrac{T_2 W}{T_2 - T_1} = q$

$T_2 W = qT_2 - qT_1$

$qT_2 - T_2 W = qT_1$

$T_2(q - W) = qT_1$

$T_2 = \dfrac{qT_1}{q - W}$

27. $\dfrac{s - s_o}{v_o + gt} = t$

$s - s_o = v_o t + gt^2$

$v_o t = s - s_o - gt^2$

$v_o = \dfrac{s - s_o - gt^2}{t}$

29. $\dfrac{1.98V}{1.96V_o} = 0.983 + 5.936(T - T_o)$

$\dfrac{1.98V}{1.96V_o} = 0.983 + 5.936T - 5.936T_o$

$5.936T = 5.936T_o + \dfrac{1.98V}{1.96V_o} - 0.983$

(continued)

29. (continued)

$T \approx T_o + 0.1702\left(\dfrac{V}{V_o}\right) - 0.1656$

31. $\dfrac{3}{55} = \dfrac{6.5}{x}$

$3x = 357.5$

$x = 119.1\overline{6}$

The two cities are approximately 119.2 kilometers apart.

33. $\dfrac{60}{88} = \dfrac{x}{80}$

$88x = 4800$

$x = 54.\overline{54}$

The speed limit is approximately 54.55 mph.

35. $\dfrac{35}{x} = \dfrac{22}{50}$

$22x = 1750$

$x = 79.\overline{54}$

The number of grizzly bears, to the nearest whole number, is 80.

37. x = number of officers

$\dfrac{2}{7} = \dfrac{x}{117 - x}$

$234 - 2x = 7x$

$9x = 234$

$x = 26$

$117 - 26 = 91$

There are 26 officers and 91 seamen aboard the Russian Navy ship.

39. $x =$ number of people in marketing

$$\frac{4}{13} = \frac{x}{187 - x}$$

$$748 - 4x = 13x$$

$$17x = 748$$

$$x = 44$$

$$187 - 44 = 143$$

There are 44 people in marketing and 143 people in sales.

41. $\dfrac{w}{l} = \dfrac{3}{5} = \dfrac{w}{\dfrac{115.2 - 2w}{2}} = \dfrac{2w}{115.2 - 2w}$

$$345.6 - 6w = 10w$$

$$16w = 345.6$$

$$w = 21.6$$

$$\frac{115.2 - 2w}{2} = 36$$

The width is 21.6 inches and the length is 36 inches.

43. $x =$ number preferring new software

$$\frac{3}{11} = \frac{x}{280 - x}$$

$$840 - 3x = 11x$$

$$14x = 840$$

$$x = 60$$

Sixty people prefer the new software.

45. $x =$ number of miles he runs

$$\frac{2}{7} = \frac{x}{63 - x}$$

$$126 - 2x = 7x$$

$$9x = 126$$

$$x = 14$$

$$63 - x = 49$$

He should run 14 miles and walk 49 miles.

47. $\dfrac{x}{8} = \dfrac{12}{15}$

$$15x = 96$$

$$x = 6.4$$

The wall is 6.4 feet high.

49. $\dfrac{F}{M} = \dfrac{7}{4} = \dfrac{F}{112}$

$$4F = 112(7) = 784$$

$$F = 196$$

She should eat fish 196 times.

51. $\dfrac{t}{6} + \dfrac{t}{9} = 1$

$$\frac{5}{18}t = 1$$

$$t = \frac{18}{5} = 3.6$$

It will take them 3.6 hours to do the work together.

53. $t =$ time it takes Fred to do the work

$$\frac{1}{4} \cdot 3 + \frac{1}{t} \cdot 3 = 1$$

$$\frac{3}{t} = \frac{1}{4}$$

$$t = 12$$

It would take Fred 12 hours if he worked without the lumberjack.

55. $\dfrac{b}{a} = \dfrac{x + 8}{c}$

$$\frac{5}{2} = \frac{x + 8}{116}$$

$$2x + 16 = 580$$

$$2x = 564$$

$$x = 282$$

The width of the river is 282 feet.

Cumulative Review Problems

57. $8x^2 - 6x - 5 = (4x - 5)(2x + 1)$

59. $25x^2 - 90xy + 81y^2 = (5x - 9y)^2$

61. F = value of full scholarship

$$3750F + 3750\left(\frac{F}{2}\right) = 50,000,000$$

$$5625F = 50,000,000$$

$$F = 8888.\overline{8}$$

$$\frac{F}{2} = 4444.\overline{4}$$

Full scholarships were $8888.89 per student and partial scholarships were $4444.44 per student.

Putting Your Skills to Work

Summarize the given information in tabular form.

year	1985	1992	2000
budget	$10,000 \div 1.085$ =9216.59	10,000	$10,000(1.085)$ =10,850

year	Italy	Spain	France
1985	0.43(9216.59) =3963.13	0.20(9216.59) =1843.32	0.37(9216.59) = 3410.14
1992	1.18(3963.13) =4676.49	10,000 − 4676.49 −4126.27 = 1197.24	1.21(3410.14) = 4126.27
2000	1340.91+1000 =2340.91	1.12(1197.24) = 1340.91	

From the table,

1. Stahl's 1985 budget was $9216.59

2. Stahl's 2000 budget was $10,850

3. In Italy in 1985 the Stahl's spent

$$\$3963.13 \text{ or } \$3963.13\left(\frac{2159 \text{ lire}}{\text{USD}}\right)$$

$$= 8,556,397.67 \text{ lire}$$

In Italy in 1992 the Stahl's spent

$$\$4676.49 \text{ or } \$4676.49\left(\frac{1064 \text{ lire}}{\text{USD}}\right)$$

$$= 4,975,785.36 \text{ lire}$$

In Italy in 2000 the Stahl's spent

$$\$2340.91 \text{ or } \$2340.91\left(\frac{2158 \text{ lire}}{\text{USD}}\right)$$

$$= 5,051,683.78 \text{ lire}$$

4. In Spain in 1985 the Stahl's spent

$$\$1843.32 \text{ or } \$1843.32\left(\frac{191.57 \text{ pesetas}}{\text{USD}}\right)$$

$$= 353,124.81 \text{ pesetas}$$

In Spain in 1992 the Stahl's spent

$$\$1197.24 \text{ or } \$1197.24\left(\frac{90.35 \text{ pesetas}}{\text{USD}}\right)$$

$$= 108,170.63 \text{ pesetas}$$

In Spain in 2000 the Stahl's spent

$$\$1340.91 \text{ or } \$1340.91\left(\frac{185.45 \text{ pesetas}}{\text{USD}}\right)$$

$$= 248,671.76 \text{ pesetas}$$

Chapter 6 Review Problems

1. $\dfrac{6x^3 - 9x^2}{12x^2 - 18x} = \dfrac{3x^2(2x-3)}{6x(2x-3)} = \dfrac{x}{2}$

2. $\dfrac{15x^4}{5x^2 - 20x} = \dfrac{15x^4}{5x(x-4)} = \dfrac{3x^3}{x-4}$

3. $\dfrac{26x^3 y^2}{39xy^4} = \dfrac{13 \cdot 2x \cdot x^2 y^2}{13 \cdot 3xy^2 y^2} = \dfrac{2x^2}{3y^2}$

4. $\dfrac{x^2 - 2x - 35}{x^2 - x - 42} = \dfrac{(x-7)(x+5)}{(x-7)(x+6)} = \dfrac{x+5}{x+6}$

5. $\dfrac{2x^2 - 5x + 3}{3x^2 + 2x - 5} = \dfrac{(2x-3)(x-1)}{(3x+5)(x-1)} = \dfrac{2x-3}{3x+5}$

6. $\dfrac{ax + 2a - bx - 2b}{3x^2 - 12} = \dfrac{a(x+2) - b(x+2)}{3(x^2 - 4)}$

7. $\dfrac{4x^2 - 1}{x^2 - 4} \cdot \dfrac{2x^2 + 4x}{4x + 2}$

$= \dfrac{(2x-1)(2x+1)}{(x+2)(x-2)} \cdot \dfrac{2x(x+2)}{2(2x+1)}$

$= \dfrac{x(2x-1)}{x-2}$

8. $\dfrac{3y}{4xy - 6y^2} \cdot \dfrac{2x - 3y}{12xy}$

$= \dfrac{3y}{2y(2x-3y)} \cdot \dfrac{(2x-3y)}{12xy}$

$= \dfrac{1}{8xy}$

9. $\dfrac{y^2 + 8y - 20}{y^2 + 6y - 16} \cdot \dfrac{y^2 + 3y - 40}{y^2 + 6y - 40}$

$= \dfrac{(y+10)(y-2)}{(y+8)(y-2)} \cdot \dfrac{(y+8)(y-5)}{(y+10)(y-4)}$

$= \dfrac{y-5}{y-4}$

10. $\dfrac{3x^3 y}{x^2 + 7x + 12} \cdot \dfrac{x^2 + 8x + 15}{6xy^2}$

$= \dfrac{3xx^2 y}{(x+4)(x+3)} \cdot \dfrac{(x+5)(x+3)}{3 \cdot 2xyy}$

$= \dfrac{x^2(x+5)}{2y(x+4)}$

11. $\dfrac{2x + 12}{3x - 15} \div \dfrac{2x^2 - 6x - 20}{x^2 - 10x + 25}$

$= \dfrac{2(x+6)}{3(x-5)} \cdot \dfrac{(x-5)(x-5)}{2(x-5)(x+2)}$

$= \dfrac{x+6}{3(x+2)}$

12. $\dfrac{6x^2 - 6a^2}{3x^2 + 3} \div \dfrac{x^4 - a^4}{a^2 x^2 + a^2}$

$= \dfrac{6(x+a)(x-a)}{3(x^2+1)} \cdot \dfrac{a^2(x^2+1)}{(x^2 - a^2)(x^2 + a^2)}$

$= \dfrac{2(x+a)(x-a)}{(x^2+1)} \cdot \dfrac{a^2(x^2+1)}{(x-a)(x+a)(x^2 + a^2)}$

$= \dfrac{a^2}{x^2 + a^2}$

13. $\dfrac{9y^2 - 3y - 2}{6y^2 - 13y - 5} \div \dfrac{3y^2 + 10y - 8}{2y^2 + 13y + 20}$

$= \dfrac{(3y+1)(3y-2)}{(3y+1)(2y-5)} \cdot \dfrac{(y+4)(2y+5)}{(y+4)(3y-2)}$

$= \dfrac{2y+5}{2y-5}$

14. $\dfrac{4a^2+12a+5}{2a^2-7a-13} \div (4a^2+2a)$

$= \dfrac{(2a+1)(2a+5)}{2a^2-7a-13} \cdot \dfrac{1}{2a(2a+1)}$

$= \dfrac{2a+5}{2a(2a^2-7a-13)}$

15. $\dfrac{x-5}{2x+1} - \dfrac{x+1}{x-2}$

$= \dfrac{(x-5)(x-2)-(2x+1)(x+1)}{(2x+1)(x-2)}$

$= \dfrac{x^2-7x+10-2x^2-3x-1}{(2x+1)(x-2)}$

$= \dfrac{-x^2-10x+9}{(2x+1)(x-2)} = -\dfrac{x^2+10x-9}{(2x+1)(x-2)}$

16. $\dfrac{5}{4x} + \dfrac{-3}{x+4} = \dfrac{5(x+4)-3(4x)}{4x(x+4)}$

$= \dfrac{5x+20-12x}{4x(x+4)}$

$= \dfrac{-7x+20}{4x(x+4)}$

17. $\dfrac{(2y-1)(3)}{12y(3)} - \dfrac{(3y+2)(4)}{9y(4)}$

$= \dfrac{6y-3-12y-8}{36y}$

$= \dfrac{-6y-11}{36y}$

18. $\dfrac{4}{y+5} + \dfrac{3y+2}{y^2-25}$

$= \dfrac{4(y-5)}{(y+5)(y-5)} + \dfrac{3y+2}{(y+5)(y-5)}$

$= \dfrac{4y-20+3y+2}{(y+5)(y-5)}$

(continued)

18. (continued)

$= \dfrac{7y-18}{(y+5)(y-5)}$

19. $\dfrac{4y}{y^2+2y+1} + \dfrac{3}{y^2-1}$

$= \dfrac{4y(y-1)}{(y+1)^2(y-1)} + \dfrac{3(y+1)}{(y+1)^2(y-1)}$

$= \dfrac{4y^2-4y+3y+3}{(y+1)^2(y-1)}$

$= \dfrac{4y^2-y+3}{(y+1)^2(y-1)}$

20. $\dfrac{y^2-4y-19}{y^2+8y+15} - \dfrac{2y-3}{y+5}$

$= \dfrac{y^2-4y-19}{(y+5)(y+3)} - \dfrac{(2y-3)(y+3)}{(y+5)(y+3)}$

$= \dfrac{y^2-4y-19-2y^2-3y+9}{(y+5)(y+3)}$

$= \dfrac{-y^2-7y-10}{(y+5)(y+3)} = -\dfrac{(y+5)(y+2)}{(y+5)(y+3)}$

$= -\dfrac{y+2}{y+3}$

21. $\dfrac{a}{5-a} - \dfrac{2}{a+3} + \dfrac{2a^2-2a}{a^2-2a-15}$

$= \dfrac{-a(a+3)}{(a-5)(a+3)} - \dfrac{2(a-5)}{(a+3)(a-5)} + \dfrac{2a(a-1)}{(a-5)(a+3)}$

$= \dfrac{-a^2-3a-2a+10+2a^2-2a}{(a+3)(a-5)}$

$= \dfrac{a^2-7a+10}{(a+3)(a-5)} = \dfrac{(a-2)(a-5)}{(a+3)(a-5)}$

$= \dfrac{a-2}{a-5}$

22. $\dfrac{5}{a^2+3a+2}+\dfrac{6}{a^2+4a+3}-\dfrac{7}{a^2+5a+6}$

$=\dfrac{5(a+3)}{(a+2)(a+1)(a+3)}$

$\quad+\dfrac{6(a+2)}{(a+3)(a+1)(a+2)}$

$\quad\quad-\dfrac{7(a+1)}{(a+3)(a+2)(a+1)}$

$=\dfrac{5a+15+6a+12-7a-7}{(a+1)(a+2)(a+3)}$

$=\dfrac{4a+20}{(a+1)(a+2)(a+3)}$

23. $4a+3-\dfrac{2a+1}{a+4}$

$=\dfrac{(4a+3)(a+4)-(2a+1)}{a+4}$

$=\dfrac{4a^2+19a+12-2a-1}{a+4}$

$=\dfrac{4a^2+17a+11}{a+4}$

24. $\dfrac{1}{x}+\dfrac{3}{2x}+3+2x$

$=\dfrac{2}{2x}+\dfrac{3}{2x}+\dfrac{6x}{2x}+\dfrac{4x^2}{2x}$

$=\dfrac{2+3+6x+4x^2}{2x}$

$=\dfrac{4x^2+6x+5}{2x}$

25. $\dfrac{\dfrac{5}{x}+1}{1-\dfrac{25}{x^2}}\cdot\dfrac{x^2}{x^2}=\dfrac{5x+x^2}{x^2-25}$

$=\dfrac{x(x+5)}{(x-5)(x+5)}=\dfrac{x}{x-5}$

26. $\dfrac{\dfrac{4}{x+3}}{\dfrac{2}{x-2}-\dfrac{1}{x^2+x-6}}$

$=\dfrac{\dfrac{4}{x+3}}{\dfrac{2}{(x-2)}-\dfrac{1}{(x+3)(x-2)}}\cdot\dfrac{(x+3)(x-2)}{(x+3)(x-2)}$

$=\dfrac{4(x-2)}{2(x+3)-1}$

$=\dfrac{4(x-2)}{2x+6-1}$

$=\dfrac{4(x-2)}{2x+5}$

27. $\dfrac{\dfrac{y}{y+1}+\dfrac{1}{y}}{\dfrac{y}{y+1}-\dfrac{1}{y}}\cdot\dfrac{y(y+1)}{y(y+1)}=\dfrac{y^2+y+1}{y^2-y-1}$

28. $\dfrac{\dfrac{10}{a+2}-5}{\dfrac{4}{a+2}-2}\cdot\dfrac{a+2}{a+2}=\dfrac{10-5a-10}{4-2a-4}=\dfrac{5}{2}$

29. $\dfrac{\dfrac{2}{x+4}-\dfrac{1}{x^2+4x}}{\dfrac{3}{2x+8}}$

$=\dfrac{\dfrac{2}{x+4}-\dfrac{1}{x(x+4)}}{\dfrac{3}{2(x+4)}}\cdot\dfrac{2x(x+4)}{2x(x+4)}$

$\pm\dfrac{4x-2}{3x}$

$=\dfrac{2(2x-1)}{3x}$

30. $\dfrac{\dfrac{y^2}{y^2-x^2}-1}{x+\dfrac{xy}{x-y}}$

$=\dfrac{\dfrac{-y^2}{(x-y)(x+y)}-1}{x+\dfrac{xy}{x-y}}\cdot\dfrac{(x-y)(x+y)}{(x-y)(x+y)}$

$=\dfrac{-y^2-x^2+y^2}{x(x-y)(x+y)+xy(x+y)}$

$=\dfrac{-x^2}{x(x+y)(x-y+y)}$

$=\dfrac{-x}{(x+y)x}$

$=\dfrac{-1}{x+y}$

31. $\dfrac{\dfrac{2x+1}{x-1}}{1+\dfrac{x}{x+1}}\cdot\dfrac{(x+1)(x-1)}{(x+1)(x-1)}$

$=\dfrac{(2x+1)(x+1)}{(x+1)(x-1)+x(x-1)}$

$=\dfrac{2x^2+3x+1}{x^2-1+x^2-x}$

$=\dfrac{(2x+1)(x+1)}{2x^2-x-1}$

$=\dfrac{(2x+1)(x+1)}{(2x+1)(x-1)}$

$=\dfrac{x+1}{x-1}$

32. $\dfrac{\dfrac{3}{x}-\dfrac{2}{x+1}}{\dfrac{5}{x^2+5x+4}-\dfrac{1}{x+4}}$

(continued)

32. (continued)

$=\dfrac{\dfrac{3}{x}-\dfrac{2}{x+1}}{\dfrac{5}{(x+4)(x+1)}-\dfrac{1}{x+4}}\cdot\dfrac{x(x+1)(x+4)}{x(x+1)(x+4)}$

$=\dfrac{3(x+1)(x+4)-2x(x+4)}{5x-x(x+1)}$

$=\dfrac{3x^2+15x+12-2x^2-8x}{5x-x^2-x}$

$=\dfrac{x^2+7x+12}{-x^2+4x}$

$=\dfrac{(x+4)(x+3)}{-x(x-4)}$

$=\dfrac{-(x+3)(x+4)}{x(x-4)}$

33. $\dfrac{3}{2}=1-\dfrac{1}{x-1}$

$3(x-1)=2(x-1)-2$

$3x-3=2x-2-2$

$x=-1$

check:

$\dfrac{3}{2}\overset{?}{=}1-\dfrac{1}{-1-1}$

$\dfrac{3}{2}=\dfrac{3}{2}$

34. $\dfrac{3}{7}+\dfrac{4}{x+1}=1$

$3(x+1)+28=7(x+1)$

$3x+3+28=7x+7$

$4x=24$

$x=6$

check:

$\dfrac{3}{7}+\dfrac{4}{6+1}\overset{?}{=}1$

$1=1$

35.
$$\frac{3}{x-2}+\frac{8}{x+3}=\frac{6}{x-2}$$
$$3(x+3)+8(x-2)=6(x+3)$$
$$3x+9+8x-16=6x+18$$
$$5x=25$$
$$x=5$$
check:
$$\frac{3}{5-2}+\frac{8}{5+3}\overset{?}{=}\frac{6}{5-2}$$
$$2=2$$

36.
$$\frac{1}{x+2}-\frac{1}{x}=\frac{-2}{x}$$
$$x-(x+2)=-2(x+2)$$
$$x-x-2=-2x-4$$
$$2x=-2$$
$$x=-1$$
check:
$$\frac{1}{-1+2}-\frac{1}{-1}\overset{?}{=}\frac{-2}{-1}$$
$$2=2$$

37.
$$\frac{5}{2a}=\frac{2}{a}-\frac{1}{12}$$
$$30=24-a$$
$$a=-6$$
check:
$$\frac{5}{2(-6)}\overset{?}{=}\frac{2}{-6}-\frac{1}{12}$$
$$-\frac{5}{12}=-\frac{5}{12}$$

38.
$$\frac{1}{2a}=\frac{2}{a}-\frac{3}{10}$$
$$5=20-3a$$
$$3a=15$$
$$a=5$$

(continued)

38. (continued)
check:
$$\frac{1}{2(5)}\overset{?}{=}\frac{2}{5}-\frac{3}{10}$$
$$\frac{1}{10}=\frac{1}{10}$$

39.
$$\frac{1}{y}+\frac{1}{2y}=2$$
$$2+1=4y$$
$$y=\frac{3}{4}$$
check:
$$\frac{1}{\frac{3}{4}}+\frac{1}{2\cdot\frac{3}{4}}\overset{?}{=}2$$
$$2=2$$

40.
$$\frac{5}{y^2}+\frac{7}{y}=\frac{6}{y^2}$$
$$5+7y=6$$
$$7y=1$$
$$y=\frac{1}{7}$$
check:
$$\frac{5}{\left(\frac{1}{7}\right)^2}+\frac{7}{\frac{1}{7}}\overset{?}{=}\frac{6}{\left(\frac{1}{7}\right)^2}$$
$$294=294$$

41.
$$\frac{a+2}{2a+6}=\frac{3}{2}-\frac{3}{a+3}$$
$$a+2=3(a+3)-3(2)$$
$$a+2=3a+9-6$$
$$2a=-1$$
$$a=-\frac{1}{2}\text{ (continued)}$$

41. (continued)
check:

$$\frac{-\dfrac{1}{2}+2}{2\left(-\dfrac{1}{2}\right)+6} \overset{?}{=} \frac{3}{2}-\frac{3}{-\dfrac{1}{2}+3}$$

$$\frac{3}{10}=\frac{3}{10}$$

42. $\dfrac{5}{a+5}+\dfrac{a+4}{2a+10}=\dfrac{3}{2}$

$$\frac{5}{a+5}+\frac{a+4}{2(a+5)}=\frac{3}{2}$$

$$10+a+4=3(a+5)$$

$$14+a=3a+15$$

$$2a=-1$$

$$a=-\frac{1}{2}$$

check:

$$\frac{5}{-\dfrac{1}{2}+5}+\frac{-\dfrac{1}{2}+4}{2\left(-\dfrac{1}{2}+10\right)} \overset{?}{=} \frac{3}{2}$$

$$\frac{3}{2}=\frac{3}{2}$$

43. $\dfrac{1}{x+2}-\dfrac{5}{x-2}=\dfrac{-15}{x^2-4}$

$$\frac{1}{x+2}-\frac{5}{x-2}=\frac{-15}{(x+2)(x-2)}$$

$$x-2-5(x+2)=-15$$

$$x-2-5x-10=-15$$

$$-4x=-3$$

$$x=\frac{3}{4}$$

(continued)

43. (continued)
check:

$$\frac{1}{\dfrac{3}{4}+2}-\frac{5}{\dfrac{3}{4}-2} \overset{?}{=} \frac{-15}{\left(\dfrac{3}{4}\right)^2-4}$$

$$\frac{48}{11}=\frac{48}{11}$$

44. $\dfrac{y+1}{y^2+2y-3}-\dfrac{1}{y+3}=\dfrac{1}{y-1}$

$$\frac{y+1}{(y+3)(y-1)}-\frac{1}{y+3}=\frac{1}{y-1}$$

$$y+1-(y-1)=y+3$$

$$y+1-y+1=y+3$$

$$y=-1$$

check:

$$\frac{-1+1}{(-1)^2+2(-1)-3}-\frac{1}{-1+3} \overset{?}{=} \frac{1}{-1-1}$$

$$-\frac{1}{2}=-\frac{1}{2}$$

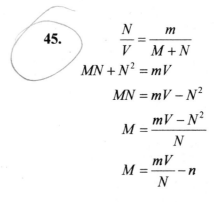

45. $\dfrac{N}{V}=\dfrac{m}{M+N}$

$$MN+N^2=mV$$

$$MN=mV-N^2$$

$$M=\frac{mV-N^2}{N}$$

$$M=\frac{mV}{N}-n$$

46. $m=\dfrac{y-y_o}{x-x_o}$

$$mx-mx_o=y-y_o$$

$$mx=y-y_o+mx_o$$

$$x=\frac{y-y_o+mx_o}{m}$$

47.
$$\frac{1}{f} = \frac{1}{a} + \frac{1}{b}$$
$$ab = fb + fa$$
$$ab - fa = fb$$
$$a(b - f) = fb$$
$$a = \frac{fb}{b - f}$$

48.
$$S = \frac{V_1 t + V_2 t}{2}$$
$$t(V_1 + V_2) = 2S$$
$$t = \frac{2S}{V_1 + V_2}$$

49.
$$d = \frac{LR_2}{R_2 + R_1}$$
$$dR_2 + dR_1 = LR_2$$
$$R_2(L - d) = dR_1$$
$$R_2 = \frac{dR_1}{L - d}$$

50.
$$\frac{S - P}{\Pr} = t$$
$$\Pr t = S - P$$
$$r = \frac{S - P}{Pt}$$

51.
$$\frac{7}{4} = \frac{S}{G} = \frac{S}{253 - S}$$
$$1771 - 7S = 4S$$
$$11S = 1771$$
$$S = 161$$
$$G = 253 - 161 = 92$$

The campus bookstore at Boston University ordered 161 scientific calculators and 92 graphing calculators for the spring semester.

52. $x =$ number of one-story homes
$$\frac{3}{13} = \frac{x}{112 - x}$$
$$336 - 3x = 13x$$
$$16x = 336$$
$$x = 21$$
$$112 - x = 91$$

Walter Johnson built 21 one-story homes and 91 two-story homes.

53.
$$\frac{5}{7} = \frac{w}{l} = \frac{w}{\dfrac{168 - 2w}{2}} = \frac{2w}{168 - 2w}$$
$$840 - 10w = 14w$$
$$24w = 840$$
$$w = 35$$
$$l = \frac{168 - 2w}{2} = 49$$

The enlarged photograph will be 35 inches wide and 49 inches long.

54.
$$\frac{4}{3500} = \frac{x}{4900}$$
$$3500x = 19,600$$
$$x = 5.6$$

The pump can empty the 4900-gallon pool in 5.6 hours.

55.
$$\frac{100}{P} = \frac{8}{40}$$
$$8P = 4000$$
$$P = 500$$

There is a population of 500 rabbits in the sanctuary.

56. x = number of officers

$$\frac{2}{9} = \frac{x}{154 - x}$$

$$308 - 2x = 9x$$

$$11x = 308$$

$$x = 28$$

There are 28 officers on the force.

57. x = number of nautical miles

$$\frac{7}{2} = \frac{x}{3.5}$$

$$2x = 24.5$$

$$x = 12.25$$

The course is 12.25 nautical miles.

58. x = height of building

$$\frac{7}{6} = \frac{x}{156}$$

$$6x = 1092$$

$$x = 182$$

The building is 182 feet tall.

59. t = time if both faucets are open

$$\frac{t}{15} + \frac{t}{10} = 1$$

$$\frac{t}{6} = 1$$

$$t = 6$$

It would take 6 minutes to fill if both faucets are left open.

60. t = time for both to paint barn

$$\frac{t}{12} + \frac{t}{18} = 1$$

$$\frac{5}{36}t = 1$$

$$t = 7.2$$

It would take 7.2 hours to paint the barn if they worked together.

61. x = number of messages in July 2001

$$\frac{651 - 430}{10} = \frac{x - 651}{17}$$

$$3757 = 10x - 6510$$

$$10x = 10,267$$

$$x = 1026.7$$

The expected number of daily messages in July 2001 would be 1027 million.

62. x = number of messages in Dec. 2000

$$\frac{651 - 430}{10} = \frac{x - 651}{10}$$

$$x - 651 = 221$$

$$x = 872$$

The expected number of daily messages in December 2000 would be 872 million.

63. x = number of messages in Dec. 2000

$$\frac{651 - 430}{10} = \frac{1}{2} \cdot \frac{x - 651}{10}$$

$$x = 2(221) + 651$$

$$x = 1093$$

The expected number of daily messages in December 2000 would be 1093 million.

64. x = number of messages in April 2001

$$\frac{430 - 94}{24} = 2 \cdot \frac{x - 430}{24}$$

$$\frac{336}{2} = x - 430$$

$$x = 168 + 430$$

$$x = 598$$

The expected number of daily messages in April 2001 would be 598 million.

Chapter 6 Test

1. $\dfrac{x^3+3x^2+2x}{x^3-2x^2-3x}=\dfrac{x(x^2+3x+2)}{x(x-2x-3)}$

$=\dfrac{(x+2)(x+1)}{(x-3)(x+1)}$

$=\dfrac{x+2}{x-3}$

2. $\dfrac{y^3+8}{y^2-4}=\dfrac{(y+2)(y^2-2y+4)}{(y+2)(y-2)}$

$=\dfrac{(y^2-2y+4)}{(y-2)}$

3. $\dfrac{2y^2+7y-4}{y^2+2y-8}\cdot\dfrac{2y^2-8}{3y^2+11y+10}$

$=\dfrac{(2y-1)(y+4)}{(y+4)(y-2)}\cdot\dfrac{2(y+2)(y-2)}{(3y+5)(y+2)}$

$=\dfrac{2(2y+1)}{3y+5}$

4. $\dfrac{4-2x}{3x^2-2x-8}\div\dfrac{2x^2+x-1}{9x+12}$

$=\dfrac{-2(x-2)}{(3x+4)(x-2)}\cdot\dfrac{3(3x+4)}{(2x-1)(x+1)}$

$=\dfrac{-6}{(2x-1)(x+1)}$

5. $\dfrac{3x+8}{x^2-25}-\dfrac{5}{x-5}$

$=\dfrac{3x+8}{(x+5)(x-5)}-\dfrac{5}{x-5}$

$=\dfrac{3x+8-5(x+5)}{(x+5)(x-5)}$

$=\dfrac{3x+8-5x-25}{(x+5)(x-5)}=\dfrac{-2x-17}{(x+5)(x-5)}$

6. $\dfrac{2}{x^2+5x+6}+\dfrac{3x}{x^2+6x+9}$

$=\dfrac{2(x+3)}{(x+3)(x+2)(x+3)}$

$\quad+\dfrac{3x(x+2)}{(x+3)(x+3)(x+2)}$

$=\dfrac{2x+6+3x^2+6x}{(x+3)(x+2)(x+3)}$

$=\dfrac{3x^2+8x+6}{(x+3)^2(x+2)}$

7. $\dfrac{\dfrac{4}{y+2}-2}{5-\dfrac{10}{y+2}}\cdot\dfrac{y+2}{y+2}$

$=\dfrac{4-2(y+2)}{5(y+2)-10}$

$=\dfrac{4-2y-4}{5y+10-10}$

$=-\dfrac{2}{5}$

8. $\dfrac{\dfrac{1}{x}-\dfrac{3}{x+2}}{\dfrac{2}{x^2+2x}}=\dfrac{\dfrac{1}{x}-\dfrac{3}{x+2}}{\dfrac{2}{x(x+2)}}\cdot\dfrac{x(x+2)}{x(x+2)}$

$=\dfrac{x+2-3x}{2}$

$=\dfrac{-2x+2}{2}$

$=\dfrac{-2(x-1)}{2}$

$=-x+1$

9. $\dfrac{7}{4} = \dfrac{x+4}{x}$

$7x = 4x + 16$

$3x = 16$

$x = \dfrac{16}{3}$

check:

$\dfrac{7}{4} \overset{?}{=} \dfrac{\frac{16}{3}+4}{\frac{16}{3}}$

$\dfrac{7}{4} = \dfrac{7}{4}$

10. $2 + \dfrac{x}{x+4} = \dfrac{3x}{x-4}$

$2(x+4)(x-4) + x(x-4) = 3x(x+4)$

$2x^2 - 32 + x^2 - 4x = 3x^2 + 12x$

$16x = -32$

$x = -2$

check:

$2 + \dfrac{-2}{-2+4} \overset{?}{=} \dfrac{3(-2)}{-2-4}$

$1 = 1$

11. $\dfrac{1}{2y+4} - \dfrac{1}{6} = \dfrac{-2}{3y+6}$

$\dfrac{1}{2(y+2)} - \dfrac{1}{6} = \dfrac{-2}{3(y+2)}$

$3 - (y+2) = -4$

$3 - y - 2 = -4$

$y = 5$

check:

$\dfrac{1}{2(5)+4} - \dfrac{1}{6} \overset{?}{=} \dfrac{-2}{3(5)+6}$

$-\dfrac{2}{21} = -\dfrac{2}{21}$

12. $\dfrac{3}{2x+3} - \dfrac{1}{2x-3} = \dfrac{2}{4x^2-9}$

$\dfrac{3}{2x+3} - \dfrac{1}{2x-3} = \dfrac{2}{(2x+3)(2x-3)}$

$3(2x-3) - (2x+3) = 2$

$6x - 9 - 2x - 3 = 2$

$4x = 14$

$x = \dfrac{7}{2}$

check:

$\dfrac{3}{2\left(\frac{7}{2}\right)+3} - \dfrac{1}{2\left(\frac{7}{2}\right)-3} \overset{?}{=} \dfrac{2}{4\left(\frac{7}{2}\right)^2 - 9}$

$\dfrac{1}{20} = \dfrac{1}{20}$

13. $h = \dfrac{S - 2WL}{2W + 2L}$

$2hW + 2hL = S - 2WL$

$(2h + 2L)W = S - 2hL$

$W = \dfrac{S - 2hL}{2h + 2L}$

14. $\dfrac{4}{a} = \dfrac{3}{b} + \dfrac{2}{c}$

$4bc = 3ac + 2ab$

$(4c - 2a)b = 3ac$

$b = \dfrac{3ac}{4c - 2a}$

15. $\dfrac{3}{19} = \dfrac{x}{286 - x}$

$858 - 3x = 19x$

$22x = 858$

$x = 39$

$286 - 39 = 247$

39 got the bonus, 247 did not.

16.

$$\frac{500}{850} = \frac{W}{\dfrac{8100 - 2W}{2}}$$

$$\frac{500}{850} = \frac{2W}{8100 - 2W}$$

$$4{,}050{,}000 - 1000W = 1700W$$

$$2700W = 4{,}050{,}000$$

$$W = 1500$$

$$\frac{8100 - 2W}{2} = 2550$$

The width is 1500 feet and the length is 2550 feet.

Cumulative Test for Chapters 1-6

1. $\left(\dfrac{3x^{-2}y^3}{z^4}\right)^{-2} = \dfrac{3^{-2}x^4 y^{-6}}{z^{-8}} = \dfrac{x^4 z^8}{9y^6}$

2. $\dfrac{2}{3}(3x - 1) = \dfrac{2}{5}x + 3$

$$30x - 10 = 6x + 45$$

$$24x = 55$$

$$x = \frac{55}{24}$$

3. $-6x + 2y = -12$

x	y
0	−6
1	−3
2	0

4. $5x - 6y = 8$

$$6y = 5x - 8$$

$$y = \frac{5}{6}x - \frac{4}{3} \Rightarrow m = \frac{5}{6}, \; m_{\parallel} = \frac{5}{6}$$

$$y - y_1 = m(x - x_1)$$

$$y - (-3) = \frac{5}{6}(x - (-1))$$

$$6y + 18 = 5x + 5$$

$$5x - 6y = 13$$

5. $x =$ amount at 5%

$$0.05x + 0.08(7000 - x) = 539$$

$$0.05x + 560 - 0.08x = 539$$

$$-0.03x = -21$$

$$x = 700$$

$$7000 - x = 6300$$

$700 was invested at 5% and $6300 was invested at 8%.

6. $3(2 - 6x) > 4(x + 1) + 24$

$$6 - 18x > 4x + 4 + 24$$

$$-22x > 22$$

$$x < -1$$

7. $2x^2 - 3x - 4y^2 = 2(-2)^2 - 3(-2) - 4(3)^2$

$$= 2(4) + 6 - 4(9)$$

$$= 8 + 6 - 36$$

$$= 14 - 36$$

$$= -22$$

8. $|3x - 4| \le 10$

$$-10 \le 3x - 4 \le 10$$

$$-6 \le 3x \le 14$$

$$-2 \le x \le \frac{14}{3}$$

9. $8x^3 - 125y^3$

$= (2x - 5y)(4x^2 + 10xy - 25y^2)$

10. $81x^3 - 90x^2y + 25xy^2$

$= x(9x^2 - 90xy + 25y^2)$

$= x(3y - 5y)^2$

11. $x^2 + 20x + 36 = 0$

$(x + 18)(x + 2) = 0$

$x + 18 = 0$ or $x + 2 = 0$

$x = -18 \qquad\qquad x = -2$

12. $3x^2 - 11x - 4 = 0$

$(3x + 1)(x - 4) = 0$

$3x + 1 = 0$ or $x - 4 = 0$

$3x + 1 = 0 \qquad x - 4 = 0$

$x = -\dfrac{1}{3} \qquad\qquad x = 4$

13. $\dfrac{7x^2 - 28}{x^2 + 6x + 8} = \dfrac{7(x + 2)(x - 2)}{(x + 4)(x + 2)}$

$= \dfrac{7(x - 2)}{x + 4}$

14. $\dfrac{2x^2 + x - 1}{2x^2 - 9x + 4} \cdot \dfrac{3x^2 - 12x}{6x + 15}$

$= \dfrac{(2x - 1)(x + 1)}{(2x - 1)(x - 4)} \cdot \dfrac{3x(x - 4)}{3(2x + 5)}$

$= \dfrac{x(x + 10}{2x + 5}$

15. $\dfrac{x^3 + 27}{x^2 + 7x + 12} \div \dfrac{x^2 - 6x + 9}{2x^2 + 13x + 20}$

$= \dfrac{(x + 3)(x^2 - 3x + 9)}{(x + 3)(x + 4)} \cdot \dfrac{(2x + 5)(x + 4)}{(x - 3)(x - 3)}$

(continued)

15.(continued)

$= \dfrac{(x^2 - 3x + 9)(2x + 5)}{(x - 3)^2}$

16. $\dfrac{5}{2x - 8} - \dfrac{3x}{x^2 - 9x + 20}$

$= \dfrac{5(x - 5)}{2(x - 4)(x - 5)} - \dfrac{3x(2)}{(x - 4)(x - 5)(2)}$

$= \dfrac{5x - 25 - 6x}{2(x - 4)(x - 5)}$

$= -\dfrac{x + 25}{2(x - 4)(x - 5)}$

17. $\dfrac{\dfrac{1}{2x + 1} + 1}{4 - \dfrac{3}{4x^2 - 1}}$

$= \dfrac{\dfrac{1}{2x + 1} + 1}{4 - \dfrac{3}{(2x - 1)(2x + 1)}} \cdot \dfrac{(2x - 1)(2x + 1)}{(2x - 1)(2x + 1)}$

$= \dfrac{2x - 1 + (2x - 1)(2x + 1)}{4(2x - 1)(2x + 1) - 3}$

$= \dfrac{(2x - 1)(1 + 2x + 1)}{16x^2 - 4 - 3}$

$= \dfrac{(2x - 1)(2x + 2)}{16x^2 - 7}$

$= \dfrac{2(x + 1)(2x - 1)}{16x^2 - 7}$

18. $\dfrac{3}{x - 6} + \dfrac{4}{x + 4} = \dfrac{3(x + 4) + 4(x - 6)}{(x - 6)(x + 4)}$

$= \dfrac{3x + 12 + 4x - 24}{(x - 6)(x + 4)}$

$= \dfrac{7x - 12}{(x - 6)(x + 4)}$

19. $\dfrac{1}{2x+3} - \dfrac{4}{4x^2-9} = \dfrac{3}{2x-3}$

$\dfrac{1}{2x+3} - \dfrac{4}{(2x+3)(2x-3)} = \dfrac{3}{2x-3}$

$2x - 3 - 4 = 3(2x+3)$

$2x - 7 = 6x + 9$

$4x = -16$

$x = -4$

check:

$\dfrac{1}{2(-4)+3} - \dfrac{4}{4(-4)^2-9} \overset{?}{=} \dfrac{3}{2(-4)-3}$

$-\dfrac{3}{11} = -\dfrac{3}{11}$

20. $\dfrac{1}{4x} - \dfrac{3}{2x} = \dfrac{5}{8}$

$2 - 12 = 5x$

$5x = -10$

$x = -2$

check:

$\dfrac{1}{4(-2)} - \dfrac{3}{2(-2)} \overset{?}{=} \dfrac{5}{8}$

$\dfrac{5}{8} = \dfrac{5}{8}$

21. $H = \dfrac{3b+2x}{5-4b}$

$5H - 4bH = 3b + 2x$

$3b + 4bH = 5H - 2x$

$b(3 + 4H) = 5H - 2x$

$b = \dfrac{5H - 2x}{3 + 4H}$

22. x = number patrolling on foot

$\dfrac{3}{11} = \dfrac{x}{3234 - x}$

$9702 - 3x = 11x$

$14x = 9702$

$x = 693$

$3234 - x = 2541$

693 are patrolling on foot and 2541 are patrolling in squad cars.

Chapter 7

Pretest Chapter 7

1. $(-3x^{\frac{1}{4}}y^{\frac{1}{2}})(-2x^{-\frac{1}{2}}y^{\frac{1}{3}}) = 6x^{\frac{1}{4}-\frac{1}{2}}y^{\frac{1}{2}+\frac{1}{3}}$

$$= 6x^{-\frac{1}{4}}y^{\frac{5}{6}}$$

$$= \frac{6y^{\frac{5}{6}}}{x^{\frac{1}{4}}}$$

2. $(-4x^{-\frac{1}{4}}y^{\frac{1}{3}})^3 = (-4)^3 x^{-\frac{3}{4}}y$

$$= -\frac{64y}{x^{\frac{3}{4}}}$$

3. $\dfrac{-18x^{-2}y^2}{-3x^{-5}y^{\frac{1}{3}}} = 6x^{-2+5}y^{2-\frac{1}{3}}$

$$= 6x^3 y^{\frac{5}{3}}$$

4. $\left(\dfrac{27x^2 y^{-5}}{x^{-4}y^4}\right)^{\frac{2}{3}} = \left(\dfrac{3^3 x^6}{y^9}\right)^{\frac{2}{3}}$

$$= \left(\dfrac{3^2 x^4}{y^6}\right)$$

$$= \dfrac{9x^4}{y^6}$$

5. $27^{-\frac{4}{3}} = \dfrac{1}{(3^3)^{\frac{4}{3}}}$

$$= \dfrac{1}{3^4}$$

$$= \dfrac{1}{81}$$

6. $\sqrt{169} + \sqrt[3]{-64} = 13 - 4 = 9$

7. $\sqrt[3]{27a^{12}b^6 c^{15}} = \sqrt[3]{27}\sqrt[3]{a^{12}}\sqrt[3]{b^6}\sqrt[3]{c^{15}}$

$$= 3a^4 b^2 c^5$$

8. $\sqrt[4]{32x^8 y^{15}} = \sqrt[4]{16 \cdot 2x^8 y^{12} y^3}$

$$= \sqrt[4]{16}\sqrt[4]{x^8}\sqrt[4]{y^{12}}\sqrt[4]{2y^3}$$

$$= 2x^2 y^3 \sqrt[4]{2y^3}$$

9. $3\sqrt{48y^3} - 2\sqrt[3]{16} + 3\sqrt[3]{54} - 5y\sqrt{12y}$

$= 3\sqrt{16y^2 \cdot 3y} - 2\sqrt[3]{8 \cdot 2} + 3\sqrt[3]{27 \cdot 2} - 5y\sqrt{4 \cdot 3y}$

$= 12y\sqrt{3y} - 4\sqrt[3]{2} + 9\sqrt[3]{2} - 10y\sqrt{3y}$

$= 2y\sqrt{3y} + 5\sqrt[3]{2}$

10. $(3\sqrt{3} - 5\sqrt{6})(\sqrt{12} - 3\sqrt{6})$

$= 3\sqrt{36} - 9\sqrt{18} - 5\sqrt{72} + 15(6)$

$= 3(6) - 9\sqrt{9 \cdot 2} - 5\sqrt{36 \cdot 2} + 90$

$= 108 - 27\sqrt{2} - 30\sqrt{2}$

$= 108 - 57\sqrt{2}$

11. $\dfrac{6}{\sqrt[3]{9x}} \cdot \dfrac{\sqrt[3]{3x^2}}{\sqrt[3]{3x^2}} = \dfrac{6\sqrt[3]{3x^2}}{\sqrt[3]{27x^3}}$

$$= \dfrac{6\sqrt[3]{3x^2}}{3x}$$

$$= \dfrac{2\sqrt[3]{3x^2}}{x}$$

12. $\dfrac{\sqrt{2}+\sqrt{3}}{\sqrt{2}-\sqrt{3}} \cdot \dfrac{\sqrt{2}+\sqrt{3}}{\sqrt{2}+\sqrt{3}} = \dfrac{2+2\sqrt{6}+3}{2-3}$

$$= -5 - 2\sqrt{6}$$

180

13. $\sqrt{3x+4}+2=x$

$\sqrt{3x+4}=x-2$

$3x+4=x^2-4x+4$

$x^2-7x=0$

$x(x-7)=0$

$\qquad x=0 \ \text{ or } x=7$

check:

$\sqrt{3(0)+4}+2\overset{?}{=}0$

$\qquad 4\neq 0$

$\sqrt{3(7)+4}+2\overset{?}{=}7$

$\qquad 7=7$

$x=7$ is the solution.

14. $\sqrt{2x+3}-\sqrt{x-2}=2$

$\sqrt{2x+3}=2+\sqrt{x-2}$

$2x+3=4+4\sqrt{x-2}+x-2$

$x+1=4\sqrt{x-2}$

$x^2+2x+1=16x-32$

$x^2-14x+33=0$

$(x-11)(x-3)=0$

$\qquad x=11 \ \text{ or } x=3$

check:

$\sqrt{2(11)+3}-\sqrt{11-2}\overset{?}{=}2$

$\qquad 5-3=2$

$\sqrt{2(3)+3}-\sqrt{3-2}\overset{?}{=}2$

$\qquad 3-1=2$

$x=3,\ x=11$ are solutions.

15. $(3-2i)-(-1+3i)=3-2i+1-3i$

$\qquad =4-5i$

16. $i^{15}+\sqrt{-25}=i^{4(3)+3}+i\sqrt{25}$

$=(i^4)^3\cdot i^3+5i$

$=(1)^3\cdot(-i)+5i$

$=4i$

17. $(3+5i)^2=9+30i+25i^2$

$=9+30i-25$

$=-16+30i$

18. $\dfrac{3+2i}{2+3i}\cdot\dfrac{2-3i}{2-3i}=\dfrac{6-5i-6i^2}{4+9}$

$=\dfrac{6-5i+6}{13}$

$=\dfrac{12-5i}{13}$

19. $y=kx^2$

$18=k(3)^2$

$k=2$

$y=2x^2$

$y=2(5)^2=50$

20. $y=\dfrac{k}{x}$

$12=\dfrac{k}{6}$

$k=72$

$y=\dfrac{72}{x^2}$

$y=\dfrac{72}{10^2}=7.2$

7.1 Exercises

1. $\left(\dfrac{4x^2y^{-3}}{x}\right)^2=\dfrac{4^2x^4y^{-6}}{x^2}=\dfrac{16x^2}{y^6}$

3. $\left(\dfrac{2a^{-1}b^3}{-3b^2}\right)^3 = \dfrac{8a^{-3}b^9}{-27b^6} = \dfrac{8b^3}{-27a^3}$

5. $(x^{\frac{3}{4}})^2 = x^{\frac{6}{4}} = x^{\frac{3}{2}}$

7. $(y^{12})^{\frac{2}{3}} = y^{12 \cdot \frac{2}{3}} = \bcancel{} \; y^8$

9. $\dfrac{x^{\frac{7}{12}}}{x^{\frac{1}{12}}} = x^{\frac{7}{12} - \frac{1}{12}} = x^{\frac{6}{12}} = x^{\frac{1}{2}}$

11. $\dfrac{x^3}{x^{\frac{1}{2}}} = x^{3 - \frac{1}{2}} = x^{\frac{6}{2} - \frac{1}{2}} = x^{\frac{5}{2}}$

13. $x^{\frac{1}{7}} \cdot x^{\frac{3}{7}} = x^{\frac{1}{7} + \frac{3}{7}} = x^{\frac{4}{7}}$

15. $y^{\frac{3}{5}} \cdot y^{-\frac{1}{10}} = y^{\frac{6}{10} - \frac{1}{10}} = y^{\frac{1}{2}}$

17. $x^{-\frac{3}{4}} = \dfrac{1}{x^{\frac{3}{4}}}$

19. $a^{-\frac{5}{6}}b^{\frac{1}{3}} = \dfrac{b^{\frac{1}{3}}}{a^{\frac{5}{6}}}$

21. $6^{-\frac{1}{2}} = \dfrac{1}{6^{\frac{1}{2}}}$

23. $2a^{-\frac{1}{4}} = \dfrac{2}{a^{\frac{1}{4}}}$

25. $(x^{\frac{1}{2}}y^{\frac{1}{3}})(x^{\frac{1}{3}}y^{\frac{2}{3}}) = x^{\frac{3}{6} + \frac{2}{6}}y^{\frac{1}{3} + \frac{2}{3}} = x^{\frac{5}{6}}y$

27. $(7x^{\frac{1}{3}}y^{\frac{1}{4}})(-2x^{\frac{1}{4}}y^{-\frac{1}{6}}) = -14x^{\frac{7}{12}}y^{\frac{1}{12}}$

29. $6^2 \cdot 6^{-\frac{2}{3}} = 6^{\frac{6}{3} - \frac{2}{3}} = 6^{\frac{4}{3}}$

31. $\dfrac{2x^{\frac{1}{5}}}{x^{-\frac{1}{2}}} = 2x^{\frac{2}{10} + \frac{5}{10}} = 2x^{\frac{7}{10}}$

33. $\dfrac{-20x^2 y^{-\frac{1}{5}}}{5x^{-\frac{1}{2}}y} = -\dfrac{4x^{\frac{5}{2}}}{y^{\frac{6}{5}}}$

35. $\left(\dfrac{8a^2b^6}{a^{-1}b^3}\right)^{\frac{1}{3}} = \left(8a^3b^3\right)^{\frac{1}{3}} = 2ab$

37. $(-3x^{\frac{2}{5}}y^{\frac{3}{2}}z^{\frac{1}{3}})^2 = (-3)^2 x^{\frac{4}{5}}y^3 z^{\frac{2}{3}} = 9x^{\frac{4}{5}}y^3 z^{\frac{2}{3}}$

39. $x^{\frac{2}{3}}(x^{\frac{4}{3}} - x^{\frac{1}{5}}) = x^{\frac{6}{3}} - x^{\frac{2}{3} + \frac{1}{5}} = x^2 - x^{\frac{13}{15}}$

41. $m^{\frac{7}{8}}(m^{-\frac{1}{2}} + 2m) = m^{\frac{7}{8} - \frac{4}{8}} + 2m^{\frac{7}{8} + \frac{8}{8}}$
$= m^{\frac{3}{8}} + 2m^{\frac{15}{8}}$

43. $\dfrac{(x^2 \cdot x^{-\frac{3}{2}})^{\frac{1}{2}}}{x^{\frac{1}{2}}} = \dfrac{(x^{\frac{4}{2} - \frac{3}{2}})^{\frac{1}{2}}}{x^{\frac{1}{2}}} = \dfrac{(x^{\frac{1}{2}})^{\frac{1}{2}}}{x^{\frac{1}{2}}}$
$= \dfrac{x^{\frac{1}{4}}}{x^{\frac{1}{2}}} = \dfrac{1}{x^{\frac{1}{4}}}$

45. $(27)^{\frac{2}{3}} = (3^3)^{\frac{2}{3}} = 3^2 = 9$

47. $(4)^{\frac{3}{2}} = (2^2)^{\frac{3}{2}} = 2^3 = 8$

49. $(81)^{\frac{3}{4}} + (25)^{\frac{1}{2}} = (3^4)^{\frac{3}{4}} + (5^2)^{\frac{1}{2}}$
$= 3^3 + 5 = 27 + 5 = 32$

182

51. $2y^{\frac{1}{3}} + y^{-\frac{2}{3}} = 2y^{\frac{1}{3}} + \dfrac{1}{y^{\frac{2}{3}}}$

$$= \dfrac{2y^{\frac{1}{3}+\frac{2}{3}} + 1}{y^{\frac{2}{3}}}$$

$$= \dfrac{2y + 1}{y^{\frac{2}{3}}}$$

53. $5^{-\frac{1}{4}} + x^{-\frac{1}{2}} = \dfrac{1}{5^{\frac{1}{4}}} + \dfrac{1}{x^{\frac{1}{2}}}$

$$= \dfrac{x^{\frac{1}{2}} + 5^{\frac{1}{4}}}{5^{\frac{1}{4}} x^{\frac{1}{2}}}$$

55. $6a^{\frac{4}{3}} - 8a^{\frac{3}{2}} = 2a(3a^{\frac{1}{3}} - 4a^{\frac{1}{2}})$

57. $x^b \div x^{\frac{1}{3}} = x^{\frac{1}{12}}$

$$x^{b - \frac{1}{3}} = x^{\frac{1}{12}}$$

$$b - \dfrac{1}{3} = \dfrac{1}{12}$$

$$b = \dfrac{1}{4}$$

59. $r = 0.62(V)^{\frac{1}{3}}$

$r = 0.62(64)^{\frac{1}{3}}$

$r = 0.62(4) = 2.48$ m

61. $r = \left(\dfrac{3V}{\pi h}\right)^{\frac{1}{2}}$

$r = \left(\dfrac{3(3140)}{3.14(30)}\right)^{\frac{1}{2}} \approx 10$ feet

Cumulative Review Problems

63. $A = \dfrac{h}{2}(a+b)$

$2A = ha + hb$

$hb = 2A - ha$

$b = \dfrac{2A - ha}{h}$

65. $y = \dfrac{ax}{a+12}$

$75 = \dfrac{a \cdot 250}{a+12}$

$75a + 900 = 250a$

$175a = 900$

$a = \dfrac{900}{175} = \dfrac{36}{7} = 5\dfrac{1}{7}$

To the nearest year, the child is 5 years old.

7.3 Exercises

1. A square root of a number is a value that when multiplied by itself is equal to the original number.

3. One answer is $\sqrt[3]{-8} = -2$ because $(-2)(-2)(-2) = -8$.

5. $\sqrt{64} = 8$

7. $\sqrt{25} + \sqrt{49} = 5 + 7 = 12$

9. $-\sqrt{\dfrac{1}{9}} = -\dfrac{1}{3}$

11. $\sqrt{36} - \sqrt{25} = 6 - 5 = 1$

13. $\sqrt{0.04} = 0.02$

15. $f(x) = \sqrt{10x+5}$

$f(0) = \sqrt{10(0)+5} = \sqrt{5} = 2.2$

$f(1) = \sqrt{10(1)+5} = \sqrt{15} = 3.9$

$f(2) = \sqrt{10(2)+5} = \sqrt{25} = 5$

$f(3) = \sqrt{10(3)+5} = \sqrt{35} = 5.9$

domain is $10x+5 \geq 0$

$$10x \geq -5$$

$$x \geq -0.5$$

17. $f(x) = \sqrt{0.5x-3}$

$f(6) = \sqrt{0.5(6)-3} = \sqrt{3-3} = \sqrt{0} = 0$

$f(8) = \sqrt{0.5(8)-3} = \sqrt{4-3} = \sqrt{1} = 1$

$f(14) = \sqrt{0.5(14)-3} = \sqrt{7-3} = \sqrt{4} = 2$

$f(16) = \sqrt{0.5(16)-3} = \sqrt{8-3} = \sqrt{5} = 2.2$

domain is $0.5x-3 \geq 0$

$$0.5x \geq 3$$

$$x \geq 6$$

19. $f(x) = \sqrt{x-3}$

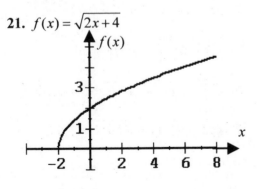

21. $f(x) = \sqrt{2x+4}$

23. $\sqrt[3]{216} = 6$

25. $\sqrt[3]{64} = 4$

27. $\sqrt[3]{-8} = -2$

29. $\sqrt[4]{81} = 3$

31. $\sqrt[5]{(8)^5} = 8$

33. $\sqrt[8]{(5)^8} = 5$

35. $\sqrt[3]{-\dfrac{1}{64}} = -\dfrac{1}{4}$

37. $\sqrt[3]{\dfrac{27}{125}} = \dfrac{\sqrt[3]{27}}{\sqrt[3]{125}} = \dfrac{3}{5}$

39. $\sqrt{a} = a^{\frac{1}{2}}$

41. $\sqrt[4]{3y} = (3y)^{\frac{1}{4}}$

43. $\sqrt[9]{(a-b)^5} = (a-b)^{\frac{5}{9}}$

45. $\sqrt[5]{\sqrt{y}} = (y^{\frac{1}{2}})^{\frac{1}{5}} = y^{\frac{1}{10}}$

47. $(\sqrt[5]{2x})^3 = \left[(2x)^{\frac{1}{5}}\right]^3 = (2x)^{\frac{3}{5}}$

49. $\sqrt[5]{(-11)^5} = -11$

51. $\sqrt[4]{a^8 b^4} = \sqrt[4]{(a^2)^4 b^4} = a^2 b$

53. $\sqrt{49x^2y^8} = \sqrt{7^2x^2(y^4)^2} = 7xy^4$

55. $\sqrt[4]{81a^{12}b^{20}} = \sqrt[4]{3^4(a^3)^4(b^5)^4} = 3a^3b^5$

57. $\sqrt[3]{8x^3y^9} = \sqrt[3]{2^3x^3(y^3)^3} = 2xy^3$

59. $x^{\frac{5}{6}} = (x^{\frac{1}{6}})^5 = (\sqrt[6]{x})^5$

61. $5^{-\frac{3}{5}} = \dfrac{1}{(5^{\frac{1}{5}})^3} = \dfrac{1}{(\sqrt[5]{5})^3}$

63. $(x+3y)^{\frac{4}{7}} = \left((x+3y)^{\frac{1}{7}}\right)^4 = \left(\sqrt[7]{x+3y}\right)^4$

65. $(-y)^{\frac{5}{7}} = \left((-y)^{\frac{1}{7}}\right)^5 = \left(\sqrt[7]{-y}\right)^5$

67. $(3ab)^{\frac{2}{7}} = \sqrt[7]{(3ab)^2} = \sqrt[7]{9a^2b^2}$

69. $27^{\frac{2}{3}} = (\sqrt[3]{27})^2 = 3^2 = 9$

71. $\left(\dfrac{4}{25}\right)^{\frac{1}{2}} = \sqrt{\dfrac{4}{25}} = \dfrac{2}{5}$

73. $(-125)^{\frac{2}{3}} = (\sqrt[3]{-125})^2 = (-5)^2 = 25$

75. $(36y^8)^{-\frac{1}{2}} = \dfrac{1}{(6^2(y^4)^2)^{\frac{1}{2}}} = \dfrac{1}{6y^4}$

77. $\sqrt{49x^8} = 7x^4$

79. $\sqrt{25a^{14}b^{18}} = 4a^7b^9$

81. $\sqrt{100x^{10}y^{12}z^2} = 10x^5y^{12}z$

83. $\sqrt[3]{-125a^6b^{15}c^{21}} = -5a^2b^5c^7$

85. $\sqrt{100x^2} = 10|x|$

87. $\sqrt[3]{-27x^9} = -3x^3$

89. $\sqrt[4]{x^{16}y^{40}} = x^4y^{10}$

91. $\sqrt[4]{a^4b^{20}} = \sqrt[4]{(ab^5)^4} = |ab^5|$

93. $\sqrt{49a^{12}b^4} = 7a^6b^2$

95. $C = 120\sqrt[3]{n} + 375$
$C = 120\sqrt[3]{216} + 375 = 120(6) + 375$
$C = 1095$
Cost is \$1095 per day.

Cumulative Review Problems

97. $\dfrac{1.24(0.30)}{1.52} = \dfrac{93}{380} = 0.2447368421...$

24.47% of the world's energy in 2020 will be produced in North America.

7.3 Exercises

1. $\sqrt{8} = \sqrt{4 \cdot 2} = \sqrt{4}\sqrt{2} = 2\sqrt{2}$

3. $\sqrt{18} = \sqrt{9 \cdot 2} = \sqrt{9}\sqrt{2} = 3\sqrt{2}$

5. $\sqrt{120} = \sqrt{4 \cdot 30} = \sqrt{4}\sqrt{30} = 2\sqrt{30}$

7. $\sqrt{44} = \sqrt{4 \cdot 11} = \sqrt{4}\sqrt{11} = 2\sqrt{11}$

9. $\sqrt{9x^3} = \sqrt{9}\sqrt{x^2}\sqrt{x} = 3x\sqrt{x}$

11. $\sqrt{60a^4b^5} = \sqrt{4 \cdot 15a^4b^4b}$
$\qquad = 2a^2b^2\sqrt{15b}$

13. $\sqrt{98x^5y^6z} = \sqrt{49 \cdot 2x^4xy^6z}$
$\qquad = 7x^2y^3\sqrt{2xz}$

15. $\sqrt[3]{8} = 2$

17. $\sqrt[3]{108} = \sqrt[3]{27 \cdot 4} = 3\sqrt[3]{4}$

19. $\sqrt[3]{56y} = \sqrt[3]{8 \cdot 7y} = 2\sqrt[3]{7y}$

21. $\sqrt[3]{8a^3b^8} = \sqrt[3]{2^3a^3b^6b^2} = 2ab^2\sqrt[3]{b^2}$

23. $\sqrt[3]{24x^6y^{11}} = \sqrt[3]{8 \cdot 3x^6y^9y^2}$
$\qquad = 2x^2y^3\sqrt[3]{3y^2}$

25. $\sqrt[4]{81kp^{23}} = \sqrt[4]{3^4kp^{20}p^3}$
$\qquad = 3p^5\sqrt[4]{kp^3}$

27. $\sqrt[5]{-32x^5y^6} = \sqrt[5]{(-2)^5x^5y^5y}$
$\qquad = -2xy\sqrt[5]{y}$

29. $\quad \sqrt[4]{1792} = a\sqrt[4]{7}$
$\sqrt[4]{256 \cdot 7} = a\sqrt[4]{7}$
$\sqrt[4]{4^4 \cdot 7} = a\sqrt[4]{7}$
$\quad 4\sqrt[4]{7} = a\sqrt[4]{7}$
$\qquad a = 4$

31. $\sqrt{49} + \sqrt{100} = 7 + 10$
$\qquad = 17$

33. $\sqrt{3} + 7\sqrt{3} - 2\sqrt{3} = 6\sqrt{3}$

35. $3\sqrt{18} - \sqrt{2} = 3\sqrt{9 \cdot 2} - \sqrt{2}$
$\qquad = 9\sqrt{2} - \sqrt{2}$
$\qquad = 8\sqrt{2}$

37. $-2\sqrt{50} + \sqrt{32} - 3\sqrt{8}$
$\qquad = -2\sqrt{25 \cdot 2} + \sqrt{16 \cdot 2} - 3\sqrt{4 \cdot 2}$
$\qquad = -10\sqrt{2} + 4\sqrt{2} - 6\sqrt{2}$
$\qquad = -12\sqrt{2}$

39. $-5\sqrt{45} + 6\sqrt{20} + 3\sqrt{5}$
$\qquad = -5\sqrt{9 \cdot 5} + 6\sqrt{4 \cdot 5} + 3\sqrt{5}$
$\qquad = -15\sqrt{4} + 12\sqrt{4} + 3\sqrt{5}$
$\qquad = 0$

41. $\sqrt{44} - 3\sqrt{63x} + 4\sqrt{28x}$
$\qquad = \sqrt{4 \cdot 11} - 3\sqrt{9 \cdot 7x} + 4\sqrt{4 \cdot 7x}$
$\qquad = 2\sqrt{11} - 9\sqrt{7x} + 8\sqrt{7x}$
$\qquad = 2\sqrt{11} - \sqrt{7x}$

43. $\sqrt{200x^3} - x\sqrt{32x}$
$\qquad = \sqrt{100x^2 \cdot 2x} - x\sqrt{16 \cdot 2x}$
$\qquad = 10x\sqrt{2x} - 4x\sqrt{2x}$
$\qquad = 6x\sqrt{2x}$

45. $\sqrt[3]{16} + 3\sqrt[3]{54} = \sqrt[3]{2^3 \cdot 2} + 3\sqrt[3]{3^3 \cdot 2}$
$\qquad = 2\sqrt[3]{2} + 9\sqrt[3]{2}$
$\qquad = 11\sqrt[3]{2}$

47. $-2\sqrt[3]{125x^3y^4} + 3y^2\sqrt[3]{8x^3}$
$\qquad = -2\sqrt[3]{5^3x^3y^3y} + 3y^2 \cdot 2x$
$\qquad = -10xy\sqrt[3]{y} + 6xy^2$

49. $\sqrt{48} + \sqrt{27} + \sqrt{75}$
   $= 6.92820323$
   $\qquad + 5.196152423$
   $\qquad\qquad + 8.660254038 = 20.78460969$
   $12\sqrt{3} = 20.78460969$ which shows
   $\sqrt{48} + \sqrt{27} + \sqrt{75} = 12\sqrt{3}$

51. $I = \sqrt{\dfrac{P}{R}} = \sqrt{\dfrac{500}{10}} = \sqrt{50} \approx 7.071$

The current is approximately 7.071 amps.

53. $T = 2\pi\sqrt{\dfrac{L}{32}} = 2(3.14)\sqrt{\dfrac{8}{32}} \approx 3.14$

The period of the pendulum is approximately 3.14 seconds.

Cumulative Review Problems

55. $81x^2 y - 25y = y(81x^2 - 25)$
   $\qquad\qquad = y(9x + 5)(9x - 5)$

57. For scallops $\dfrac{1}{0.2} = \dfrac{x}{1} \Rightarrow x = 5$

For skim milk $\dfrac{1}{0.25} = \dfrac{x}{1} \Rightarrow x = 4$
Five small servings of scallops(thirty scallops) would meet the requirement. Four small servings of skim milk (4 cups) would meet the requirement.

59. $\dfrac{250 - 307}{307} = -\dfrac{57}{307} = -0.1856677524...$
The percent decrease in population during the 50 years from 2000 to 2050 is approximately 18.6%

61. $\dfrac{280 - 307}{307} = -\dfrac{27}{307} = -0.0879478827...$
The percent decrease in population during the 50 years from 2000 to 2050 is approximately 8.8%

7.4 Exercises

1. $(2\sqrt{6})(-3\sqrt{2}) = -6\sqrt{12}$
   $\qquad\qquad\qquad = -6\sqrt{4 \cdot 3}$
   $\qquad\qquad\qquad = -12\sqrt{3}$

3. $(-3\sqrt{y})(\sqrt{5x}) = -3\sqrt{5xy}$

5. $7\sqrt{x}(2\sqrt{3} - 5\sqrt{x}) = 14\sqrt{3x} - 35x$

7. $(3 - \sqrt{2})(8 + \sqrt{2})$
   $= 24 - 5\sqrt{2} - 2$
   $= 22 - 5\sqrt{2}$

9. $(2\sqrt{3} + \sqrt{2})(2\sqrt{3} - 4\sqrt{2})$
   $= 12 - 6\sqrt{6} - 8$
   $= 4 - 6\sqrt{6}$

11. $(\sqrt{7} + 4\sqrt{5x})(2\sqrt{7} + 3\sqrt{5x})$
   $= 7 + 11\sqrt{35x} + 12(5x)$
   $= 7 + 11\sqrt{35x} + 60x$

13. $(\sqrt{3} + 2\sqrt{2})(\sqrt{5} + \sqrt{3})$
   $= \sqrt{15} + 3 + 2\sqrt{10} + 2\sqrt{6}$

15. $(\sqrt{x} - 2\sqrt{3x})(\sqrt{x} + 2\sqrt{3x})$
   $= (\sqrt{x})^2 - (2\sqrt{3x})^2$
   $= x - 12x$
   $= -11x$

17. $(\sqrt{5} - 2\sqrt{6})^2$
$$= (\sqrt{5})^2 - 4\sqrt{5}\sqrt{6} + (2\sqrt{6})^2$$

19. $(\sqrt{3x+4} + 3)^2$
$$= (\sqrt{3x+4})^2 + 6\sqrt{3x+4} + 9$$
$$= 3x + 4 + 6\sqrt{3x+4} + 9$$
$$= 3x + 13 + 6\sqrt{3x+4}$$

21. $(6 - 5\sqrt{a})^2 = 36 - 60\sqrt{a} + 25a$

23. $(\sqrt[3]{x^2})(3\sqrt[3]{4x} - 4\sqrt[3]{x^5})$
$$= 3\sqrt[3]{4x^3} - 4\sqrt[3]{x^7}$$
$$= 3x\sqrt[3]{4} - 4\sqrt[3]{x^6 \cdot x}$$
$$= 3x\sqrt[3]{4} - 4x^2\sqrt[3]{x}$$

25. $\sqrt{\dfrac{49}{25}} = \dfrac{\sqrt{49}}{\sqrt{25}} = \dfrac{7}{5}$

27. $\sqrt{\dfrac{12x}{49y^6}} = \dfrac{\sqrt{4 \cdot 3x}}{\sqrt{49y^6}} = \dfrac{2\sqrt{3x}}{7y^3}$

29. $\sqrt[3]{\dfrac{8x^5y^6}{27}} = \dfrac{\sqrt[3]{8x^3y^6 \cdot x^2}}{\sqrt[3]{27}} = \dfrac{2xy^2\sqrt[3]{x^2}}{3}$

31. $\dfrac{\sqrt[3]{24x^3y^5}}{\sqrt[3]{3y^2}} = \sqrt[3]{\dfrac{24x^3y^5}{3y^2}} = \sqrt[3]{8x^3y^3} = 2xy$

33. $\dfrac{3}{\sqrt{2}} \cdot \dfrac{\sqrt{2}}{\sqrt{2}} = \dfrac{3\sqrt{2}}{2}$

35. $\sqrt{\dfrac{4}{3}} = \dfrac{\sqrt{4}}{\sqrt{3}} \cdot \dfrac{\sqrt{3}}{\sqrt{3}} = \dfrac{2\sqrt{3}}{3}$

37. $\dfrac{1}{\sqrt{5y}} \cdot \dfrac{\sqrt{5y}}{\sqrt{5y}} = \dfrac{\sqrt{5y}}{5y}$

39. $\dfrac{x}{\sqrt{5} - \sqrt{2}} \cdot \dfrac{\sqrt{5} + \sqrt{2}}{\sqrt{5} + \sqrt{2}} = \dfrac{x(\sqrt{5} + \sqrt{2})}{5 - 2}$
$$= \dfrac{x(\sqrt{5} + \sqrt{2})}{3}$$

41. $\dfrac{\sqrt{3}}{\sqrt{5} - 2} \cdot \dfrac{\sqrt{5} + 2}{\sqrt{5} + 2} = \dfrac{\sqrt{15} + 2\sqrt{3}}{5 - 4}$
$$= \sqrt{15} + 2\sqrt{3}$$

43. $\dfrac{\sqrt{x}}{\sqrt{3x} + \sqrt{2}} \cdot \dfrac{\sqrt{3x} - \sqrt{2}}{\sqrt{3x} - \sqrt{2}} = \dfrac{x\sqrt{3} - \sqrt{2x}}{3x - 2}$

45. $\dfrac{\sqrt{5} + \sqrt{3}}{\sqrt{5} - \sqrt{3}} \cdot \dfrac{\sqrt{5} + \sqrt{3}}{\sqrt{5} + \sqrt{3}} = \dfrac{5 + 2\sqrt{15} + 3}{5 - 3}$
$$= \dfrac{8 + 2\sqrt{15}}{2}$$
$$= 4 + \sqrt{15}$$

47. $\dfrac{\sqrt{3x} - 2\sqrt{y}}{\sqrt{3x} + \sqrt{y}} \cdot \dfrac{\sqrt{3x} - \sqrt{y}}{\sqrt{3x} - \sqrt{y}}$
$$= \dfrac{3x - 3\sqrt{3xy} + 2y}{3x - y}$$

49. $\dfrac{x\sqrt{5} + 1}{\sqrt{5} + 2} \cdot \dfrac{\sqrt{5} - 2}{\sqrt{5} - 2} = \dfrac{5x - 2x\sqrt{5} + \sqrt{5} - 2}{5 - 4}$
$$= 5x - 2x\sqrt{5} + \sqrt{5} - 2$$

51. $\dfrac{5}{\sqrt{2} - \sqrt{3}} \cdot \dfrac{\sqrt{2} + \sqrt{3}}{\sqrt{2} + \sqrt{3}} = \dfrac{5(\sqrt{2} + \sqrt{3})}{2 - 3}$
$$= -5(\sqrt{2} + \sqrt{3})$$

53. $\dfrac{\sqrt[3]{x^2}}{\sqrt[3]{7x^2}} \cdot \dfrac{\sqrt[3]{7^2 x}}{\sqrt[3]{7^2 x}} = \dfrac{x\sqrt[3]{49}}{7x}$

$$= \dfrac{\sqrt[3]{49}}{7}$$

55. $\dfrac{\sqrt{6}}{2\sqrt{3} - \sqrt{2}} = 1.194938299...$

$\dfrac{\sqrt{3} + 3\sqrt{2}}{5} = 1.194938299...$

The decimal approximations are the same. The student worked correctly.

57. $\dfrac{\sqrt{3} + 2\sqrt{7}}{8} \cdot \dfrac{\sqrt{3} - 2\sqrt{7}}{\sqrt{3} - 2\sqrt{7}} = \dfrac{3 - 28}{8(\sqrt{3} + 2\sqrt{7})}$

$$= \dfrac{-25}{\sqrt{3} + 2\sqrt{7}}$$

59. $C = 0.18\dfrac{\sqrt{21}\sqrt{50}}{2} = 2.916333314...$

$2.92 is the cost to fertilize the lawn.

61. $A = LW$

$A = (\sqrt{x} + 5)(\sqrt{x} + 3)$

$A = x + 8\sqrt{x} + 15 \text{ mm}^2$

Cumulative Review Problems

63. $2x + 3y = 13 \rightarrow 4x + 6y = 26$

$\underline{5x - 2y = 4 \rightarrow 15x - 6y = 12}$

$\qquad\qquad\qquad 19x = 38$

$\qquad\qquad\qquad\quad x = 2$

$2(2) + 3y = 13$

$3y = 9$

$y = 3$

$x = 2$, $y = 3$ is the solution.

65. From the table, she reaches her goal on January 11th.

day	consumption
2nd	$280 \div 2 = 140$
5th	$140 \div 2 = 70$
8th	$70 \div 2 = 35$
11th	$35 \div 2 = 17.5$

67. $15\% + 31\% + 26\% = 72\%$

7.5 Exercises

1. Isolate one of the radicals on one side of the equation.

3. $\sqrt{8x + 1} = 5$

$8x + 1 = 25$

$8x = 24$

$x = 3$

check: $\sqrt{8(3) + 1} \overset{?}{=} 5$

$\qquad\qquad\qquad 5 = 5$

5. $2x = \sqrt{x + 3}$

$4x^2 = x + 3$

$4x^2 - x - 3 = 0$

$(4x + 3)(x - 1) = 0$

$x = -\dfrac{3}{4}$ or $x = 1$

check: $2\left(-\dfrac{3}{4}\right) \overset{?}{=} \sqrt{-\dfrac{3}{4} + 3}$

$-\dfrac{3}{2} \neq \dfrac{3}{2}$

$2(1) \overset{?}{=} \sqrt{1 + 3}$

$2 = 2$

$x = 1$ is the solution.

7. $y - \sqrt{y-3} = 5$

$$y - 5 = \sqrt{y-3}$$

$$y^2 - 10y + 25 = y - 3$$

$$y^2 - 11y + 28 = 0$$

$$(y-7)(y-4) = 0$$

$$y = 7 \ \text{ or } \ y = 4$$

check:

$$4 - \sqrt{4-3} \overset{?}{=} 5$$

$$3 \neq 5$$

$$7 - \sqrt{7-3} \overset{?}{=} 5$$

$$5 = 5$$

$x = 7$ is the solution.

9. $\sqrt{y+1} - 1 = y$

$$\sqrt{y+1} = y + 1$$

$$y + 1 = y^2 + 2y + 1$$

$$y^2 + y = 0$$

$$y(y+1) = 0$$

$$y = 0 \ \text{ or } \ y = -1$$

check: $\sqrt{0+1} - 1 \overset{?}{=} 0$

$$0 = 0$$

$$\sqrt{-1+1} - 1 \overset{?}{=} -1$$

$$-1 = -1$$

$y = 0, \ y = -1$ is the solution.

11. $x - 2\sqrt{x-3} = 3$

$$x - 3 = 2\sqrt{x-3}$$

$$x^2 - 6x + 9 = 4x - 12$$

$$x^2 - 10x + 21 = 0$$

$$(x-7)(x-3) = 0$$

$$x = 7 \ \text{ or } \ x = 3$$

(continued)

11. (continued)

check:

$$7 - 2\sqrt{7-3} \overset{?}{=} 3$$

$$3 = 3$$

$$3 - 2\sqrt{3-3} \overset{?}{=} 3$$

$$3 = 3$$

$x = 7, \ x = 3$ is the solution.

13. $\sqrt{3x^2 - x} = x$

$$3x^2 - x = x^2$$

$$2x^2 - x = 0$$

$$x(2x-1) = 0$$

$$x = 0 \ \text{ or } \ x = \frac{1}{2}$$

check:

$$\sqrt{3(0)^2 - 0} \overset{?}{=} 0$$

$$0 = 0$$

$$\sqrt{3\left(\frac{1}{2}\right)^2 - \frac{1}{2}} \overset{?}{=} \frac{1}{2}$$

$$\frac{1}{2} = \frac{1}{2}$$

$x = 0, \ x = \dfrac{1}{2}$ is the solution.

15. $\sqrt[3]{2x+3} = 2$

$$2x + 3 = 8$$

$$2x = 5$$

$$x = \frac{5}{2}$$

check: $\sqrt[3]{2 \cdot \dfrac{5}{2} + 3} \overset{?}{=} 2$

$$2 = 2$$

$x = \dfrac{5}{2}$ is the solution.

17. $\sqrt[3]{4x-1}=3$

$4x-1=27$

$4x=28$

$x=7$

check:

$\sqrt[3]{4(7)-1}\overset{?}{=}3$

$3=3$

$x=7$ is the solution.

19. $\sqrt{x+4}=1+\sqrt{x-3}$

$x+4=1+2\sqrt{x-3}+x-3$

$3=\sqrt{x-3}$

$9=x-3$

$x=12$

check:

$\sqrt{12+4}\overset{?}{=}1+\sqrt{12-3}$

$4=4$

$x=12$ is the solution.

21. $\sqrt{x+6}=1+\sqrt{x+2}$

$x+6=1+2\sqrt{x+2}+x+2$

$9=4(x+2)$

$4x+8=9$

$4x=1$

$x=\dfrac{1}{4}$

check:

$\sqrt{\dfrac{1}{4}+6}\overset{?}{=}1+\sqrt{\dfrac{1}{4}+2}$

$\dfrac{5}{2}=\dfrac{5}{2}$

$x=\dfrac{1}{4}$ is the solution.

23. $\sqrt{6x+6}=1+\sqrt{4x+5}$

$6x+6=1+2\sqrt{4x+5}+4x+5$

$x^2=4x+5$

$x^2-4x-5=0$

$(x-5)(x+1)=0$

$x=5$ or $x=-1$

check:

$\sqrt{6(5)+6}\overset{?}{=}1+\sqrt{4(5)+5}$

$6=6$

$\sqrt{6(-1)+6}\overset{?}{=}1+\sqrt{4(-1)+5}$

$0\neq 2$

$x=5$ is the solution.

25. $\sqrt{2x+9}-\sqrt{x+1}=2$

$\sqrt{2x+9}=2+\sqrt{x+1}$

$2x+9=4+4\sqrt{x+1}+x+1$

$x+4=4\sqrt{x+1}$

$x^2+8x+16=16x+16$

$x^2-8x=0$

$x(x-8)=0$

$x=0$ or $x=8$

check:

$\sqrt{2(0)+9}-\sqrt{0+1}\overset{?}{=}2$

$2=2$

$\sqrt{2(8)+9}-\sqrt{8+1}\overset{?}{=}2$

$2=2$

$x=0,\ x=8$ is the solution.

27. $\sqrt{3x+4} + \sqrt{x+5} = \sqrt{7-2x}$

$3x+4+2\sqrt{3x+4}\sqrt{x+5}+x+5 = 7-2x$

$\sqrt{3x+4}\sqrt{x+5} = -1-3x$

$3x^2+19x+20 = 1+6x+9x^2$

$6x^2-13x-19 = 0$

$(6x-19)(x+1) = 0$

$x = \dfrac{19}{6}$ or $x = -1$

check:

$\sqrt{3\left(\dfrac{19}{6}\right)+4} + \sqrt{\dfrac{19}{6}+5} \overset{?}{=} \sqrt{7-2\left(\dfrac{19}{6}\right)}$

$6.53197264... \neq 0.8164965809...$

$\sqrt{3(-1)+4} + \sqrt{-1+5} \overset{?}{=} \sqrt{7-2(-1)}$

$3 = 3$

$x = -1$ is the solution.

29. $2\sqrt{x} - \sqrt{x-5} = \sqrt{2x-2}$

$4x - 4\sqrt{x}\sqrt{x-5} + x - 5 = 2x-2$

$4\sqrt{x}\sqrt{x-5} = 3x-3$

$16x^2-80x = 9x^2-18x+9$

$7x^2-62x-9 = 0$

$(7x+1)(x-9) = 0$

$x = -\dfrac{1}{7}$ or $x = 9$

$x = -\dfrac{1}{7}$ does not check since it gives

the square root of a negative.

$2\sqrt{9} - \sqrt{9-5} \overset{?}{=} \sqrt{2(9)-2}$

$4 = 4$

$x = 9$ is the solution.

31. $x = \sqrt{5.326x-1.983}$

$x^2 = 5.326x-1.983$

$x^2-5.326x+1.983 = 0$

$x = 0.4027856296$ or $x = 4.92321437$

check:

$0.4027856296 \overset{?}{=} \sqrt{5.326(0.4027856296)-1.983}$

$0.4027856296 \overset{?}{=} 0.4027856294$

$0.4028 = 0.4028$

$4.92321437 \overset{?}{=} \sqrt{5.326(4.92321437)-1.983}$

$4.92321437 \overset{?}{=} 4.92321437$

$4.9232 = 4.9232$

$x = 0.4028,\ x = 4.9232$ is the solution.

33. (a) $V = 2\sqrt{3S}$

$V^2 = 4(3S)$

$S = \dfrac{V^2}{12}$

(b) $S = \dfrac{18^2}{12} = 27$ feet

35. $0.11y+1.25 = \sqrt{3.7625+0.22x}$

$0.0121y^2+0.275y+1.5625 = 3.7625+0.22x$

$0.22x = 0.0121y^2+0.275y-2.2$

$x = 0.055y^2+1.25y-10$

37. $\sqrt{x^2-4x+c} = x-1$

$\sqrt{4^2-4(4)+c} = 4-1$

$\sqrt{c} = 3$

$c = 9$

Cumulative Review Problems

39. $(4^3 x^6)^{\frac{2}{3}} = 4^{3 \cdot \frac{2}{3}} x^{6 \cdot \frac{2}{3}} = 4^2 x^4 = 16x^4$

41. $\sqrt[3]{-216 x^6 y^9} = -6x^2 y^3$

43. $V = Ah = (4x^2 + 2x + 9)(2x + 3)$
$V = 8x^3 + 12x^2 + 4x^2 + 6x + 18x + 27$
$V = 8x^3 + 16x^2 + 24x + 27 \text{ cm}^3$

45. $w =$ speed of current
$(12 + w) \cdot 3 = (12 - w) \cdot 5$
$36 + 3w = 60 - 5w$
$8w = 24$
$w = 3$

The current flows at 3 miles per hour.

7.6 Exercises

1. No. There is no real number that, when squared, will equal -9.

3. No. To be equal, the real number parts must be equal, and the imaginary number parts must be equal. $2 \neq 3$ and $3i \neq 2i$.

5. $\sqrt{-36} = \sqrt{36}\sqrt{-1} = 6i$

7. $\sqrt{-50} = \sqrt{25 \cdot 2}\sqrt{-1} = 5i\sqrt{2}$

9. $\sqrt{-\dfrac{1}{4}} = \sqrt{\dfrac{1}{4}}\sqrt{-1} = \dfrac{1}{2}i$

11. $-\sqrt{-81} = -\sqrt{81}\sqrt{-1} = -9i$

13. $2 + \sqrt{-3} = 2 + \sqrt{3}\sqrt{-1} = 2 + i\sqrt{3}$

15. $-3 + \sqrt{-24} = -3 + \sqrt{4 \cdot 6}\sqrt{-1}$
$\qquad = -3 + 2i\sqrt{6}$

17. $x - 3i = 5 + yi$
$x = 5$
$y = -3$

19. $1.3 - 2.5yi = x - 5i$
$x = 1.3$
$-2.5y = -5$
$y = 2$

21. $23 + yi = 17 - x + 3i$
$23 = 17 - x$
$x = -6$
$y = 3$

23. $\left(-\dfrac{3}{2} + \dfrac{1}{2}i\right) + \left(\dfrac{5}{2} - \dfrac{3}{2}i\right)$
$= -\dfrac{3}{2} + \dfrac{5}{2} + \left(\dfrac{1}{2} - \dfrac{3}{2}\right)i$
$= 1 - i$

25. $(2.8 - 0.7i) - (1.6 - 2.8i)$
$= 2.8 - 1.6 + (-0.7 + 2.8)i$
$= 1.2 + 2.1i$

27. $(2 + 3i)(2 - i) = 4 - 2i + 6i - 3i^2$
$\qquad = 4 + 4i + 3$
$\qquad = 7 + 4i$

29. $5i - 2(-4 + i) = 5i + 8 - 2i = 8 + 3i$

31. $2i(5i - 6) = 10i^2 - 12i = -10 - 12i$

33. $\left(\dfrac{1}{2} + i\right)^2 = \dfrac{1}{4} + i + i^2 = \dfrac{1}{4} + i - 1 = -\dfrac{3}{4} + i$

35. $(i\sqrt{3})(i\sqrt{7}) = i^2\sqrt{21}$
$$= -\sqrt{21}$$

37. $(\sqrt{-3})(\sqrt{-2}) = (i\sqrt{3})(i\sqrt{2})$
$$= i^2\sqrt{6}$$
$$= -\sqrt{6}$$

39. $(\sqrt{-36})(\sqrt{-4}) = (i\sqrt{36})(i\sqrt{4})$
$$= (6i)(2i)$$
$$= 12i^2$$
$$= -12$$

41. $(3 + \sqrt{-2})(4 + \sqrt{-5})$
$$= (3 + i\sqrt{2})(4 + i\sqrt{5})$$
$$= 12 + 3i\sqrt{5} + 4i\sqrt{2} + i^2\sqrt{10}$$
$$= 12 + 3i\sqrt{5} + 4i\sqrt{2} - \sqrt{10}$$
$$= 12 - \sqrt{10} + (3\sqrt{5} + 4\sqrt{2})i$$

43. $i^{17} = (i^4)^4 \cdot i$
$$= 1^4 \cdot i$$
$$= i$$

45. $i^{24} = (i^4)^6$
$$= 1^6$$
$$= 1$$

47. $i^{46} = (i^4)^{11} \cdot i^2$
$$= 1^{11}(-1)$$
$$= -1$$

49. $i^{30} + i^{28} = (i^4)^7 \cdot i^2 + (i^4)^7$
$$= 1^7(-1) + 1^7$$
$$= -1 + 1$$
$$= 0$$

51. $\dfrac{2+i}{3-i} \cdot \dfrac{3+i}{3+i} = \dfrac{6 + 5i + i^2}{9 + 1}$
$$= \dfrac{6 + 5i - 1}{10}$$
$$= \dfrac{5(1+i)}{10}$$
$$= \dfrac{1+i}{2}$$

53. $\dfrac{3i}{4+2i} \cdot \dfrac{4-2i}{4-2i} = \dfrac{12i - 6i^2}{16 + 4}$
$$= \dfrac{6 + 12i}{20}$$
$$= \dfrac{3 + 6i}{10}$$

55. $\dfrac{5-2i}{6i} \cdot \dfrac{-6i}{-6i} = \dfrac{-30i + 12i^2}{36}$
$$= \dfrac{-12 - 30i}{36}$$
$$= -\dfrac{2 + 5i}{6}$$

57. $\dfrac{7}{5-6i} \cdot \dfrac{5+6i}{5+6i} = \dfrac{35 + 42i}{25 + 36}$
$$= \dfrac{35 + 42i}{61}$$

59. $\dfrac{5-2i}{3+2i} \cdot \dfrac{3-2i}{3-2i}$
$$= \dfrac{15 - 10i - 6i + 4i^2}{9 + 4}$$
$$= \dfrac{15 - 16i - 4}{13}$$
$$= \dfrac{11 - 16i}{13}$$

61. $\dfrac{2-3i}{2+i} \cdot \dfrac{2-i}{2-i}$

$= \dfrac{4-2i-6i+3i^2}{4+1}$

$= \dfrac{4-8i-3}{5}$

$= \dfrac{1-8i}{5}$

63. $(29.3+56.2i)^2$

$= 858.49+3293.32i+3158.44i^2$

$= 858.49+3293.32i-3158.44$

$= -2299.95+3293.32i$

65. $Z = \dfrac{V}{I}$

$Z = \dfrac{3+2i}{3i} \cdot \dfrac{-3i}{-3i}$

$Z = \dfrac{-9i-6i^2}{9}$

$Z = \dfrac{6-9i}{9}$

$Z = \dfrac{2-3i}{3}$

Cumulative Review Problems

67. $x+3+2x-5+4x+2 = 105$

$7x = 105$

$x = 15$

$x+3 = 18$

$2x-5 = 25$

$4x+2 = 62$

18 hours producing juice in glass bottles, 25 hours producing juice in cans, and 62 hours producing juice in plastic bottles.

7.7 Exercises

1. Answers will vary. A person's weekly paycheck varies as the number of hours worked, $y = kx$ where y is the weekly salary, k is the hourly salary, and x is the number of hours worked.

3. $y = \dfrac{k}{x}$

5. $y = kx,\ 15 = k \cdot 40,\ k = \dfrac{3}{8}$

$y = \dfrac{3}{8}x,\ y = \dfrac{3}{8} \cdot 64 = 24$

7. $p = kd,\ 21 = k \cdot 50,\ k = \dfrac{21}{50}$

$p = \dfrac{21}{50}d,\ p = \dfrac{21}{50} \cdot 170 = 71.4$

The pressure would be 71.4 psi.

9. $d = kt^2,\ 1 = k\left(\dfrac{1}{4}\right)^2,\ k = 16$

$d = 16t^2 = 16(1)^2 = 16$

$d = 16(2)^2 = 64$

It falls 16 feet in 1 second and 64 feet in 2 seconds.

11. $y = \dfrac{k}{x^2},\ 10 = \dfrac{k}{2^2},\ k = 40$

$y = \dfrac{40}{x^2},\ y = \dfrac{40}{0.5^2} = 160$

13. $h = \dfrac{k}{t},\ 2000 = \dfrac{k}{6},\ k = 12,000$

$h = \dfrac{12,000}{t},\ h = \dfrac{12,000}{9} = 1333.\bar{3}$

The heat loss is approximately 1333.3 Btu per hour.

15. $s = \dfrac{k}{t}$, $45 = \dfrac{k}{6}$, $k = 270$

$s = \dfrac{270}{t}$, $s = \dfrac{270}{9} = 30$

The speed is 30 miles per hour.

17. $w = \dfrac{k}{l}$, $900 = \dfrac{k}{8}$, $k = 7200$

$w = \dfrac{7200}{l}$, $w = \dfrac{7200}{18} = 400$

400 pounds can be safely supported.

19. $i = \dfrac{kf}{s}$, $4 = \dfrac{k(700)}{200}$, $k = \dfrac{8}{7} \approx 1.1$

$i \approx \dfrac{1.1f}{s}$, $i \approx \dfrac{1.1(500)}{250} = 2.2$

The intensity is approximately 2.2 oersteds.

21. $d = kav$, $222 = k(37.8)(45)$, $k \approx 0.1305$

$d \approx 0.1305av$

$450 \approx 0.1305(55)v$

$v \approx 62.7$

The Caravan must travel at approximately 62.7 miles per hour.

Cumulative Review Problems

23. $3x^2 - 8x + 4 = 0$

$(3x - 2)(x - 2) = 0$

$x = \dfrac{2}{3}$ or $x = 2$

25. $488.75 = 1.0625p$

$p = 460$

The original price was \$460.

27. x = number of gold leaf frames

$140x + 95(110 - x) = 13,375$

$140x + 10,450 - 95x = 13,375$

$45x = 2907$

$x = 64.6$

$110 - x = 45.4$

He can buy 65 gold and 45 silver.

Putting Your Skills to Work

1. $D = \sqrt{\dfrac{3H}{2}}$

$D = \sqrt{\dfrac{3(10)}{2}} = 3.9$ miles

$D = \sqrt{\dfrac{3(50)}{2}} = 8.7$ miles

$D = \sqrt{\dfrac{3(100)}{2}} = 1.22$ miles

2. $D = \sqrt{\dfrac{3(150)}{2}} = 15$ miles

$D = \sqrt{\dfrac{3(250)}{2}} = 19.4$ miles

H	D
10	3.9
50	8.7
100	12.2
150	15.0
250	19.4

(continued)

2. (continued)

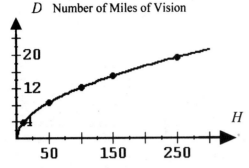

D   Number of Miles of Vision

20

12

4

50   150   250

H

Number of Feet Above Sea Level

3. $D = \sqrt{\dfrac{3H}{2}}$

$$D^3 = \frac{3H}{2}$$

$$H = \frac{2D^2}{3}$$

$$H = \frac{2(20)^2}{3} = 266.\overline{6}$$

A sailor must be 266.7 feet above sea level to see a distance of 20 miles.

4. $\sqrt{\dfrac{3(210)}{2}} - \sqrt{\dfrac{3(150)}{2}} = 2.748239349\ldots$

The viewing distance increased by approximately 2.7 miles.

Chapter 7 Review Problems

1. $(3xy^{\frac{1}{2}})(5x^2 y^{-3}) = 15x^3 y^{-\frac{5}{2}} = \dfrac{15x^3}{y^{\frac{5}{2}}}$

2. $\dfrac{3x^{\frac{2}{3}}}{6x^{\frac{1}{6}}} = \dfrac{x^{\frac{2}{3}-\frac{1}{6}}}{2} = \dfrac{x^{\frac{1}{6}}}{2} = \dfrac{1}{2x^{\frac{1}{6}}}$

3. $(25a^3 b^4)^{\frac{1}{2}} = 25^{\frac{1}{2}} a^{3\cdot\frac{1}{2}} b^{4\cdot\frac{1}{2}}$

$$= 5a^{\frac{3}{2}} b^2$$

4. $5^{\frac{1}{4}} \cdot 5^{\frac{1}{2}} = 5^{\frac{1}{4}+\frac{1}{2}} = 5^{\frac{3}{4}}$

5. $(2a^{\frac{1}{3}} b^{\frac{1}{4}})(-3a^{\frac{1}{2}} b^{\frac{1}{2}}) = -6a^{\frac{1}{3}+\frac{1}{2}} b^{\frac{1}{4}+\frac{1}{2}}$

$$= -6a^{\frac{5}{6}} b^{\frac{3}{4}}$$

6. $\dfrac{6x^{\frac{2}{3}} y^{\frac{1}{10}}}{12x^{\frac{1}{6}} y^{-\frac{1}{5}}} = \dfrac{x^{\frac{2}{3}-\frac{1}{6}} y^{\frac{1}{10}+\frac{1}{5}}}{2}$

$$= \dfrac{x^{\frac{1}{2}} y^{\frac{3}{10}}}{2}$$

7. $(2x^{-\frac{1}{5}} y^{\frac{1}{10}} z^{\frac{4}{5}})^{-5} = 2^{-5} x^{-\frac{1}{5}(-5)} y^{\frac{1}{10}(-5)} z^{\frac{4}{5}(-5)}$

$$= \dfrac{xy^{-\frac{1}{2}} z^{-4}}{2^5}$$

$$= \dfrac{x}{32y^{\frac{1}{2}} z^4}$$

8. $\left(\dfrac{49a^3 b^6}{a^{-7} b^4}\right)^{\frac{1}{2}} = \left(49a^{10} b^2\right)^{\frac{1}{2}}$

$$= 49^{\frac{1}{2}} a^{10(\frac{1}{2})} b^{2(\frac{1}{2})}$$

$$= 7a^5 b$$

9. $\dfrac{(x^{\frac{3}{4}} y^{\frac{2}{5}})^{\frac{1}{2}}}{x^{-\frac{1}{8}}} = x^{\frac{3}{4}\cdot\frac{1}{2}} x^{\frac{1}{8}} y^{\frac{2}{5}\cdot\frac{1}{2}}$

$$= x^{\frac{3}{8}+\frac{1}{8}} y^{\frac{1}{5}}$$

$$= x^{\frac{1}{2}} y^{\frac{1}{5}}$$

10. $\left(\dfrac{27x^{5n}}{x^{2n-3}}\right)^{\frac{1}{3}} = \left(27x^{5n-(2n-3)}\right)^{\frac{1}{3}}$

$ = 27^{\frac{1}{3}}(x^{3n+3})^{\frac{1}{3}}$

$ = 3x^{(3n+3)(\frac{1}{3})}$

$ = 3x^{n+1}$

11. $(5^{\frac{6}{5}})^{\frac{10}{7}} = 5^{\frac{6}{5}\cdot\frac{10}{7}} = 5^{\frac{12}{7}}$

12. $2x^{\frac{1}{3}} + x^{-\frac{2}{3}} = 2x^{\frac{1}{3}} + \dfrac{1}{x^{\frac{2}{3}}}$

$ = \dfrac{2x^{\frac{1}{3}+\frac{2}{3}}+1}{x^{\frac{2}{3}}}$

$ = \dfrac{2x+1}{x^{\frac{2}{3}}}$

13. $6x^{\frac{3}{2}} - 9x^{\frac{1}{2}} = 3x(2x^{\frac{1}{2}} - 3x^{-\frac{1}{2}})$

14. $\sqrt{\sqrt[5]{2x}} = \sqrt{(2x)^{\frac{1}{5}}} = ((2x)^{\frac{1}{5}})^{\frac{1}{2}} = (2x)^{\frac{1}{10}}$

15. $(2x+3y)^{\frac{4}{9}} = ((2x+3y)^{\frac{1}{9}})^4$

$ = (\sqrt[9]{2x+3y})^4$

16. $\sqrt[3]{125} + \sqrt[4]{81} = 5 + 3 = 8$

17. $-\sqrt[6]{64} = -\sqrt[6]{2^6} = -2$

$\sqrt[6]{-64}$ is not a real number.

18. $27^{-\frac{4}{3}} = \dfrac{1}{(27^{\frac{1}{3}})^4} = \dfrac{1}{3^4} = \dfrac{1}{81}$

19. $\left(\dfrac{4}{9}\right)^{\frac{3}{2}} = \left(\left(\dfrac{4}{9}\right)^{\frac{1}{2}}\right)^3 = \left(\dfrac{2}{3}\right)^3 = \dfrac{8}{27}$

20. $\sqrt{99x^3y^6z^{10}} = \sqrt{9x^2y^6z^{10}\cdot 11x}$

$ = 3xy^2z^5\sqrt{11x}$

21. $\sqrt[3]{-56a^8b^{10}c^{12}} = \sqrt[3]{-8a^6b^9c^{12}\cdot 7a^2b}$

$ = -2a^2b^3c^4\sqrt[3]{7a^2b}$

22. $\sqrt{144x^{10}y^{12}z^0} = 12x^5y^6$

23. $\sqrt[3]{125a^9b^6c^{300}} = 5a^3b^2c^{100}$

24. $\sqrt[3]{y^3} = y$

25. $\sqrt{y^2} = |y|$

26. $\sqrt[4]{x^4y^4} = \sqrt[4]{(xy)^4} = |xy|$

27. $\sqrt[5]{x^{10}} = \sqrt[5]{(x^2)^5} = x^2$

28. $\sqrt[3]{x^{21}} = \sqrt[3]{(x^7)^3} = x^7$

29. $\sqrt{x^8} = \sqrt{(x^4)^2} = x^4$

30. $\sqrt{50} + 2\sqrt{32} - \sqrt{8} = 5\sqrt{2} + 8\sqrt{2} - 2\sqrt{2}$

$ = 11\sqrt{2}$

31. $\sqrt{28} - 4\sqrt{7} + 5\sqrt{63}$

$ = \sqrt{4\cdot 7} - 4\sqrt{7} + 5\sqrt{9\cdot 7}$

$ = 2\sqrt{7} - 4\sqrt{7} + 15\sqrt{7}$

$ = 13\sqrt{7}$

32. $\sqrt[3]{8} + 3\sqrt[3]{16} - 4\sqrt[3]{54}$

$= 2 + 3\sqrt[3]{8 \cdot 2} - 4\sqrt[3]{27 \cdot 2}$

$= 2 + 6\sqrt[3]{2} - 12\sqrt[3]{2}$

$= 2 - 6\sqrt[3]{2}$

33. $2\sqrt{32x} - 5x\sqrt{2} + \sqrt{18x} + 2\sqrt{8x^2}$

$= 2\sqrt{16 \cdot 2x} - 5x\sqrt{2} + \sqrt{9 \cdot 2x} + 2\sqrt{4x^2 \cdot 2}$

$= 8\sqrt{2x} - 5x\sqrt{2} + 3\sqrt{2x} + 4x\sqrt{2}$

$= 11\sqrt{2x} - x\sqrt{2}$

34. $(5\sqrt{12})(3\sqrt{6}) = 15\sqrt{72}$

$= 15\sqrt{36 \cdot 2}$

$= 90\sqrt{2}$

35. $3\sqrt{x}(2\sqrt{8x} - 3\sqrt{48}$

$= 3\sqrt{x}(4\sqrt{2x} - 12\sqrt{3})$

$= 12x\sqrt{2} - 36\sqrt{3x}$

36. $(5\sqrt{2} + \sqrt{3})(\sqrt{2} - 2\sqrt{3})$

$= 10 - 9\sqrt{6} - 6 = 4 - 9\sqrt{6}$

37. $(5\sqrt{6} - 2\sqrt{2})(\sqrt{6} - \sqrt{2})$

$= 30 - 7\sqrt{12} + 4$

$= 34 - 7\sqrt{4 \cdot 3}$

$= 34 - 14\sqrt{3}$

38. $(2\sqrt{5} - 3\sqrt{6})^2 = 20 - 6\sqrt{30} + 54$

$= 74 - 6\sqrt{30}$

39. $(\sqrt[3]{2x} + \sqrt[3]{6})(\sqrt[3]{4x^2} - \sqrt[3]{y})$

$= \sqrt[3]{8x^3} - \sqrt[3]{2xy} + \sqrt[3]{24x^2} - \sqrt[3]{6y}$

$= 2x - \sqrt[3]{2xy} + 2\sqrt[3]{3x^2} - \sqrt[3]{6y}$

40. $f(x) = \sqrt{5x + 20}$

  (a) $f(16) = \sqrt{5(16) + 20}$

           $= \sqrt{100} = 10$

  (b) domain: $5x + 20 \geq 0$

               $5x \geq -20$

                $x \geq -4$

41. $f(x) = \sqrt{36 - 4x}$

  (a) $f(5) = \sqrt{36 - 4(5)} = \sqrt{16} = 4$

  (b) domain: $36 - 4x \geq 0$

               $4x \leq 36$

                $x \leq 9$

42. $f(x) = \sqrt{\dfrac{3}{4}x - \dfrac{1}{2}}$

  (a) $f(1) = \sqrt{\dfrac{3}{4} \cdot 1 - \dfrac{1}{2}} = \sqrt{\dfrac{1}{4}} = \dfrac{1}{2}$

  (b) domain: $\dfrac{3}{4}x - \dfrac{1}{2} \geq 0$

               $3x - 2 \geq 0$

               $3x \geq 2$

               $x \geq \dfrac{2}{3}$

43. $\sqrt{\dfrac{3x^2}{y} \cdot \dfrac{y}{y}} = \sqrt{\dfrac{3x^2 y}{y^2}} = \dfrac{x\sqrt{3y}}{\sqrt{y^2}}$

$= \dfrac{x\sqrt{3y}}{y}$

44. $\dfrac{2}{\sqrt{3y}} \cdot \dfrac{\sqrt{3y}}{\sqrt{3y}} = \dfrac{2\sqrt{3y}}{3y}$

45. $\dfrac{3\sqrt{7x}}{\sqrt{21x}} = \dfrac{3\sqrt{7x}}{\sqrt{3} \cdot \sqrt{7x}} \cdot \dfrac{\sqrt{3}}{\sqrt{3}} = \dfrac{3\sqrt{3}}{3} = \sqrt{3}$

46. $\dfrac{2}{\sqrt{6}-\sqrt{5}} \cdot \dfrac{\sqrt{6}+\sqrt{5}}{\sqrt{6}+\sqrt{5}}$

$\quad = \dfrac{2\sqrt{6}+2\sqrt{5}}{6-5}$

$\quad = 2\sqrt{6}+2\sqrt{5}$

47. $\dfrac{\sqrt{x}}{3\sqrt{x}+\sqrt{y}} \cdot \dfrac{3\sqrt{x}-\sqrt{y}}{3\sqrt{x}-\sqrt{y}}$

$\quad = \dfrac{3x-\sqrt{xy}}{9x-y}$

48. $\dfrac{\sqrt{5}}{\sqrt{7}-3} \cdot \dfrac{\sqrt{7}+3}{\sqrt{7}+3} = \dfrac{\sqrt{35}+3\sqrt{5}}{7-9}$

$\quad\quad\quad = \dfrac{-(\sqrt{35}+3\sqrt{5})}{2}$

49. $\dfrac{2\sqrt{3}+\sqrt{6}}{\sqrt{3}+2\sqrt{6}} \cdot \dfrac{\sqrt{3}-2\sqrt{6}}{\sqrt{3}-2\sqrt{6}}$

$\quad = \dfrac{6-4\sqrt{18}+\sqrt{18}-12}{3-24}$

$\quad = \dfrac{-6-3\sqrt{9\cdot 2}}{-21}$

$\quad = \dfrac{-6-9\sqrt{2}}{-21}$

$\quad = \dfrac{2+3\sqrt{2}}{7}$

50. $\dfrac{5\sqrt{2}-\sqrt{3}}{\sqrt{6}-\sqrt{3}} \cdot \dfrac{\sqrt{6}+\sqrt{3}}{\sqrt{6}+\sqrt{3}}$

$\quad = \dfrac{5\sqrt{12}+5\sqrt{6}-\sqrt{18}-3}{6-3}$

$\quad = \dfrac{5\sqrt{4\cdot 3}+5\sqrt{6}-\sqrt{9\cdot 2}-3}{3}$

$\quad = \dfrac{10\sqrt{3}+5\sqrt{6}-3\sqrt{2}-3}{3}$

51. $\dfrac{3\sqrt{x}+\sqrt{y}}{\sqrt{x}-\sqrt{y}} \cdot \dfrac{\sqrt{x}+\sqrt{y}}{\sqrt{x}+\sqrt{y}} = \dfrac{3x+4\sqrt{xy}+y}{x-y}$

52. $\dfrac{2xy}{\sqrt[3]{16xy^5}} \cdot \dfrac{\sqrt[3]{4x^2 y}}{\sqrt[3]{4x^2 y}} = \dfrac{2xy\sqrt[3]{4x^2 y}}{4xy^2}$

$\quad\quad\quad = \dfrac{\sqrt[3]{4x^2 y}}{2y}$

53. $\sqrt{-16}+\sqrt{-45} = \sqrt{16}\sqrt{-1}+\sqrt{45}\sqrt{-1}$

$\quad\quad\quad = 4i+i\sqrt{9\cdot 5}$

$\quad\quad\quad = 4i+3i\sqrt{5}$

54. $2x-3i+5 = yi-2+\sqrt{6}$

$\quad\quad 2x+5 = -2+\sqrt{6}$

$\quad\quad\quad 2x = -7+\sqrt{6}$

$\quad\quad\quad\quad x = \dfrac{-7+\sqrt{6}}{2}$

$\quad\quad\quad\quad y = -3$

55. $(-12-6i)+(3-5i) = -12-6i+3-5i$

$\quad\quad\quad\quad = -9-11i$

56. $(2-i)-(12-3i) = 2-i-12+3i$

$\quad\quad\quad\quad = -10+2i$

57. $(7+3i)(2-5i) = 14-29i-15i^2$

$\quad\quad\quad\quad = 14-29i+15$

$\quad\quad\quad\quad = 29-29i$

58. $(8-4i)^2 = 64-64i+16i^2$

$\quad\quad\quad = 64-64i-16$

$\quad\quad\quad = 48-64i$

59. $2i(3+4i) = 6i+8i^2$

$\quad\quad\quad = -8+6i$

60. $3 - 4(2 + i) = 3 - 8 - 4i = -5 - 4i$

61. $i^{34} = (i^4)^8 \cdot i^2 = 1^8(-1) = -1$

62. $i^{65} = (i^4)^{16} \cdot i = 1^{16} \cdot i = i$

63. $\dfrac{7 - 2i}{3 + 4i} \cdot \dfrac{3 - 4i}{3 - 4i} = \dfrac{21 - 6i - 28i + 8i^2}{9 + 16}$

$= \dfrac{21 - 34i - 8}{25}$

$= \dfrac{13 - 34i}{25}$

64. $\dfrac{5 - 2i}{1 - 3i} \cdot \dfrac{1 + 3i}{1 + 3i} = \dfrac{5 - 2i + 15i - 6i^2}{1 + 9}$

$= \dfrac{5 + 13i + 6}{10}$

$= \dfrac{11 + 13i}{10}$

65. $\dfrac{4 - 3i}{5i} \cdot \dfrac{-5i}{-5i} = \dfrac{-20i + 15i^2}{25} = \dfrac{-15 - 20i}{25}$

$= -\dfrac{3 + 4i}{5}$

66. $\dfrac{12}{3 - 5i} \cdot \dfrac{3 + 5i}{3 + 5i} = \dfrac{36 + 60i}{9 + 25} = \dfrac{36 + 60i}{34}$

$= \dfrac{18 + 30i}{17}$

67. $\dfrac{10 - 4i}{2 + 5i} \cdot \dfrac{2 - 5i}{2 - 5i} = \dfrac{20 - 58i + 20i^2}{4 + 25}$

$= \dfrac{20 - 58i - 20}{29}$

$= \dfrac{-58i}{29}$

$= -2i$

68. $2\sqrt{6x + 1} = 10$

$\sqrt{6x + 1} = 5$

$6x + 1 = 25$

$6x = 24$

$x = 4$

check:

$2\sqrt{6(4) + 1} \overset{?}{=} 10$

$10 = 10$

69. $\sqrt[3]{3x - 1} = \sqrt[3]{5x + 1}$

$3x - 1 = 5x + 1$

$2x = -2$

$x = -1$

check:

$\sqrt[3]{3(-1) - 1} \overset{?}{=} \sqrt[3]{5(-1) + 1}$

$\sqrt[3]{-4} = \sqrt[3]{-4}$

70. $\sqrt{2x + 1} = 2x - 5$

$2x + 1 = 4x^2 - 20x + 25$

$4x^2 - 22x + 24 = 0$

$2x - 11x + 12 = 0$

$(x - 4)(2x - 3) = 0$

$x = 4$ or $x = -\dfrac{3}{2}$

check:

$\sqrt{2(4) + 1} \overset{?}{=} 2(4) - 5$

$3 = 3$

$\sqrt{2\left(-\dfrac{3}{2}\right) + 1} \overset{?}{=} 2\left(-\dfrac{3}{2}\right) - 5$

$\sqrt{-2} \neq -8$

$x = 4$ is the solution.

71. $1+\sqrt{3x+1}=x$

$\sqrt{3x+1}=x-1$

$3x+1=x^2-2x+1$

$x^2-5x=0$

$x(x-5)=0$

$x=0$ or $x=5$

check:

$1+\sqrt{3(0)+1}\overset{?}{=}0$

$2\neq 0$

$1+\sqrt{3(5)+1}\overset{?}{=}5$

$5=5$

$x=5$ is the solution.

72. $\sqrt{3x+1}-\sqrt{2x-1}=1$

$\sqrt{3x+1}=\sqrt{2x-1}+1$

$3x+1=2x-1+2\sqrt{2x-1}+1$

$x+1=2\sqrt{2x-1}$

$x^2+2x+1=8x-4$

$x^2-6x+5=0$

$(x-5)(x-1)=0$

$x=5$ or $x=1$

check:

$\sqrt{3(5)+1}-\sqrt{2(5)-1}\overset{?}{=}1$

$1=1$

$\sqrt{3(1)+1}-\sqrt{2(1)-1}\overset{?}{=}1$

$1=1$

$x=5$, $x=1$ is the solution.

73. $\sqrt{7x+2}=\sqrt{x+3}+\sqrt{2x-1}$

$7x+2=x+3+2\sqrt{x+3}\sqrt{2x-1}+2x-1$

$4x=2\sqrt{x+3}\sqrt{2x-1}$

$2x=\sqrt{x+3}\sqrt{2x-1}$

$4x^2=2x^2-5x-3$

$2x^2+5x+3=0$

$(x+1)(2x+3)=0$

$x=-1$ or $x=-\dfrac{3}{2}$

check:

$\sqrt{7(1)+2}\overset{?}{=}\sqrt{1+3}+\sqrt{2(1)-1}$

$3=3$

$\sqrt{7\left(\dfrac{3}{2}\right)+2}\overset{?}{=}\sqrt{\left(\dfrac{3}{2}\right)+3}+\sqrt{2\left(\dfrac{3}{2}\right)-1}$

$\dfrac{5}{\sqrt{2}}=\dfrac{5}{\sqrt{2}}$

$x=1$, $x=\dfrac{3}{2}$ is the solution.

74. $y=kx$, $16=k\cdot 5$, $k=\dfrac{16}{5}$

$y=\dfrac{16}{5}x$, $y=\dfrac{16}{5}\cdot 3=\dfrac{48}{5}=9.6$

75. $y=kx$, $5=k\cdot 20$, $k=\dfrac{1}{4}$

$y=\dfrac{1}{4}x$, $y=\dfrac{1}{4}\cdot 50=\dfrac{50}{4}=12.5$

76. $d=kv^2$, $50=k\cdot 30^2$, $k=\dfrac{1}{18}$

$d=\dfrac{1}{18}v^2$, $d=\dfrac{1}{18}\cdot 55^2\approx 168$

The car stops in approximately 168 ft.

77. $t = k\sqrt{d}$, $2 = k\sqrt{64}$, $k = \dfrac{1}{4}$

$t = \dfrac{1}{4}\sqrt{d}$, $t = \dfrac{1}{4}\sqrt{196} = \dfrac{14}{4} = 3.5$

Object drops 196 ft. in 3.5 seconds.

78. $y = \dfrac{k}{x}$, $8 = \dfrac{k}{3}$, $k = 24$

$y = \dfrac{24}{x}$, $y = \dfrac{24}{48} = 0.5$

79. $V = \dfrac{k}{P}$, $70 = \dfrac{k}{24}$, $k = 1680$

$V = \dfrac{1680}{P}$, $100 = \dfrac{1680}{P}$, $P = 16.8$ lb

80. $y = k\dfrac{x}{z^2}$, $1 = k \cdot \dfrac{8}{4^2}$, $k = 2$

$y = 2 \cdot \dfrac{x}{z^2}$, $y = 2 \cdot \dfrac{6}{3^2} = \dfrac{4}{3}$

81. $V = khr^2$, $50 = k(5)(3)^2$, $k = \dfrac{10}{9}$

$V = \dfrac{10}{9}r^2 h$, $V = \dfrac{10}{9}(4)^2(9) = 160$ cm^3

Chapter 7 Test

1. $(2x^{\frac{1}{2}}y^{\frac{1}{3}})(-3x^{\frac{1}{3}}y^{\frac{1}{6}}) = -6x^{\frac{1}{2}+\frac{1}{3}}y^{\frac{1}{3}+\frac{1}{6}}$

$\qquad\qquad\qquad\qquad = -6x^{\frac{5}{6}}y^{\frac{1}{2}}$

2. $\dfrac{7x^3}{4x^{\frac{3}{4}}} = \dfrac{7x^{\frac{9}{4}}}{4}$

3. $(8x^3)^{\frac{1}{2}} = 8^{\frac{3}{2}}x^{\frac{3}{3}\cdot\frac{1}{2}} = 8^{\frac{3}{2}}x^2$

4. $6^{\frac{1}{5}} \cdot 6^{\frac{3}{5}} = 6^{\frac{1}{5}+\frac{3}{5}} = 6^{\frac{4}{5}}$

5. $8^{-\frac{2}{3}} = \dfrac{1}{\left(8^{\frac{1}{3}}\right)^2} = \dfrac{1}{2^2} = \dfrac{1}{4}$

6. $16^{\frac{5}{4}} = \left(16^{\frac{1}{4}}\right)^5 = 2^5 = 32$

7. $\sqrt{75a^4b^9} = \sqrt{25a^4b^8 \cdot 3b}$

$\qquad\qquad = 5a^2b^4\sqrt{3b}$

8. $\sqrt{64x^6y^5} = \sqrt{64x^6y^4 \cdot y}$

$\qquad\qquad = 8x^3y^2\sqrt{y}$

9. $\sqrt[3]{250x^4y^6} = \sqrt[3]{125x^3y^6 \cdot 2x}$

$\qquad\qquad = 5xy^2\sqrt[3]{2x}$

10. $3\sqrt{48} - \sqrt[3]{54x^5} + 2\sqrt{27} + 2x\sqrt[3]{16x^2}$

$= 3\sqrt{16 \cdot 3} - \sqrt[3]{27x^3 \cdot 2x^2}$

$\qquad\qquad + 2\sqrt{9 \cdot 3} + 2x\sqrt[3]{8 \cdot 2x^2}$

$= 12\sqrt{3} - 3x\sqrt[3]{2x^2} + 6\sqrt{3} + 4x\sqrt[3]{2x^2}$

$= 18\sqrt{3} + x\sqrt[3]{2x^2}$

11. $\sqrt{32} - 3\sqrt{8} + 2\sqrt{72}$

$= \sqrt{16 \cdot 2} - 3\sqrt{4 \cdot 2} + 2\sqrt{36 \cdot 2}$

$= 4\sqrt{2} - 6\sqrt{2} + 12\sqrt{2}$

$= 10\sqrt{2}$

12. $2\sqrt{3}(3\sqrt{6} - 5\sqrt{2}) = 6\sqrt{18} - 10\sqrt{6}$

$\qquad\qquad\qquad\qquad = 6\sqrt{9 \cdot 2} - 10\sqrt{6}$

$\qquad\qquad\qquad\qquad = 18\sqrt{2} - 10\sqrt{6}$

13. $(5\sqrt{3}-\sqrt{6})(2\sqrt{3}+3\sqrt{6})$

$\quad = 30+15\sqrt{18}-2\sqrt{18}-18$

$\quad = 12+13\sqrt{9\cdot 2}$

$\quad = 12+39\sqrt{2}$

14. $\dfrac{8}{\sqrt{20x}}\cdot\dfrac{\sqrt{20x}}{\sqrt{20x}}=\dfrac{8\sqrt{4\cdot 5x}}{20x}$

$\qquad\qquad\qquad = \dfrac{16\sqrt{5x}}{20x}$

$\qquad\qquad\qquad = \dfrac{4\sqrt{5x}}{5x}$

15. $\sqrt{\dfrac{xy}{3}}=\dfrac{\sqrt{xy}}{\sqrt{3}}\cdot\dfrac{\sqrt{3}}{\sqrt{3}}$

$\qquad\quad = \dfrac{\sqrt{3xy}}{3}$

16. $\dfrac{5+2\sqrt{3}}{4-\sqrt{3}}\cdot\dfrac{4+\sqrt{3}}{4+\sqrt{3}}=\dfrac{20+13\sqrt{3}+6}{16-3}$

$\qquad\qquad\qquad\qquad = \dfrac{26+13\sqrt{3}}{13}$

$\qquad\qquad\qquad\qquad = 2+\sqrt{3}$

17. $\qquad \sqrt{3x-2}=x$

$\qquad\qquad 3x-2=x^2$

$\qquad x^2-3x+2=0$

$\qquad (x-2)(x-1)=0$

$\qquad\qquad\qquad x=2 \ \text{ or } \ x-1$

check:

$\qquad \sqrt{3(2)-2}\overset{?}{=}2$

$\qquad\qquad\quad 2=2$

$\qquad \sqrt{3(1)-2}\overset{?}{=}1$

$\qquad\qquad\quad 1=1$

$x=2, \ x=1$ is the solution.

18. $\qquad 5+\sqrt{x+15}=x$

$\qquad\qquad \sqrt{x+15}=x-5$

$\qquad\qquad x+15=x^2-10x+25$

$\qquad x^2-11x+10=0$

$\qquad (x-10)(x-11)=0$

$\qquad\qquad\qquad x=10 \ \text{ or } \ x=11$

check:

$\qquad 5+\sqrt{10+15}\overset{?}{=}10$

$\qquad\qquad\qquad 10=10$

$\qquad 5+\sqrt{11+15}\overset{?}{=}11$

$\qquad\qquad 5+\sqrt{26}\neq 11$

$x=10$ is the solution.

19. $\qquad\qquad 5-\sqrt{x-2}=\sqrt{x+3}$

$\qquad 25-10\sqrt{x-2}+x-2=x+3$

$\qquad\qquad\qquad 20=10\sqrt{x-2}$

$\qquad\qquad\qquad\quad 2=\sqrt{x-2}$

$\qquad\qquad\qquad\quad 4=x-2$

$\qquad\qquad\qquad\quad x=6$

check:

$\qquad 5-\sqrt{6-2}\overset{?}{=}\sqrt{6+3}$

$\qquad\qquad\qquad 3=3$

$x=6$ is the solution.

20. $(8+2i)-3(2-4i)=8+2i-6+12i$

$\qquad\qquad\qquad\qquad = 2+14i$

21. $i^{18}+\sqrt{-16}=(i^4)^4\cdot i^2+\sqrt{16}\sqrt{-1}$

$\qquad\qquad\qquad = 1^4(-1)+4i$

$\qquad\qquad\qquad = -1+4i$

22. $(3-2i)(4+3i)=12-i-6i^2=12-i+6$

$\qquad\qquad\qquad\qquad = 18-i$

23. $\dfrac{2+5i}{1-3i}\cdot\dfrac{1+3i}{1+3i}=\dfrac{2+11i+15i^2}{1+9}$

$\qquad\qquad\qquad =\dfrac{-13+11i}{10}$

24. $(6+3i)^2=36+36i+9i^2$

$\qquad\qquad =27+36i$

25. $i^{43}=(i^4)^{10}\cdot i^3=1^{10}\cdot(-i)=-i$

26. $y=\dfrac{k}{x}$, $9=\dfrac{k}{2}$, $k=18$

$\quad y=\dfrac{18}{x}$, $y=\dfrac{18}{6}=3$

27. $y=k\cdot\dfrac{x}{z^2}$, $3=k\cdot\dfrac{8}{4^2}$, $k=6$

$\quad y=6\cdot\dfrac{x}{z^2}$, $y=6\cdot\dfrac{5}{6^2}=\dfrac{5}{6}$

28. $d=kv^2$, $30=k\cdot30^2$, $k=\dfrac{1}{30}$

$\quad d=\dfrac{1}{30}\cdot v^2$, $d=\dfrac{1}{30}\cdot50^2\approx83.3$ feet

Cumulative Test for Chapters 1-7

1. Associative property of addition.

2. $2a(3a^3-4)-3a^2(a-5)$

$\quad=6a^4-8a-3a^3+15a^2$

$\quad=6a^4-3a^3+15a^2-8a$

3. $7(12-14)^3-7+3\div(-3)$

$\quad=7(-2)^3-7-1$

$\quad=7(-8)-8$

$\quad=-56-8$

$\quad=-64$

4. $y=-\dfrac{3}{4}x+2$

$\quad 4y=-3x+8$

$\quad 3x=-4y+8$

$\quad x=\dfrac{-4y+8}{3}$

5. $3x-5y=15$

x	y
5	0
0	-3

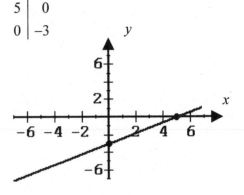

6. $16x^2+24x-16=8(2x^2+3x-2)$

$\qquad\qquad\qquad\quad =8(2x-1)(x+2)$

7. Multiply first equation by -3 and add to second equation; then multiply first equation by -2 and add to third equation.

x	$+4y$	$-z$	$=$	10
$3x$	$+2y$	$+z$	$=$	4
$2x$	$-3y$	$+2z$	$=$	-7

x	$+4y$	$-z$	$=$	10
	$-10y$	$+4z$	$=$	-26
	$-11y$	$+4z$	$=$	-27

Multiply the second equation by $-\dfrac{11}{10}$ and add to the third equation.

(continued)

7. (continued)

$$x \quad +4y \quad -z \quad = \quad 10$$
$$-10y \quad +4z \quad = \quad -26$$
$$-\frac{2}{5}z \quad = \quad \frac{8}{5}$$

$$-\frac{2}{5}z = \frac{8}{5}$$

$z = -4$, substitute in second equation

$$-10y + 4(-4) = -26$$

$y = 1$, substitute in first equation

$$x + 4(1) - (-4) = 10$$

$$x = 2$$

$x = 2$, $y = 1$, $z = -4$ is the solution.

8. $\dfrac{7x}{x^2 - 2x - 15} - \dfrac{2}{x-5}$

$$= \dfrac{7x}{(x-5)(x+3)} - \dfrac{2(x+3)}{(x-5)(x+3)}$$

$$= \dfrac{7x - 2x - 6}{(x-5)(x+3)}$$

$$= \dfrac{5x - 6}{(x-5)(x+3)}$$

9.
$$P = 2L + 2W$$
$$48 = 2(2W + 3) + 2W$$
$$24 = 2W + 3 + W$$
$$3W = 21$$
$$W = 7$$
$$2W + 3 = 17$$

The length is 7 meters and the width is 17 meters.

10. $56x + 2 = 8b + 4x$
$$4b = 26x + 1$$
$$b = \dfrac{26x + 1}{4}$$

11. $\dfrac{2x^{-3}y^{-4}}{4x^{\frac{5}{2}}y^2} = \dfrac{1}{2}x^{-3-\left(\frac{5}{2}\right)}y^{-4-\frac{7}{2}} = \dfrac{1}{2}x^{-\frac{1}{2}}y^{-\frac{15}{2}}$

$$= \dfrac{1}{2x^{\frac{1}{2}}y^{\frac{15}{2}}}$$

12. $(3x^{-\frac{1}{2}}y^2)^{-\frac{1}{3}} = 3^{-\frac{1}{3}}x^{-\frac{1}{2}\left(-\frac{1}{3}\right)}y^{2\left(-\frac{1}{3}\right)}$

$$= \dfrac{x^{\frac{1}{6}}}{3^{\frac{1}{3}}y^{\frac{2}{3}}}$$

13. $64^{-\frac{1}{3}} = \dfrac{1}{64^{\frac{1}{3}}} = \dfrac{1}{4}$

14. $\sqrt[3]{40x^5y^9} = \sqrt[3]{8 \cdot 5x^3y^9x^2} = 2xy^3\sqrt[3]{5x^2}$

15. $\sqrt{80x} + 2\sqrt{45x} - 3\sqrt{20x}$
$$= \sqrt{16 \cdot 5x} + 2\sqrt{9 \cdot 5x} - 3\sqrt{4 \cdot 5x}$$
$$= 4\sqrt{5x} + 6\sqrt{5x} - 6\sqrt{5x}$$
$$= 4\sqrt{5x}$$

16. $(2\sqrt{3} - 5\sqrt{2})(\sqrt{3} + 4\sqrt{2})$
$$= 6 + 3\sqrt{6} - 40 = -34 + 3\sqrt{6}$$

17. $\dfrac{\sqrt{3} + 2}{2\sqrt{3} - 5} \cdot \dfrac{2\sqrt{3} + 5}{2\sqrt{3} + 5} = \dfrac{6 + 9\sqrt{3} + 10}{12 - 25}$
$$= -\dfrac{16 + 9\sqrt{3}}{13}$$

18. $i^{21} + \sqrt{-16} + \sqrt{-49}$
$$= (i^4)^5 \cdot i + \sqrt{16}\sqrt{-1} + \sqrt{49}\sqrt{-1}$$
$$= 1^5 \cdot i + 4i + 7i$$
$$= 12i$$

19. $(3-4i)^2 = 3^2 - 2(3)(4i) + (4i)^2$

$= 9 - 24i + 16i^2 = 9 - 24i - 16$

$= -7 - 24i$

20. $\dfrac{1+4i}{1+3i} \cdot \dfrac{1-3i}{1-3i} = \dfrac{1+i-12i^2}{1^2+3^2} = \dfrac{13+i}{10}$

21.
$$x - 3 = \sqrt{3x+1}$$
$$x^2 - 6x + 9 = 3x + 1$$
$$x^2 - 9x + 8 = 0$$
$$(x-1)(x-8) = 0$$
$$x = 1 \text{ or } x = 8$$

check:

$1 - 3 \overset{?}{=} \sqrt{3(1)+1}$

$-2 \neq 2$

$8 - 3 \overset{?}{=} \sqrt{3(8)+1}$

$5 = 5$

$x = 8$ is the solution.

22.
$$1 + \sqrt{x+1} = \sqrt{x+2}$$
$$1 + 2\sqrt{x+1} + x + 1 = x + 2$$
$$\sqrt{x+1} = 0$$
$$x + 1 = 0$$
$$x = -1$$

check:

$1 + \sqrt{-1+1} \overset{?}{=} \sqrt{-1+2}$

$1 = 1$

$x = 1$ is the solution.

23. $y = kx^2$, $12 = k \cdot 2^2$, $k = 3$

$y = 3x^2$, $y = 3 \cdot 5^2 = 75$

24. $I = \dfrac{k}{d^2}$, $120 = \dfrac{k}{10^2}$, $k = 12{,}000$

$I = \dfrac{12{,}000}{d^2}$, $I = \dfrac{12{,}000}{15^2} \approx 53.3$ lumens

Chapter 8

Pretest Chapter 8

1. $2x^2 + 3 = 39$

$$2x^2 = 36$$

$$x^2 = 18$$

$$x = \pm\sqrt{18}$$

$$x = \pm\sqrt{9 \cdot 2}$$

$$x = \pm 3\sqrt{2}$$

2. $2x^2 - 4x - 3 = 0$

$$(x-1)^2 = \frac{5}{2}$$

$$x - 1 = \pm\sqrt{\frac{5}{2} \cdot \frac{2}{2}}$$

$$x - 1 = \pm\sqrt{\frac{10}{4}}$$

$$x - 1 = \pm\frac{\sqrt{10}}{2}$$

$$x = 1 \pm \frac{\sqrt{10}}{2}$$

$$x = \frac{2 \pm \sqrt{10}}{2}$$

3. $8x^2 - 2x - 7 = 0$

$$x = \frac{-b \pm \sqrt{b^2 - 4ac}}{2a}$$

$$x = \frac{-(-2) \pm \sqrt{(-2)^2 - 4(8)(-7)}}{2(8)}$$

$$x = \frac{2 \pm \sqrt{228}}{16} = \frac{2 \pm \sqrt{4 \cdot 57}}{16}$$

$$x = \frac{2 \pm 2\sqrt{57}}{16} = \frac{1 \pm \sqrt{57}}{8}$$

4. $(x-1)(x+5) = 2$

$$x^2 + 4x - 5 = 2$$

$$x^2 + 4x - 7 = 0$$

$$x = \frac{-b \pm \sqrt{b^2 - 4ac}}{2a}$$

$$x = \frac{-4 \pm \sqrt{4^2 - 4(1)(-7)}}{2(1)}$$

$$x = \frac{-4 \pm \sqrt{44}}{2} = \frac{-4 \pm \sqrt{4 \cdot 11}}{2}$$

$$x = \frac{-4 \pm 2\sqrt{11}}{2}$$

$$x = -2 \pm \sqrt{11}$$

5. $\qquad 4x^2 = -12x - 17$

$$4x^2 + 12x + 17 = 0$$

$$x = \frac{-b \pm \sqrt{b^2 - 4ac}}{2a}$$

$$x = \frac{-12 \pm \sqrt{12^2 - 4 \cdot 4 \cdot 17}}{2 \cdot 4}$$

$$x = \frac{-12 \pm \sqrt{-128}}{8}$$

$$x = \frac{-12 \pm \sqrt{64 \cdot 2}\sqrt{-1}}{8}$$

$$x = \frac{-12 \pm 8i\sqrt{2}}{8}$$

$$x = \frac{-3 \pm 2i\sqrt{2}}{2}$$

6. $5x^2 + 4x - 12 = 0$

$$(x+2)(5x-6) = 0$$

$$x = -2, \; x = \frac{6}{5}$$

7. $7x^2 + 9x = 14x^2 - 3x$

$7x^2 - 12x = 0$

$x(7x - 12) = 0$

$x = 0$ or $7x - 12 = 0$

$x = 0$　　　　$x = \dfrac{12}{7}$

8. $\dfrac{18}{x} + \dfrac{12}{x+1} = 9$

$18(x+1) + 12x = 9x(x+1)$

$18x + 18 + 12x = 9x^2 + 9x$

$9x^2 - 21x - 18 = 0$

$3x^2 - 7x - 6 = 0$

$(x - 3)(3x + 2) = 0$

$x - 3 = 0$ or $3x + 2 = 0$

$x = 3$　　　　$x = -\dfrac{2}{3}$

9. $x^6 - 7x^3 - 8 = 0$

$(x^3)^2 - 7x^3 - 8 = 0$

$(x^3 - 8)(x^3 + 1) = 0$

$x^3 - 8 = 0$ or $x^3 + 1 = 0$

$x^3 = 8$　　　$x^3 = -1$

$x = 2$　　　　$x = -1$

10. $w^{\frac{4}{3}} - 6w^{\frac{2}{3}} + 8 = 0$

$(w^{\frac{2}{3}})^2 - 6w^{\frac{2}{3}} + 8 = 0$

$(w^{\frac{2}{3}} - 4)(w^{\frac{2}{3}} - 2) = 0$

$w^{\frac{2}{3}} - 4 = 0$ or $w^{\frac{2}{3}} - 2 = 0$

$w^2 = 64$　　　$w^2 = 8$

$w = \pm 8$　　　$w^2 = 8$

　　　　　　　$w = \pm 2\sqrt{2}$

11. $3x^2 + 2wx + 8w = 0$

$x = \dfrac{-2w \pm \sqrt{(2w)^2 - 4(3)(8w)}}{2(3)}$

$x = \dfrac{-2w \pm \sqrt{4w^2 - 96w}}{6}$

$x = \dfrac{-2w \pm \sqrt{4(w^2 - 24w)}}{6}$

$x = \dfrac{-2w \pm 2\sqrt{w^2 - 24w}}{6}$

$x = \dfrac{-w \pm \sqrt{w^2 - 24w}}{3}$

12. $A = LW = (3W + 1)W = 52$

$3W^2 + W - 52 = 0$

$(3W + 13)(W - 4) = 0$

$W - 4 = 0$ or $3W + 13 = 0$

$W = 4$　　　　$W = -\dfrac{13}{3}$

　　　　　　　　reject, $W > 0$

$L = 3W + 1 = 13$

The width is 4 m and the length is 13 m.

13. $f(x) = 3x^2 + 6x - 9$

$x_{\text{vertex}} = \dfrac{-b}{2a}$

$x_{\text{vertex}} = \dfrac{-6}{2(3)} = -1$

$y_{\text{vertex}} = 3(-1)^2 + 6(-1) - 9$

$y_{\text{vertex}} = -12$

Vertex $(-1, -12)$

$f(0) = 3(0)^2 + 6(0) - 9 = -9$

y-intecept $(0, -9)$

(continued)

13. (continued)

$$3x^2 + 6x - 9 = 0$$

$$x^2 + 2x - 3 = 0$$

$$(x+3)(x-1) = 0$$

$$x + 3 = 0 \text{ or } x - 1 = 0$$

$$x = -3 \qquad x = 1$$

x-intercepts $(1,0), \ (-3,0)$

14. $g(x) = -x^2 + 6x - 5$

$$x_{\text{vertex}} = \frac{-b}{2a}$$

$$x_{\text{vertex}} = \frac{-6}{2(-1)} = 3$$

$$y_{\text{vertex}} = -(3)^2 + 6(3) - 5$$

$$y_{\text{vertex}} = 4$$

Vertex $(3,4)$

$$f(0) = -(0)^2 + 6(0) - 5 = -5$$

y-intecept $(0,-5)$

$$-x^2 + 6x - 5 = 0$$

$$x^2 - 6x + 5 = 0$$

$$(x-5)(x-1) = 0$$

$$x - 5 = 0 \text{ or } x - 1 = 0$$

$$x = 5 \qquad x = 1$$

x-intercepts $(5,0), \ (1,0)$

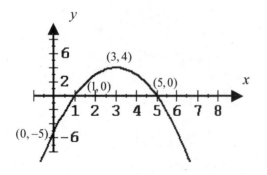

15. $x^2 - x - 6 > 0$

$$(x-3)(x+2) = 0$$

$$x - 3 = 0 \text{ or } x + 2 = 0$$

$$x = 3 \qquad x = -2$$

Critical values: $x = -2, 3$

Test: $x^2 - x - 6$

Region	Test	Result
I	$x = -3$	$6 > 0$
II	$x = 0$	$-6 < 0$
III	$x = 4$	$6 > 0$

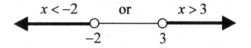

16. $2x^2 + 9x \le -9$

$$2x^2 + 9x + 9 = 0$$

$$(x+3)(2x+3) = 0$$

$$x + 3 = 0 \text{ or } 2x + 3 = 0$$

$$x = -3 \qquad x = -\frac{3}{2}$$

Critical values: $x = -3, -\frac{3}{2}$

Test: $2x^2 + 9x$

Region	Test	Result
I	$x = -4$	$-4 > -9$
II	$x = -2$	$-10 < -9$
III	$x = 0$	$0 > -9$

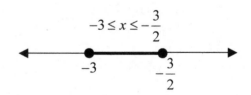

17. $4x^2 - 3x - 5 < 0$

$4x^2 - 3x - 5 = 0$

$$x = \frac{-(-3) \pm \sqrt{(-3)^2 - 4(4)(-5)}}{2(4)}$$

$x = -0.8,\ 1.6$

Critical values: $x = -3, -\dfrac{3}{2}$

Test: $4x^2 - 3x - 5$

Region	Test	Result
I	-1	$2 > 0$
II	0	$-5 < 0$
III	2	$5 > 0$

The solution is $-0.8 < x < 1.6$.

8.1 Exercises

1. $x^2 = 100$

$x = \pm\sqrt{100}$

$x = \pm 10$

3. $x^2 + 81 = 0$

$x^2 = -81$

$x = \pm 9i$

5. $3x^2 - 45 = 0$

$3x^2 = 45$

$x^2 = 15$

$x = \pm\sqrt{15}$

7. $5x^2 - 10 = 0$

$5x^2 = 10$

$x^2 = 2$

$x = \pm\sqrt{2}$

9. $x^2 = -81$

$x = \pm 9i$

11. $6x^2 + 4 = 4x^2$

$2x^2 = -4$

$x^2 = -2$

$x = \pm i\sqrt{2}$

13. $(x+2)^2 = 18$

$x + 2 = \pm\sqrt{18} = \pm 3\sqrt{2}$

$x = -2 \pm 3\sqrt{2}$

15. $(3x+2)^2 = 5$

$3x + 2 = \pm\sqrt{5}$

$3x = -2 \pm \sqrt{5}$

$x = \dfrac{-2 \pm \sqrt{5}}{3}$

17. $(5x-2)^2 = 25$

$5x - 2 = \pm 5$

$5x = 2 \pm 5$

$x = \dfrac{2 \pm 5}{5} = \dfrac{7}{5},\ -\dfrac{3}{5}$

19. $\left(\dfrac{x}{3} - 1\right)^2 = 45$

$\dfrac{x}{3} - 1 = \pm\sqrt{45}$

$\dfrac{x}{3} - 1 = \pm\sqrt{9 \cdot 5}$

$\dfrac{x}{3} - 1 = \pm 3\sqrt{5}$

$\dfrac{x}{3} = 1 \pm 3\sqrt{5}$

$x = 3 \pm 9\sqrt{5}$

21. $x^2 + 6x + 2 = 0$

$x^2 + 6x + 9 = -2 + 9$

$(x+3)^2 = 7$

$x + 3 = \pm\sqrt{7}$

$x = -3 \pm \sqrt{7}$

23. $x^2 - 12x = 4$

$x^2 - 12x + 36 = 4 + 36$

$(x-6)^2 = 40$

$x - 6 = \pm\sqrt{40} = \pm 2\sqrt{10}$

$x = 6 \pm 2\sqrt{10}$

25. $\dfrac{x^2}{3} - \dfrac{x}{3} = 3$

$x^2 - x = 9$

$x^2 - x + \dfrac{1}{4} = 9 + \dfrac{1}{4}$

$(x - \dfrac{1}{2})^2 = \dfrac{37}{4}$

$x - \dfrac{1}{2} = \pm\dfrac{\sqrt{37}}{2}$

$x = \dfrac{1 \pm \sqrt{37}}{2}$

27. $7x^2 + 4x - 5 = 0$

$x^2 + \dfrac{4}{7}x + \dfrac{4}{49} = \dfrac{5}{7} + \dfrac{4}{49}$

$\left(x + \dfrac{2}{7}\right)^2 = \dfrac{39}{49}$

$x + \dfrac{2}{7} = \pm\dfrac{\sqrt{39}}{7}$

$x = \dfrac{-2 \pm \sqrt{39}}{7}$

29. $5x^2 + 4x - 3 = 0$

$x^2 + \dfrac{4}{5}x + \dfrac{4}{25} = \dfrac{3}{5} + \dfrac{4}{25}$

$\left(x + \dfrac{2}{5}\right)^2 = \dfrac{19}{25}$

$x + \dfrac{2}{5} = \pm\dfrac{\sqrt{19}}{5}$

$x = \dfrac{-2 \pm \sqrt{19}}{5}$

31. $2y^2 - y = 15$

$y^2 - \dfrac{1}{2}y + \dfrac{1}{16} = \dfrac{15}{2} + \dfrac{1}{16}$

$\left(y - \dfrac{1}{4}\right)^2 = \dfrac{121}{16}$

$y - \dfrac{1}{4} = \pm\dfrac{11}{4}$

$y = \dfrac{1 \pm 11}{4}$

$y = 3, \; y = -\dfrac{5}{2}$

33. $2x^2 + 2 = 3x$

$x^2 - \dfrac{3}{2}x + \dfrac{9}{16} = -1 + \dfrac{9}{16}$

$\left(x - \dfrac{3}{4}\right)^2 = -\dfrac{7}{16}$

$x - \dfrac{3}{4} = \pm\dfrac{i\sqrt{7}}{4}$

$x = \dfrac{3 \pm i\sqrt{7}}{4}$

35. $x^2 + 2x - 5 = 0$

$(-1 + \sqrt{6})^2 + 2(-1 + \sqrt{6}) - 5 \overset{?}{=} 0$

$1 - 2\sqrt{6} + 6 - 2 + 2\sqrt{6} - 5 \overset{?}{=} 0$

$0 = 0$

37. $(x-7)^2(8) = 648$

$$(x-7)^2 = 81$$
$$x - 7 = 9$$
$$x = 16$$

39. $L = 4t^2$

$$t = \sqrt{\frac{L}{4}}$$
$$t = \sqrt{\frac{3.1}{4}} \approx 0.88 \text{ second}$$

41. $D = 16t^2$

$$t = \sqrt{\frac{D}{16}}$$
$$t = \sqrt{\frac{3600}{16}} = 15 \text{ seconds}$$

Cumulative Review Problems

43. $\sqrt{b^2 - 4ac} = \sqrt{4^2 - 4(3)(-4)}$
$$= \sqrt{64}$$
$$= 8$$

45. $5x^2 - 6x + 8 = 5(-2)^2 - 6(-2) + 8$
$$= 20 + 12 + 8$$
$$= 40$$

8.2 Exercises

1. Place the quadratic in standard form. Find a, b, and c. Substitute these values into the quadratic formula.

3. If the discriminant in the quadratic formula is zero, then the quadratic equation will have <u>one real</u> solution.

5. $x^2 - x - 3 = 0$
$$a = 1,\ b = -1, c = -3$$
$$x = \frac{-b \pm \sqrt{b^2 - 4ac}}{2a}$$
$$x = \frac{-(-1) \pm \sqrt{(-1)^2 - 4(1)(-3)}}{2(1)}$$
$$x = \frac{1 \pm \sqrt{13}}{2}$$

7. $2x^2 + x - 4 = 0$
$$a = 2,\ b = 1, c = -4$$
$$x = \frac{-b \pm \sqrt{b^2 - 4ac}}{2a}$$
$$x = \frac{-1 \pm \sqrt{1^2 - 4(2)(-4)}}{2(2)}$$
$$x = \frac{-1 \pm \sqrt{33}}{4}$$

9. $x^2 = \frac{2}{3}x$
$$x^2 - \frac{2}{3}x = 0$$
$$3x^2 - 2x = 0$$
$$a = 3,\ b = -2, c = 0$$
$$x = \frac{-b \pm \sqrt{b^2 - 4ac}}{2a}$$
$$x = \frac{-(-2) \pm \sqrt{(-2)^2 - 4(3)(0)}}{2(3)}$$
$$x = \frac{2 \pm \sqrt{4}}{6} = \frac{2 \pm 2}{6}$$
$$x = 0,\ x = \frac{2}{3}$$

11. $6x^2 - x - 1 = 0$

$a = 6,\ b = -1,\ c = -1$

$$x = \frac{-b \pm \sqrt{b^2 - 4ac}}{2a}$$

$$x = \frac{-(-1) \pm \sqrt{(-1)^2 - 4(6)(-1)}}{2(6)}$$

$$x = \frac{1 \pm \sqrt{25}}{12}$$

$$x = \frac{1 \pm 5}{12}$$

$$x = \frac{1}{2},\ x = -\frac{1}{3}$$

13. $4x^2 + 3x - 2 = 0$

$a = 4,\ b = 3,\ c = -2$

$$x = \frac{-b \pm \sqrt{b^2 - 4ac}}{2a}$$

$$x = \frac{-(3) \pm \sqrt{(3)^2 - 4(4)(-2)}}{2(4)}$$

$$x = \frac{-3 \pm \sqrt{41}}{8}$$

15. $3x^2 + 1 = 8$

$3x^2 - 7 = 0$

$a = 3,\ b = 0,\ c = -7$

$$x = \frac{-b \pm \sqrt{b^2 - 4ac}}{2a}$$

$$x = \frac{0 \pm \sqrt{0^2 - 4(3)(-7)}}{2(3)}$$

$$x = \frac{\pm\sqrt{84}}{6} = \frac{\pm 2\sqrt{21}}{6}$$

$$x = \frac{\pm\sqrt{21}}{3}$$

17. $2x(x + 3) - 3 = 4x - 2$

$2x^2 + 6x - 4x - 1 = 0$

$2x^2 + 2x - 1 = 0$

$a = 2,\ b = 2,\ c = -1$

$$x = \frac{-b \pm \sqrt{b^2 - 4ac}}{2a}$$

$$x = \frac{-(2) \pm \sqrt{(2)^2 - 4(2)(-1)}}{2(2)}$$

$$x = \frac{-2 \pm \sqrt{12}}{4} = \frac{-2 \pm 2\sqrt{3}}{4}$$

$$x = \frac{-1 \pm \sqrt{3}}{2}$$

19. $3x^2 + 5x + 1 = 5x + 4$

$3x^2 - 3 = 0$

$x^2 - 1 = 0$

$a = 1,\ b = 0,\ c = -1$

$$x = \frac{-b \pm \sqrt{b^2 - 4ac}}{2a}$$

$$x = \frac{-(0) \pm \sqrt{(0)^2 - 4(1)(-1)}}{2(1)}$$

$$x = \frac{\pm\sqrt{4}}{2}$$

$$x = \frac{\pm 2}{2}$$

$$x = \pm 1$$

21. $(x - 2)(x + 1) = \frac{2x + 3}{2}$

$x^2 - x - 2 = \frac{2x + 3}{2}$

$2x^2 - 2x - 4 = 2x + 3$

$2x^2 - 4x - 7 = 0$

(continued)

21. (continued)

$$a = 2, b = -4, c = -7$$

$$x = \frac{-b \pm \sqrt{b^2 - 4ac}}{2a}$$

$$x = \frac{-(-4) \pm \sqrt{(-4)^2 - 4(2)(-7)}}{2(2)}$$

$$x = \frac{4 \pm \sqrt{72}}{4}$$

$$x = \frac{4 \pm 6\sqrt{2}}{4}$$

$$x = \frac{2 \pm 3\sqrt{2}}{2}$$

23. $\dfrac{1}{x+2} + \dfrac{1}{x} = \dfrac{1}{3}$

$$3x + 3(x + 2) = x(x + 2)$$

$$3x + 3x + 6 = x^2 + 2x$$

$$x^2 - 4x - 6 = 0$$

$$a = 1, b = -4, c = -6$$

$$x = \frac{-b \pm \sqrt{b^2 - 4ac}}{2a}$$

$$x = \frac{-(-4) \pm \sqrt{(-4)^2 - 4(1)(-6)}}{2(1)}$$

$$x = \frac{4 \pm \sqrt{40}}{2}$$

$$x = \frac{4 \pm 2\sqrt{10}}{2}$$

$$x = 2 \pm \sqrt{10}$$

25. $\dfrac{1}{y} - y = \dfrac{5}{3}$

$$3 - 3y^2 = 5y$$

$$3y^2 + 5y - 3 = 0$$

(continued)

25. (continued)

$$a = 3, b = 5, c = -3$$

$$x = \frac{-b \pm \sqrt{b^2 - 4ac}}{2a}$$

$$x = \frac{-(5) \pm \sqrt{(5)^2 - 4(3)(-3)}}{2(3)}$$

$$x = \frac{-5 \pm \sqrt{61}}{6}$$

27. $\dfrac{1}{4} + \dfrac{6}{y+2} = \dfrac{6}{y}$

$$y(y + 2) + 6(4y) = 6(4(y + 2))$$

$$y^2 + 2y + 24y = 24y + 48$$

$$y^2 + 2y - 48 = 0$$

$$a = 1, b = 2, c = -48$$

$$y = \frac{-b \pm \sqrt{b^2 - 4ac}}{2a}$$

$$y = \frac{-(2) \pm \sqrt{(2)^2 - 4(1)(-48)}}{2(1)}$$

$$y = \frac{-2 \pm \sqrt{196}}{2} = \frac{-2 \pm 14}{2}$$

$$y = 6, \ y = -8$$

29. $x^2 - 2x + 4 = 0$

$$a = 1, b = -2, c = 4$$

$$x = \frac{-b \pm \sqrt{b^2 - 4ac}}{2a}$$

$$x = \frac{-(-2) \pm \sqrt{(-2)^2 - 4(1)(4)}}{2(1)}$$

$$x = \frac{2 \pm \sqrt{-12}}{2} = \frac{2 \pm \sqrt{4(3)}\sqrt{-1}}{2}$$

$$x = \frac{2 \pm 2i\sqrt{3}}{2}$$

$$x = 1 \pm i\sqrt{3}$$

31. $5x^2 = -3$

$5x^2 + 3 = 0$

$a = 5, b = 0, c = 3$

$x = \dfrac{-b \pm \sqrt{b^2 - 4ac}}{2a}$

$x = \dfrac{-(0) \pm \sqrt{(0)^2 - 4(5)(3)}}{2(5)}$

$x = \dfrac{\pm\sqrt{-60}}{10} = \dfrac{\pm\sqrt{4(15)}\sqrt{-1}}{10}$

$x = \dfrac{\pm 2i\sqrt{15}}{10}$

$x = \dfrac{\pm i\sqrt{15}}{5}$

33. $3x^2 + 4x = 2$

$3x^2 + 4x - 2 = 0$

$a = 3, b = 4, c = -2$

$b^2 - 4ac = 4^2 - 4(3)(-2) = 40$

2 irrational roots

35. $2x^2 + 10x + 8 = 0$

$a = 2, b = 10, c = 8$

$b^2 - 4ac = 10^2 - 4(2)(8) = 36$

2 rational roots

37. $9x^2 + 4 = 12x$

$9x^2 - 12x + 4 = 0$

$a = 9, b = -12, c = 4$

$b^2 - 4ac = (-12)^2 - 4(9)(4) = 0$

1 rational root

39. $13, -2$

$x = 13, \ x = -2$

$x - 13 = 0, \ x + 2 = 0$

$(x - 13)(x + 2) = 0$

$x^2 - 11x - 26 = 0$

41. $-5, \ -12$

$x = -5, \ x = -12$

$x + 5 = 0, \ x + 12 = 0$

$(x + 5)(x + 12) = 0$

$x^2 + 17x + 60 = 0$

43. $4i, \ -4i$

$x = 4i, \ x = -4i$

$x - 4i = 0, \ x + 4i = 0$

$(x - 4i)(x + 4i) = 0$

$x^2 + 16 = 0$

45. $3, \ -\dfrac{5}{2}$

$x = 3, \ x = -\dfrac{5}{2}$

$x - 3 = 0, \ 2x + 5 = 0$

$(x - 3)(2x + 5) = 0$

$2x^2 - x - 15 = 0$

47. $3x^2 + 5x - 9 = 0$

$x = \dfrac{-5 \pm \sqrt{5^2 - 4(3)(-9)}}{2(3)}$

$x = \dfrac{-5 \pm \sqrt{133}}{6}$

$x = 1.0888, \ x = -2.7554$

49. $20.6x^2 - 73.4x + 41.8 = 0$

$x = \dfrac{-(-73.4) \pm \sqrt{(-73.4)^2 - 4(20.6)(41.8)}}{2(20.6)}$

$x = \dfrac{73.4 \pm \sqrt{1943.24}}{41.2}$

$x = 2.8515, \ x = 0.7116$

51. $p = -100x^2 + 4200x - 39,476 = 0$

$$x = \frac{-4200 \pm \sqrt{4200^2 - 4(-100)(-39,476)}}{2(-100)}$$

$$x = \frac{-4200 \pm \sqrt{1,849,600}}{-200}$$

$$x = 14.2, \ x = 27.8$$

Fourteen or twenty-eight parachutes per day will produce a zero profit.

53. $p_{max} = -100(21)^2 + 4200(21) - 39,476$

$ p_{max} = 4624$

The maximum profit is $4624 per day. 21 is the average of 14 and 28.

Cumulative Review Problems

55. $9x^2 - 6x + 3 - 4x - 12x^2 + 8$
$ = -3x^2 - 10x + 11$

57. Current:
$ 2L + 2W = 50$
$ L + W = 25 \Rightarrow L = 25 - W$
 New:
$ 2(2L) + 2(3W) = 118$
$ 4L + 6W = 118$
$ 4(25 - W) + 6W = 118$
$ 100 - 4W + 6W = 118$
$ 2W = 18$
$ W = 9$
$ L = 25 - W = 16$

The width is 9 feet and the length is 16 feet.

8.3 Exercises

1. $x^4 - 9x^2 + 20 = 0$
$ y = x^2$
$ y^2 - 9y + 20 = 0$
$ (y - 5)(y - 4) = 0$
$ y = 5, \ y = 4$
$ x^2 = 5, \ x^2 = 4$
$ x = \pm\sqrt{5}, \ x = \pm 2$

3. $x^4 + x^2 - 12 = 0$
$ y = x^2$
$ y^2 + y - 12 = 0$
$ (y - 3)(y + 4) = 0$
$ y = 3, \ y = -4$
$ x^2 = 3, \ x^2 = -4$
$ x = \pm\sqrt{3}, \ x = \pm 2i$

5. $3x^4 = 10x^2 + 8$
$ 3x^4 - 10x^2 - 8 = 0$
$ y = x^2$
$ 3y^2 - 10y - 8 = 0$
$ (3y + 2)(y - 4) = 0$
$ y = -\frac{2}{3}, \ y = 4$
$ x^2 = -\frac{2}{3}, \ x^2 = 4$
$ x = \pm\frac{i\sqrt{6}}{3}, \ x = \pm 2$

7. $x^6 - 7x^3 - 8 = 0, \ y = x^3$
$ y^2 - 7y - 8 = 0$
$ (y - 8)(y + 1) = 0$

(continued)

7. (continued)

$y = 8, \quad y = -1$

$x^3 = 8, \quad x^3 = -1$

$x = 2, \quad x = -1$

9. $x^6 - 5x^3 - 14 = 0$

$y = x^3$

$y^2 - 5y - 14 = 0$

$(y - 7)(y + 2) = 0$

$y = 7, \quad y = -2$

$x^3 = 7, \quad x^3 = -2$

$x = \sqrt[3]{7}, \quad x = \sqrt[3]{-2} = -\sqrt[3]{2}$

11. $x^8 = 3x^4 - 2$

$x^8 - 3x^4 + 2 = 0$

$y = x^4$

$y^2 - 3y + 2 = 0$

$(y - 2)(y - 1) = 0$

$y = 2, \quad y = 1$

$x^4 = 2, \quad x^4 = 1$

$x = \pm\sqrt[4]{2}, \quad x = \pm 1$

13. $3x^8 + 13x^4 = 10$

$3x^8 + 13x^4 - 10 = 0$

$y = x^4$

$3y^2 + 13y - 10 = 0$

$(3y - 2)(y + 5) = 0$

$y = \dfrac{2}{3}, \quad y = -5$

$x^4 = \dfrac{2}{3}, \quad x^4 = -5, \text{ no real roots}$

$x = \pm\dfrac{\sqrt[4]{54}}{3}$

15. $x^{\frac{2}{3}} + 2x^{\frac{1}{3}} - 8 = 0$

$y = x^{\frac{1}{3}}$

$y^2 + 2y - 8 = 0$

$(y + 4)(y - 2) = 0$

$y = -4, \quad y = 2$

$x^{\frac{1}{3}} = -4, \quad x^{\frac{1}{3}} = 2$

$x = (-4)^3 = -64$

$x = 2^3 = 8$

17. $2x^{\frac{2}{3}} - 7x^{\frac{1}{3}} - 4 = 0$

$y = x^{\frac{1}{3}}$

$2y^2 - 7y - 4 = 0$

$(2y + 1)(y - 4) = 0$

$y = -\dfrac{1}{2}, \quad y = 4$

$x^{\frac{1}{3}} = -\dfrac{1}{2}, \quad x^{\frac{1}{3}} = 4$

$x = -\dfrac{1}{8}, \quad x = 64$

19. $3x^{\frac{1}{2}} - 14x^{\frac{1}{4}} - 5 = 0$

$y = x^{\frac{1}{4}}$

$3y^2 - 14y - 5 = 0$

$(3y + 1)(y - 5) = 0$

$y = -\dfrac{1}{3}, \quad y = 5$

$x^{\frac{1}{4}} = -\dfrac{1}{3}, \quad x = \dfrac{1}{81} \text{ (extraneous)}$

$x^{\frac{1}{4}} = 5, \quad x = 5^4 = 625$

21. $2x^{\frac{1}{2}} - 14x^{\frac{1}{4}} - 5 = 0$

$y = x^{\frac{1}{4}}$

$2y^2 - 14y - 5 = 0$

$(y-2)(2y+3) = 0$

$y = 2, \ y = -\dfrac{3}{2}$

$x^{\frac{1}{4}} = 2, \ x^{\frac{1}{4}} = -\dfrac{3}{2}$

$x = \left(-\dfrac{3}{2}\right)^4 = \dfrac{81}{16}$ (extraneous)

$x = 2^4 = 16$

23. $x^{\frac{2}{5}} + x^{\frac{1}{5}} - 2 = 0$

$y = x^{\frac{1}{5}}$

$y^2 + y - 2 = 0$

$(y+2)(y-1) = 0$

$y = -2, \ y = 1$

$x^{\frac{1}{5}} = -2, \ x^{\frac{1}{5}} = 1$

$x = (-2)^5 = -32, \ x = 1^5 = 1$

25. $(x^2 + x)^2 - 5(x^2 + x) = -6$

$(x^2 + x)^2 - 5(x^2 + x) + 6 = 0$

$y = (x^2 + x)$

$y^2 - 5y + 6 = 0$

$(y-2)(y-3) = 0$

$y = 2, \ y = 3$

$(x^2 + x) = 2, \ (x^2 + x) = 3$

$x^2 + x - 2 = 0, \ x^2 + x - 3 = 0$

$(x+2)(x-1) = 0, \ x = \dfrac{-1 \pm \sqrt{1^2 - 4(1)(-3)}}{2(1)}$

(continued)

25. (continued)

$x = -2, \ x = 1, \ x = \dfrac{-1 \pm \sqrt{13}}{2}$

27. $x - 5x^{\frac{1}{2}} + 6 = 0$

$y = x^{\frac{1}{2}}$

$y^2 - 5y + 6 = 0$

$(y-3)(y-2) = 0$

$y = 3, \ y = 2$

$x^{\frac{1}{2}} = 3, \ x^{\frac{1}{2}} = 2$

$x = 9, \quad x = 4$

29. $10x^{-2} + 7x^{-1} + 1 = 0$

$y = x^{-1}$

$10y^2 - 7y + 1 = 0$

$(5y+1)(2y+1) = 0$

$y = -\dfrac{1}{5}, \ y = -\dfrac{1}{2}$

$x^{-1} = -\dfrac{1}{5}, \ x^{-1} = -\dfrac{1}{2}$

$x = -5, \quad x = -2$

31. $15 - \dfrac{2x}{x-1} = \dfrac{x^2}{x^2 - 2x + 1}$

$15 - \dfrac{2x}{x-1} = \dfrac{x^2}{(x-1)(x-1)}$

$15(x-1)^2 - 2x(x-1) = x^2$

$15x^2 - 30x + 15 - 2x^2 + 2x = x^2$

$12x^2 - 28x + 15 = 0$

$(6x-5)(2x-3) = 0$

$x = \dfrac{5}{6}, \ x = \dfrac{3}{2}$

Cumulative Review Problems

33. $\sqrt{8x} + 3\sqrt{2x} - 4\sqrt{50x}$

$= \sqrt{4 \cdot 2x} + 3\sqrt{2x} - 4\sqrt{25 \cdot 2x}$

$= 2\sqrt{2x} + 3\sqrt{2x} - 20\sqrt{2x}$

$= -15\sqrt{2x}$

35. $3\sqrt{2}(\sqrt{5} - 2\sqrt{6}) = 3\sqrt{10} - 6\sqrt{12}$

$\qquad\qquad\qquad = 3\sqrt{10} - 6\sqrt{4 \cdot 3}$

$\qquad\qquad\qquad = 3\sqrt{10} - 12\sqrt{3}$

37. $\dfrac{28.3 - 16.9}{16.9} = 0.674556213\ldots$

High school graduate: 67.5%

$\dfrac{36.4 - 24.0}{24.0} = 0.51\overline{6}$

Associate's degree: 51.7%

$\dfrac{50.1 - 30.1}{30.1} = 0.6644518272\ldots$

Bachelor's degree: 66.4%

$\dfrac{87.4 - 51.1}{51.1} = 0.71037182\ldots$

Doctorate: 71.0%

39. $\dfrac{281,500,000 - 109,000,000}{4063}$

$= 42,456.31307$

The average profit per performance
was $42,456.31.

41. $(42,456.31)(7451)$

$= 316,341,965.8$

The total profit of *Cats* was
$316,341,965.80.

8.4 Exercises

1. $S = 16t^2$

$t^2 = \dfrac{S}{16}$

$t = \pm\sqrt{\dfrac{S}{16}} = \pm\dfrac{\sqrt{S}}{4}$

3. $A = \pi\left(\dfrac{d}{2}\right)^2 = \pi\dfrac{d^2}{4}$

$d^2 = \dfrac{4A}{\pi}$

$d = \pm\sqrt{\dfrac{4A}{\pi}}$

5. $3H = \dfrac{1}{2}ax^2$

$ax^2 = 6H$

$x^2 = \dfrac{6H}{a}$

$x = \pm\sqrt{\dfrac{6H}{a}}$

7. $4(y^2 + w) - 5 = 7R$

$4(y^2 + w) = 5 + 7R$

$4y^2 + 4w = 5 + 7R$

$y^2 = \dfrac{7R - 4w + 5}{4}$

$y = \pm\dfrac{\sqrt{7R - 4w + 5}}{2}$

9. $Q = \dfrac{3mxM^2}{2c}, \quad M^2 = \dfrac{2Qc}{3mx}$

$M = \pm\sqrt{\dfrac{2Qc}{3mx}}$

11. $V = \pi(r^2 + R^2)h$

$\quad\quad V = \pi r^2 h + \pi R^2 h$

$\quad\quad \pi r^2 h = V - \pi R^2 h$

$\quad\quad r^2 = \dfrac{V - \pi R^2 h}{\pi h}$

$\quad\quad r = \pm\sqrt{\dfrac{V - \pi R^2 h}{\pi h}}$

13. $x^2 + 3bx - 10b^2 = 0$

$\quad\quad (x + 5b)(x - 2b) = 0$

$\quad\quad x = -5b, \quad x = 2b$

15. $P = EI - RI^2$

$\quad\quad RI^2 - EI + P = 0$

$\quad\quad a = R, \ b = -E, \ c = P$

$\quad\quad I = \dfrac{-(-E) \pm \sqrt{(-E)^2 - 4RP}}{2R}$

$\quad\quad I = \dfrac{E \pm \sqrt{E^2 - 4RP}}{2R}$

17. $10w^2 - 3qw - 4 = 0$

$\quad\quad a = 10, \ b = -3q, \ c = -4$

$\quad\quad w = \dfrac{-(-3q) \pm \sqrt{(-3q)^2 - 4(10)(-4)}}{2(10)}$

$\quad\quad w = \dfrac{3q \pm \sqrt{9q^2 + 160}}{20}$

19. $S = 2\pi rh + \pi r^2$

$\quad\quad \pi r^2 + 2\pi hr - S = 0$

$\quad\quad a = \pi, \ b = 2\pi h, \ c = -S$

$\quad\quad r = \dfrac{-2\pi h \pm \sqrt{(2\pi h)^2 - 4\pi(-S)}}{2\pi}$

(continued)

19. (continued)

$\quad\quad r = \dfrac{-2\pi h \pm \sqrt{4\pi^2 h^2 + 4\pi S}}{2\pi}$

$\quad\quad r = \dfrac{-2\pi h \pm \sqrt{4\pi^2 h^2 + 4\pi S}}{2\pi}$

$\quad\quad r = \dfrac{-2\pi h \pm 2\sqrt{\pi^2 h^2 + \pi S}}{2\pi}$

$\quad\quad r = \dfrac{-\pi h \pm \sqrt{\pi^2 h^2 + \pi S}}{\pi}$

21. $(a+1)x^2 + 5x + 2w = 0$

$\quad\quad x = \dfrac{-5 \pm \sqrt{5^2 - 4(a+1)(2w)}}{2(a+1)}$

$\quad\quad x = \dfrac{-5 \pm \sqrt{25 - 8aw - 8w}}{2a + 2}$

23. $c^2 = a^2 + b^2$

$\quad\quad c^2 = 7^2 + \sqrt{3}^2$

$\quad\quad c^2 = 49 + 3$

$\quad\quad c^2 = 52$

$\quad\quad c = \sqrt{52}$

$\quad\quad c = \sqrt{4 \cdot 13}$

$\quad\quad c = 2\sqrt{13}$

25. $c^2 = a^2 + b^2$

$\quad\quad \sqrt{34}^2 = a^2 + \sqrt{19}^2$

$\quad\quad a^2 = 34 - 19$

$\quad\quad a^2 = 15$

$\quad\quad a = \sqrt{15}$

27. $c^2 = a^2 + b^2$

$144 = a^2 + 4a^2 = 5a^2$

$a^2 = \dfrac{144}{5}$

$a = \sqrt{\dfrac{144}{5}}$

$a = \dfrac{12}{\sqrt{5}}$

$a = \dfrac{12\sqrt{5}}{5}$

$b = 2a = \dfrac{24\sqrt{5}}{5}$

29. $c^2 = a^2 + b^2$

$c^2 = 12^2 + (c-6)^2$

$c^2 = 144 + c^2 - 12c + 36$

$12c = 180$

$c = 15$

$c - 6 = 9$

The hypotenuse is 15 miles long and the other leg is 9 miles long.

31. $c^2 = a^2 + b^2$

$11^2 = a^2 + (a+3)^2$

$121 = a^2 + a^2 + 6a + 9$

$2a^2 + 6a - 112 = 0$

$a = 2, \ b = 6, \ c = -112$

$a = \dfrac{-6 + \sqrt{6^2 - 4(2)(-112)}}{2(2)}$

$a \approx 6.13$

$a + 3 \approx 9.13$

The first leg was about 6.13 miles long and the final leg was about 9.13 miles long.

35. $b =$ base of triangle

$A = \dfrac{1}{2}bh = \dfrac{1}{2}b(2b+2) = 72$

$2b^2 + 2b = 144$

$b^2 + b - 72 = 0$

$(b-8)(b+9) = 0$

$b - 8 = 0, \ b + 9 = 0$

$b = 8 \qquad b = -9, \ \text{reject}, \ b > 0$

$2b + 2 = 18$

The base is 8 cm and the altitude is 18 cm.

37. $v =$ speed in rain

$s = vt, \quad t = \dfrac{s}{v}$

$\dfrac{225}{v} + \dfrac{150}{v+5} = 8$

$225(v+5) + 150v = 8v(v+5)$

$225v + 1125 + 150v = 8v^2 + 40v$

$8v^2 - 335v - 1125 = 0$

$(v-45)(8v+25) = 0$

$v = 45, \quad v = -\dfrac{25}{8} \ \text{reject}, \ v > 0$

$v + 5 = 50$

The speed in the rain was 45 mph and the speed without rain was 50 mph.

39. $t =$ time from home to work

$1 \text{ hr } 16 \text{ min } = 1 + \dfrac{16}{60} = \dfrac{19}{15} \text{ hr}$

$\dfrac{19}{15} - t =$ time from work to home

(continued)

222

39. (continued)

The distance from home to work and the distance from work to home are the same.

$$50t = 45\left(\frac{19}{15} - t\right)$$

$$50t = 57 - 45t$$

$$95t = 57$$

$$t = \frac{57}{95} = \frac{3}{5}$$

$$50 \cdot \frac{3}{5} = 30$$

Bob lives 30 miles from his job.

41. $2003 - 1980 = 23$

$$N = 1.11x^2 + 33.39x + 304.09$$

$$N = 1.11(23)^2 + 33.39(23) + 304.09$$

$$N = 1659.25 \text{ thousands}$$

$$N = 1,659,250$$

To the nearest hundred, there will be 1,659,300 inmates in 2003.

43. $N = 1.11x^2 + 33.39x + 304.09$

$$1,744,800 = 1744.8 \text{ thousands}$$

$$1.11x^2 + 33.39x + 304.09 = 1744.8$$

$$1.11x^2 + 33.39x - 1440.71 = 0$$

$$x = \frac{-33.39 + \sqrt{33.39^2 - 4(1.11)(-1440.71)}}{2(1.11)}$$

$$x = 23.99988462$$

$$1980 + 24 = 2004$$

The number of inmates will reach 1,744,800 in the year 2004.

45. $w = \dfrac{12b^2}{\dfrac{5}{2}w + \dfrac{7}{2}b + \dfrac{21}{2}}$

$$\frac{5}{2}w^2 + w\left(\frac{7}{2}b + \frac{21}{2}\right) - 12b^2 = 0$$

$$5w^2 + (7b + 21)w - 24b^2 = 0$$

$$w = \frac{-(7b+21) \pm \sqrt{(7b+21)^2 - 4(5)(-24b^2)}}{2(5)}$$

$$w = \frac{-7b - 21 \pm \sqrt{49b^2 + 294b + 441 + 480b^2}}{10}$$

$$w = \frac{-7b - 21 \pm \sqrt{529b^2 + 294b + 441}}{10}$$

Cumulative Review Problems

47. $\dfrac{4}{\sqrt{3x}} \cdot \dfrac{\sqrt{3x}}{\sqrt{3x}}$

$$= \frac{4\sqrt{3x}}{3x}$$

49. $\dfrac{3}{\sqrt{x} + \sqrt{y}} \cdot \dfrac{\sqrt{x} - \sqrt{y}}{\sqrt{x} - \sqrt{y}}$

$$= \frac{3(\sqrt{x} - \sqrt{y})}{x - y}$$

51. $\dfrac{3ab}{\sqrt[3]{8ab^2}} \cdot \dfrac{\sqrt[3]{a^2b}}{\sqrt[3]{a^2b}}$

$$= \frac{3ab\sqrt[3]{a^2b}}{2ab}$$

$$= \frac{3\sqrt[3]{a^2b}}{2}$$

8.5 Exercises

1. $f(x) = x^2 - 2x - 8$

$$x_{\text{vertex}} = \frac{-b}{2a}$$

$$x_{\text{vertex}} = \frac{-(-2)}{2(1)}$$

$$x_{\text{vertex}} = 1$$

$$f(x_{\text{vertex}}) = 1^2 - 2(1) - 8$$

$$y_{\text{vertex}} = -9$$

$$V(1, -9)$$

$$f(0) = 0^2 - 2(0) - 8$$

$$f(0) = -8$$

$$x^2 - 2x - 8 = 0$$

$$(x - 4)(x + 2) = 0$$

$$x = 4, \ x = -2$$

$$I(0, -8); \ (4, 0); \ (-2, 0)$$

3. $f(x) = -x^2 - 4x + 12$

$$x_{\text{vertex}} = \frac{-b}{2a}$$

$$x_{\text{vertex}} = \frac{-(-4)}{2(-1)}$$

$$x_{\text{vertex}} = -2$$

$$f(x_{\text{vertex}}) = -(-2)^2 - 4(-2) + 12$$

$$f(x_{\text{vertex}}) = 16$$

$$V(-2, 16)$$

$$f(0) = -0^2 - 4(0) + 12 = 12$$

$$-x^2 - 4x + 12 = 0$$

$$x^2 + 4x - 12 = 0$$

$$(x - 2)(x + 6) = 0$$

$$x = 2, \ x = -6$$

$$I(0, 12); \ (2, 0); \ (-6, 0)$$

5. $p(x) = 3x^2 + 12x + 3$

$$x_{\text{vertex}} = \frac{-b}{2a} = \frac{-(12)}{2(3)} = -2$$

$$f(x_{\text{vertex}}) = 3(-2)^2 + 12(-2) + 3 = -9$$

$$V(-2, -9)$$

$$f(0) = 3(0)^2 + 12(0) + 3 = 3$$

$$3x^2 + 12x + 3 = 0$$

$$x^2 + 4x + 1 = 0$$

$$a = 1, \ b = 4, \ c = 1$$

$$x = \frac{-4 \pm \sqrt{4^2 - 4(1)(1)}}{2(1)}$$

$$x = -0.3, \ x = -3.7$$

$$I(0, 3); \ (-0.3, 0); \ (-3.7, 0)$$

7. $r(x) = x^2 - 6x + 8$

$$x_{\text{vertex}} = \frac{-b}{2a}$$

$$x_{\text{vertex}} = \frac{-(-2)}{2(-3)}$$

$$x_{\text{vertex}} = -\frac{1}{3}$$

$$f(x_{\text{vertex}}) = -3\left(-\frac{1}{3}\right)^2 - 2\left(-\frac{1}{3}\right) - 6$$

$$f(x_{\text{vertex}}) = -\frac{17}{3}$$

$$V\left(-\frac{1}{3}, -\frac{17}{3}\right)$$

$$f(0) = -3(0)^2 - 2(0) - 6 = -6$$

$$-3x^2 - 2x - 6 = 0$$

$$a = -3, b = -2, c = -6$$

$$b^2 - 4ac = -68 < 0 \Rightarrow \text{ no } x\text{-intercepts}$$

$$I(0, -6)$$

9. $f(x) = 2x^2 + 2x - 4$

$$x_{\text{vertex}} = \frac{-b}{2a}$$

$$x_{\text{vertex}} = \frac{-(2)}{2(2)}$$

$$x_{\text{vertex}} = -\frac{1}{2}$$

$$f(x_{\text{vertex}}) = 2\left(-\frac{1}{2}\right)^2 + 2\left(-\frac{1}{2}\right) - 4$$

$$y_{\text{vertex}} = -\frac{9}{2}$$

$$V\left(-\frac{1}{2}, -\frac{9}{2}\right)$$

$$f(0) = 2(0)^2 + 2(0) - 4$$

$$f(0) = -4$$

$$2x^2 + 2x - 4 = 0$$

$$x^2 + x - 2 = 0$$

$$(x-1)(x+2) = 0$$

$$x = 1, \;\; x = -2$$

$$I(0,-4); \;\; (1,0); \;\; (-2,0)$$

11. $f(x) = x^2 - 6x + 8$

$$\frac{-b}{2a} = \frac{-(-6)}{2(1)} = 3$$

$$f(3) = 3^2 - 6(3) + 8 = -1$$

$$V(3,-1)$$

$$f(0) = 0^2 - 6(0) + 8 = 8$$

y-intercept: $(0,8)$

$$x^2 - 6x + 8 = 0$$

$$(x-4)(x-2) = 0$$

$$x = 4, \;\; x = 2$$

x-intercepts: $(4,0), \;\; (2,0)$

(continued)

11. (continued)

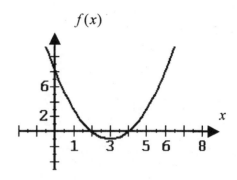

13. $g(x) = x^2 + 2x - 8$

$$\frac{-b}{2a} = \frac{-2}{2} = -1$$

$$g(-1) = (-1)^2 + 2(-1) - 8 = -9$$

$$V(-1,-9)$$

$$g(0) = 0^2 + 2(0) - 8 = -8$$

y-intercept: $(0,-8)$

$$x^2 + 2x - 8 = 0$$

$$(x+4)(x-2) = 0$$

$$x = -4, \;\; x = 2$$

x-intercepts: $(-4,0), \;\; (2,0)$

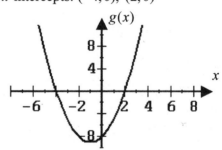

15. $p(x) = -x^2 + 4x - 3$

$$\frac{-b}{2a} = \frac{-4}{2(-1)} = 2$$

$$p(2) = -2^2 + 4(2) - 3 = 1$$

$$V(2,1)$$

(continued)

15. (continued)

$p(0) = -0^2 + 4(0) - 3 = -3$

y-intercept: $(0, -3)$

$-x^2 + 4x - 3 = 0$

$x^2 - 4x + 3 = 0$

$(x - 3)(x - 1) = 0$

$x = 3, \ x = 1$

x-intercepts: $(3, 0), \ (1, 0)$

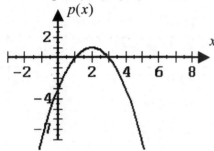

17. $r(x) = x^2 + 4x + 6$

$\dfrac{-b}{2a} = \dfrac{-4}{2(1)} = -2$

$r(-2) = (-2)^2 + 4(-2) + 6 = 2$

$V(-2, 2)$

$r(0) = 0^2 + 4(0) + 6 = 6$

y-intercept: $(0, 6)$

$x^2 + 4x + 6 = 0$

$a = 1, \ b = 4, \ c = 6$

$b^2 - 4ac = 4^2 - 4(1)(6) = -8 < 0$

x-intercepts: none

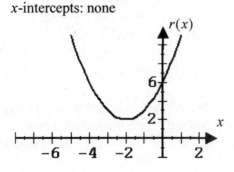

19. $f(x) = x^2 - 6x + 5$

$\dfrac{-b}{2a} = \dfrac{-(-6)}{2(1)} = 3$

$f(3) = 3^2 - 6(3) + 5 = -4$

$V(3, -4)$

$f(0) = 0^2 - 6(0) + 5 = 5$

y-intercept: $(0, 5)$

$x^2 - 6x + 5 = 0$

$(x - 5)(x - 1) = 0$

$x = 5, \ x = 1$

x-intercept: $(5, 0), \ (1, 0)$

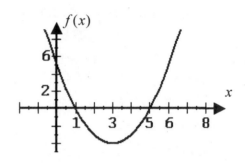

21. $g(x) = -x^2 + 6x - 9$

$\dfrac{-b}{2a} = \dfrac{-6}{2(-1)} = 3$

$g(3) = -3^2 + 6(3) - 9 = 0$

$V(3, 0)$

$g(0) = -0^2 + 6(0) - 9 = -9$

y-intercept: $(0, -9)$

$-x^2 + 6x - 9 = 0$

$x^2 - 6x + 9 = 0$

$(x - 3)(x - 3) = 0$

$x = 3$

x-intercept: $(3, 0)$

(continued)

21. (continued)

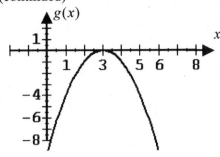

23. $N(x) = 0.18x^2 - 3.18x + 102.25$

$N(20) = 0.18(20)^2 - 3.18(20) + 102.25$

$\quad = 110.65$ thousands

$\quad = 110,650$

$N(40) = 0.18(40)^2 - 3.18(40) + 102.25$

$\quad = 263.05$ thousands

$\quad = 263,050$

$N(60) = 0.18(60)^2 - 3.18(60) + 102.25$

$\quad = 559.45$ thousands

$\quad = 559,450$

$N(80) = 0.18(80)^2 - 3.18(80) + 102.25$

$\quad = 999.85$ thousands

$\quad = 999,850$

$N(100) = 0.18(100)^2 - 3.18(100) + 102.25$

$\quad = 1584.25$ thousands

$\quad = 1,584,250$

25. From the graph, $N(70) \approx 750,000$ who scuba dive and have $70,000 mean income.

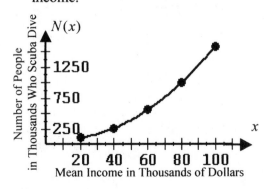

27. From the graph, $N(x) = 390$ for $x \approx 50$. This means that 390,000 people who scuba dive have a mean income of $50,000.

29. $P(x) = -6x^2 + 312x - 3672$

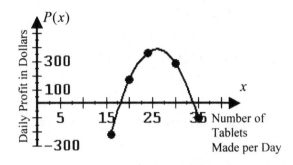

31. $P(x) = -6x^2 + 312x - 3672 = 360$

$6x^2 - 312x + 4032 = 0$

$x^2 - 52x + 672 = 0$

$(x - 24)(x - 28) = 0$

$x = 24, \quad x = 28$

24 or 28 tablets per day will give $360 profit because the parabola has vertical axis of symmetry.

33. $d(t) = -16t^2 + 32t + 40$

$\dfrac{-b}{2a} = \dfrac{-32}{2(-16)} = 1$

$d(1) = -16(1^2) + 32(1) + 40 = 56$

$-16t^2 + 32t + 40 = 0$

$t = \dfrac{-32 - \sqrt{32^2 - 4(-16)(40)}}{2(-16)}$

$t = 2.870828693...$

The maximum height is 56 feet and the ball will reach the ground about 2.9 seconds after being thrown upward.

35. $y = 2.3x^2 - 4.4x + 7.59$

$\dfrac{-b}{2a} = \dfrac{-(-4.4)}{2(1)} = 2.2$

$2.2^2 - 4.4(2.2) + 7.59 = 2.75$

$V(2.2, 2.8)$

$0^2 - 4.4(0) + 7.59$

y-intercept: $(0, 7.59)$

$a = 1,\ b = -4.4,\ c = 7.59$

$b^2 - 4ac = (-4.4)^2 - 4(2)(7.59)$

$b^2 - 4ac = -41.36 < 0$

x-intercepts: none

37. $y = 2.3x^2 - 5.4x - 1.6$

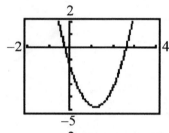

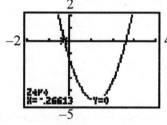

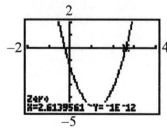

x-intercepts: $(-0.3, 0),\ (2.6, 0)$

Cumulative Review Problems

39. $9x + 5y = 6$

$\underline{2x - 5y = -17}$

$11x\quad\ \ = -11$

$x = -1$

$9(-1) + 5y = 6$

$5y = 15$

$y = 3$

$(-1, 3)$

41. $3x - y + 2z = 12$

$2x - 3y + z = 5$

$x + 3y + 8z = 22$

Multiply the first equation by $-\dfrac{2}{3}$ and add to the second equation, then multiply the first equation by $-\dfrac{1}{3}$ and add to the third equation.

$3x - y + 2z = 12$

$-\dfrac{7}{3}y - \dfrac{1}{3}z = -3$

$\dfrac{10}{3}y + \dfrac{22}{3}z = 18$

Multiply the second equation by $\dfrac{10}{7}$ and add to the third equation

$3x - y + 2z = 12$

$-\dfrac{7}{3}y - \dfrac{1}{3}z = -3$

$\dfrac{48}{7}z = \dfrac{96}{7},\quad z = 2$

(continued)

41. (continued)

Substitute $z = 2$ into second equation.

$-\dfrac{7}{3}y - \dfrac{1}{3}(2) = -3, \ \ y = 1$

Substitute $y = 1, \ z = 2$ into first equation.

$3x - 1 + 2(2) = 12, \ \ x = 3$

$(3, 1, 2)$

8.6 Exercises

1. The critical points divide the number line into regions. All values of x in a given region produce results that are greater than zero, or else all the values of x in a given region produce results that are less than zero.

3. $x^2 + x - 12 < 0$

$x^2 + x - 12 = 0$

$(x + 4)(x - 3) = 0$

$x = -4, \ \ x = 3$

Critical points: $-4, 3$

Test: $x^2 + x - 12$

Region	Test	Result
I	−5	$8 > 0$
II	0	$-12 < 0$
III	4	$8 > 0$

$-4 < x < 3$

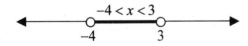

5. $2x^2 + x - 3 < 0$

$2x^2 + x - 3 = 0$

$(2x + 3)(x - 1) = 0$

$x = -\dfrac{3}{2}, \ \ x = 1$

(continued)

5. (continued)

Critical points: $-\dfrac{3}{2}, 1$

Test: $2x^2 + x - 3$

Region	Test	Result
I	−2	$3 > 0$
II	0	$-3 < 0$
III	2	$7 > 0$

$-\dfrac{3}{2} < x < 1$

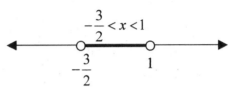

7. $x^2 \geq 4$

$x^2 - 4 \geq 0$

$x^2 - 4 = 0$

$(x + 2)(x - 2) = 0$

$x = -2, \ \ x = 2$

Critical points: $-2, 2$

Test: $x^2 - 4$

Region	Test	Result
I	−3	$5 > 0$
II	0	$-4 < 0$
III	3	$5 > 0$

$-2 \leq x$ or $x \geq 2$

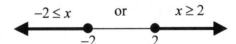

9. $5x^2 \leq 4x + 1$

$5x^2 - 4x - 1 \leq 0$

$5x^2 - 4x - 1 = 0$

$(5x + 1)(x - 1) = 0$

$x = -\dfrac{1}{5}, \ \ x = 1$

(continued)

9. (continued)

Critical points: $-\dfrac{1}{5}, 1$

Test: $5x^2 - 4x - 1$

Region	Test	Result
I	−1	$8 > 0$
II	0	$-1 < 0$
III	2	$11 > 0$

$-\dfrac{1}{5} \le x \le 1$

11. $20 - x - x^2 > 0$

$x^2 + x - 20 < 0$

$x^2 + x - 20 = 0$

$(x+5)(x-4) = 0$

$x = -5, \quad x = 4$

Critical points: $-5, 4$

Test: $20 - x - x^2$

Region	Test	Result
I	−6	$-10 < 0$
II	0	$20 > 0$
III	5	$-10 < 0$

$-5 < x < 4$

13. $6x^2 - 5x > 6$

$6x^2 - 5x - 6 > 0$

$6x^2 - 5x - 6 = 0$

$(2x-3)(3x+2) = 0$

$x = \dfrac{3}{2}, \quad x = -\dfrac{2}{3}$

Critical points: $\dfrac{3}{2}, -\dfrac{2}{3}$

Test: $6x^2 - 5x - 6$

(continued)

13. (continued)

Region	Test	Result
I	−1	$5 > 0$
II	0	$-6 < 0$
III	2	$8 > 0$

$x < -\dfrac{2}{3}$ or $x > \dfrac{3}{2}$

15. $-2x + 30 \ge x(x+5)$

$-2x + 30 \ge x^2 + 5x$

$0 \ge x^2 + 7x - 30$

$x^2 + 7x - 30 \le 0$

$x^2 + 7x - 30 = 0$

$(x+10)(x-3) = 0$

$x = -10, \quad x = 3$

Critical points: $-10, 3$

Test: $x^2 + 7x - 30$

Region	Test	Result
I	−12	$30 > 0$
II	0	$-30 < 0$
III	5	$30 > 0$

$-10 \le x \le 3$

17. $x^2 - 4x \le -4$

$x^2 - 4x + 4 \le 0$

$x^2 - 4x + 4 = 0$

$(x-2)(x-2) = 0$

Critical point: 2

Test: $x^2 - 4x + 4$

Region	Test	Result
I	0	$4 > 0$
II	3	$1 > 0$

$x = 2$

19. $x^2 - 2x > 4$

$x^2 - 2x - 4 > 0$

$x^2 - 2x - 4 = 0$

$a = 1, \ b = -2, \ c = -4$

$x = \dfrac{-(-2) \pm \sqrt{(-2)^2 - 4(1)(-4)}}{2(1)}$

$x = \dfrac{2 \pm \sqrt{20}}{2} = 1 \pm \sqrt{5}$

$x \approx 3.2, -1.2$

Approximate critical points: $3.2, -1.2$

Test: $x^2 - 2x - 4$

Region	Test	Result
I	-2	$4 > 0$
II	0	$-4 < 0$
III	4	$4 > 0$

$x < 1 - \sqrt{5} \approx -1.2$ or $x > 1 + \sqrt{5} \approx 3.2$

21. $x^2 - 6x < -7$

$x^2 - 6x + 7 < 0$

$x^2 - 6x + 7 = 0$

$a = 1, \ b = -6, \ c = 7$

$x = \dfrac{-(-6) \pm \sqrt{(-6)^2 - 4(1)(7)}}{2(1)}$

$x = \dfrac{6 \pm \sqrt{8}}{2}$

$x \approx 4.4, \ 1.6$

Approximate critical points: $4.4, 1.6$

Test: $x^2 - 6x + 7$

Region	Test	Result
I	0	$7 > 0$
II	1	$-2 < 0$
III	3	$2 > 0$

$1.6 \approx 3 - \sqrt{2} < x < 3 + \sqrt{2} \approx 4.4$

23. $2x^2 \geq x^2 - 4$

$x^2 \geq -4$

All real numbers satisfy this inequality.

25. $s = -16t^2 + 640t$

$-16t^2 + 640t > 6000$

$t^2 - 40t + 375 < 0$

$t^2 - 40t + 375 = 0$

$(t - 15)(t - 25) = 0$

$t = 15, \ t = 25$

Critical points: $15, 25$

Test: $t^2 - 40t + 375$

Region	Test	Result
I	0	$375 > 0$
II	20	$-25 < 0$
III	30	$75 > 0$

$15 < t < 25$

The height will be greater than 6000 feet for times between 15 and 24 seconds.

27. Profit $= -20(x^2 - 220x + 2400)$

$-20(x^2 - 220x + 2400) > 0$

$x^2 - 220x + 2400 < 0$

$x^2 - 220x + 2400 = 0$

$x = \dfrac{220 \pm \sqrt{38,800}}{2} \approx 208.5, \ 11.5$

Approximate critical points: $208.4, 11.5$

Test: $x^2 - 220x + 2400$

Region	Test	Result
I	0	$2400 > 0$
II	20	$-1600 < 0$
III	300	$26,400 > 0$

$\approx 11.5 < x < 208.5$

(continued)

27. (continued)

(a) The profit will be greater than zero for production levels between approximately 11.5 and 208.5 units.

(b) $-20(50^2 - 220(50) + 2400)$

$= 122,000$

The daily profit is $122,000 when 50 units are manufactured.

(c) $-20(60^2 - 220(60) + 2400)$

$= 144,000$

The daily profit is $144,000 when 60 units are manufactured.

Cumulative Review Problems

29. $\dfrac{0 + 81 + 92 + 80 + \text{E5} + \text{E6}}{6} \geq 70$

$\text{E5} + \text{E6} \geq 420 - 253$

$\text{E5} + \text{E6} \geq 167$

She must score a combined total of 167 points on the two remaining tests. Any two scores that total 167 will be sufficient to participate in synchronized swimming.

31. For the 2-hour trip,

$10(18) + 14(10) + 5(16) = \$400$

For the 3-hour trip,

$10(22) + 14(12) + 5(19) = \$483$

Putting Your Skills to Work

1. $V = 4 \cdot \pi r^2 h = 4(3.14)(28)^2(4000)$

$V = 39,388,160 \text{ ft}^3$

2. $V = 4\pi(r_{\text{outer}}^2 - r_{\text{inner}}^2)h$

$V = 4(3.14)(28^2 - 25^2)(4000)$

$V = 7,988,160 \text{ ft}^3$

3. $y = 0.00205x^2$, $y = 726$, $x = 660$

$y = 0$

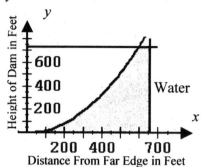

4. $t = $ thickness

$400 = 0.00205x^2$, $x \approx 441.7$

$t = 660 - x \approx 218.3$

From the equations it is approximately 218 ft thick or wide when the dam is 400 ft high which is in agreement with an estimate from the graph.

Chapter 8 Review Problems

1. $5x^2 = 100$

$x^2 = 20$

$x = \pm\sqrt{20} = \pm\sqrt{4 \cdot 5}$

$x = \pm 2\sqrt{5}$

2. $(x + 8)^2 = 81$

$x + 8 = \pm 9$

$x = -8 \pm 9$

$x = 1, \ x = -17$

3. $x^2 + 8x + 13 = 0$

$x^2 + 8x + 16 = -13 + 16$

$(x + 4)^2 = 3$

$x + 4 = \pm\sqrt{3}$

$x = -4 \pm \sqrt{3}$

4. $4x^2 - 8x + 1 = 0$

$$x^2 - 2x + 1 = -\frac{1}{4} + 1$$

$$(x-1)^2 = \frac{3}{4}$$

$$x - 1 = \pm\frac{\sqrt{3}}{2}$$

$$x = 1 \pm \frac{\sqrt{3}}{2}$$

5. $3x^2 - 10x + 6 = 0$

$a = 3,\ b = -10,\ c = 6$

$$x = \frac{-(-10) \pm \sqrt{(-10)^2 - 4(3)(6)}}{2(3)}$$

$$x = \frac{10 \pm \sqrt{28}}{6}$$

$$x = \frac{10 \pm \sqrt{4 \cdot 7}}{6}$$

$$x = \frac{5 \pm \sqrt{7}}{3}$$

6. $x^2 - 6x - 4 = 0$

$a = 1,\ b = -6,\ c = -4$

$$x = \frac{-(-6) \pm \sqrt{(-6)^2 - 4(1)(-4)}}{2(1)}$$

$$x = \frac{6 \pm \sqrt{52}}{2} = \frac{6 \pm \sqrt{4 \cdot 13}}{2}$$

$$x = 3 \pm \sqrt{13}$$

7. $4x^2 - 12x + 9 = 0$

$$(2x - 3)^2 = 0$$

$$2x - 3 = 0$$

$$x = \frac{3}{2}$$

8. $3x^2 - 8x + 6 = 0$

$a = 3,\ b = -8,\ c = 6$

$$x = \frac{-(-8) \pm \sqrt{(-8)^2 - 4(3)(6)}}{2(3)}$$

$$x = \frac{8 \pm \sqrt{-8}}{6} = \frac{8 \pm \sqrt{4 \cdot 2}\sqrt{-1}}{6} = \frac{8 \pm 2i\sqrt{2}}{6}$$

$$x = \frac{4 \pm i\sqrt{2}}{3}$$

9. $6x^2 - 23x = 4x$

$$6x^2 - 27x = 0$$

$$3x(2x - 9) = 0$$

$$3x = 0,\quad 2x - 9 = 0$$

$$x = 0,\quad x = \frac{9}{2}$$

10. $12x^2 - 29x + 15 = 0$

$$(3x - 5)(4x - 3) = 0$$

$$3x - 5 = 0,\quad 4x - 3 = 0$$

$$x = \frac{5}{3},\quad x = \frac{3}{4}$$

11. $x^2 - 3x - 23 = 5$

$$x^2 - 3x - 28 = 0$$

$$(x - 7)(x + 4) = 0$$

$$x - 7 = 0,\quad x + 4 = 0$$

$$x = 7,\quad x = -4$$

12. $3x^2 + 7x + 13 = 13$

$$3x^2 + 7x = 0$$

$$x(3x + 7) = 0$$

$$x = 0,\quad 3x + 7 = 0$$

$$x = 0,\quad x = -\frac{7}{3}$$

13. $3x^2 - 2x = 15x - 10$

$3x^2 - 17x + 10 = 0$

$(3x - 2)(x - 5) = 0$

$3x - 2 = 0, \ x - 5 = 0$

$3x - 2 = 0, \ x - 5 = 0$

$x = \dfrac{2}{3}, \ x = 5$

14. $6x^2 + 12x - 24 = 0$

$x^2 + 2x - 4 = 0$

$a = 1, \ b = 2, \ c = -4$

$x = \dfrac{-2 \pm \sqrt{2^2 - 4(1)(-4)}}{2(1)}$

$x = \dfrac{-2 \pm \sqrt{20}}{2} = \dfrac{-2 \pm \sqrt{4 \cdot 5}}{2}$

$x = -1 \pm \sqrt{5}$

15. $4x^2 - 3x + 2 = 0$

$a = 4, \ b = -3, \ c = 2$

$x = \dfrac{-(-3) \pm \sqrt{(-3)^2 - 4(4)(2)}}{2(4)}$

$x = \dfrac{3 \pm \sqrt{-23}}{8}$

$x = \dfrac{3 \pm \sqrt{23}\sqrt{-1}}{8}$

$x = \dfrac{3 \pm i\sqrt{23}}{8}$

16. $3x^2 + 5x + 1 = 0$

$a = 3, \ b = 5, \ c = 1$

$x = \dfrac{-5 \pm \sqrt{5^2 - 4(3)(1)}}{2(3)}$

$x = \dfrac{-5 \pm \sqrt{13}}{6}$

17. $3x(3x + 2) - 2 = 3x$

$9x^2 + 6x - 2 - 3x = 0$

$9x^2 + 3x - 2 = 0$

$(3x + 2)(3x - 1) = 0$

$3x + 2 = 0, \ 3x - 1 = 0$

$x = -\dfrac{2}{3}, \ x = \dfrac{1}{3}$

18. $10x(x - 2) + 10 = 2x$

$10x^2 - 20x + 10 - 2x = 0$

$10x^2 - 22x + 10 = 0$

$5x^2 - 11x + 5 = 0$

$a = 5, \ b = -11, \ c = 5$

$x = \dfrac{-(-11) \pm \sqrt{(-11)^2 - 4(5)(5)}}{2(5)}$

$x = \dfrac{11 \pm \sqrt{21}}{10}$

19. $\dfrac{5}{6}x^2 - x + \dfrac{1}{3} = 0$

$5x^2 - 6x + 2 = 0$

$a = 5, \ b = -6, \ c = 2$

$x = \dfrac{-(-6) \pm \sqrt{(-6)^2 - 4(5)(2)}}{2(5)}$

$x = \dfrac{6 \pm \sqrt{-4}}{10}$

$x = \dfrac{6 \pm \sqrt{4}\sqrt{-1}}{10}$

$x = \dfrac{6 \pm 2i}{10}$

$x = \dfrac{3 \pm i}{5}$

20. $\dfrac{4}{5}x^2 + x + \dfrac{1}{5} = 0$

$4x^2 + 5x + 1 = 0$

$(4x + 1)(x + 1) = 0$

$4x + 1 = 0, \ x + 1 = 0$

$x = -\dfrac{1}{4}, \ x = -1$

21. $y + \dfrac{5}{3y} + \dfrac{17}{6} = 0$

$6y^2 + 10 + 17y = 0$

$6y^2 + 17y + 10 = 0$

$(6y + 5)(y + 2) = 0$

$6y - 5 = 0, \ y + 2 = 0$

$y = \dfrac{5}{6}, \ y = -2$

22. $\dfrac{19}{y} - \dfrac{15}{y^2} + 10 = 0$

$19y - 15 + 10y^2 = 0$

$10y^2 + 19y - 15 = 0$

$(2y + 5)(5y - 3) = 0$

$2y + 5 = 0, \ 5y - 3 = 0$

$y = -\dfrac{5}{2}, \ y = \dfrac{3}{5}$

23. $\dfrac{15}{y^2} - \dfrac{2}{y} = 1$

$15 - 2y = y^2$

$y^2 + 2y - 15 = 0$

$(y + 5)(y - 3) = 0$

$y + 5 = 0, \ y - 3 = 0$

$y = -5, \ y = 3$

24. $y - 18 + \dfrac{81}{y} = 0$

$y^2 - 18y + 81 = 0$

$(y - 9)^2 = 0$

$y - 9 = 0$

$y = 9$

25. $(3y + 2)(y - 1) = 7(-y + 1)$

$3y^2 - y - 2 = -7y + 7$

$3y^2 + 6y - 9 = 0$

$y^2 + 2y - 3 = 0$

$(y + 3)(y - 1) = 0$

$y + 3 = 0, \ y - 1 = 0$

$y = -3, \ y = 1$

26. $y(y + 1) + (y + 2)^2 = 4$

$y^2 + y + y^2 + 4y + 4 = 4$

$2y^2 + 5y = 0$

$y(2y + 5) = 0$

$y = 0, \ 2y + 5 = 0$

$y = 0, \ y = -\dfrac{5}{2}$

27. $\dfrac{2x}{x + 3} + \dfrac{3x - 1}{x + 1} = 3$

$2x(x + 1) + (3x - 1)(x + 3) = 3(x + 3)(x + 1)$

$2x^2 + 2x + 3x^2 + 8x - 3 = 3x^2 + 12x + 9$

$2x^2 - 2x - 12 = 0$

$x^2 - x - 6 = 0$

$(x - 3)(x + 2) = 0$

$x - 3 = 0, \ x + 2 = 0$

$x = 3, \ x = -2$

28. $\dfrac{4x+1}{2x+5} + \dfrac{3x}{x+4} = 2$

$(4x+1)(x+4)+3x(2x+5) = 2(2x+5)(x+4)$

$4x^2+17x+4+6x^2+15x = 4x^2+26x+40$

$6x^2+6x-36 = 0$

$x^2+x-6 = 0$

$(x+3)(x-2) = 0$

$x+3 = 0,\ x-2 = 0$

$x = -2,\ x = 2$

29. $2x^2+5x-3 = 0$

$a = 2,\ b = 5,\ c = -3$

$b^2-4ac = 5^2-4(2)(-3) = 49 = 7^2$

Two rational solutions.

30. $3x^2-7x-12 = 0$

$a = 3, b = -7, c = -12$

$b^2-4ac = (-7)^2-4(3)(-12) = 193$

Two irrational solutions.

31. $4x^2-6x+5 = 0$

$a = 4,\ b = -6,\ c = 5$

$b^2-4ac = (-6)^2-4(4)(5) = -44$

Two complex solutions.

32. $25x^2-20x+4 = 0$

$a = 25, b = -20, c = 4$

$b^2-4ac = (-20)^2-4(25)(4) = 0$

One rational solution.

33. $5,\ -5$

$x = 5,\ x = -5$

$x-5 = 0,\ x+5 = 0$

$(x-5)(x+5) = 0$

$x^2-25 = 0$

34. $3i,\ -3i$

$x = 3i,\ x = -3i$

$x-3i = 0,\ x+3i = 0$

$(x-3i)(x+3i) = 0$

$x^2-9i^2 = 0$

$x^2+9 = 0$

35. $4\sqrt{2},\ -4\sqrt{2}$

$x = 4\sqrt{2},\ x = -4\sqrt{2}$

$x-4\sqrt{2} = 0,\ x+4\sqrt{2} = 0$

$(x-4\sqrt{2})(x+4\sqrt{2}) = 0$

$x^2-(4\sqrt{2})(4\sqrt{2}) = 0$

$x^2-32 = 0$

36. $-\dfrac{3}{4},\ -\dfrac{1}{2}$

$x = -\dfrac{3}{4},\ x = -\dfrac{1}{2}$

$4x = -3,\ 2x = -1$

$4x+3 = 0,\ 2x+1 = 0$

$(4x+3)(2x+1) = 0$

$8x^2+14x+3 = 0$

37. $x^4-6x^2+8 = 0$

$y = x^2$

$y^2-6y+8 = 0$

$(y-4)(y-2) = 0$

$y-4 = 0, y-2 = 0$

$y = 4,\ y = 2,$

$x^2 = 4,\ x^2 = 2$

$x = \pm 2,\ x = \pm\sqrt{2}$

38. $2x^6 - 5x^3 - 3 = 0$

$y = x^3$

$2y^2 - 5y - 3 = 0$

$(2y+1)(y-3) = 0$

$2y+1 = 0, y-3 = 0$

$y = -\dfrac{1}{2}, y = 3$

$x^3 = -\dfrac{1}{2}, x^2 = 3$

$x = -\sqrt[3]{\dfrac{1}{2}} = -\dfrac{\sqrt[3]{4}}{2}, x = \sqrt[3]{3}$

39. $x^{\frac{2}{3}} + 9x^{\frac{1}{3}} = -8$

$y = x^{\frac{1}{3}}$

$y^2 + 9y + 8 = 0$

$(y+8)(y+1) = 0$

$y+8 = 0, y+1 = 0$

$y = -8, y = -1$

$x^{\frac{1}{3}} = -8, x^{\frac{1}{3}} = -1$

$x = -512, x = -1$

40. $3x^{\frac{1}{2}} - 11x^{\frac{1}{4}} = 4$

$y = x^{\frac{1}{4}}$

$3y^2 - 11y - 4 = 0$

$(3y+1)(y-4) = 0$

$3y+1 = 0, y-4 = 0$

$y = -\dfrac{1}{3}, y = 4$

$x^{\frac{1}{4}} = -\dfrac{1}{3} \text{ (not a valid real root)}, x^{\frac{1}{4}} = 4$

$x = 4^4 = 256$

41. $(2x-5)^2 + 4(2x-5) + 3 = 0$

$y = 2x-5$

$y^2 + 4y + 3 = 0$

$(y+3)(y+1) = 0$

$y+3 = 0, y+1 = 0$

$y = -3, y = -1$

$2x-5 = -3, 2x-5 = -1$

$2x = 2, 2x = 4$

$x = 1, x = 2$

42. $1 + 4x^{-8} = 5x^{-4}$

$x^8 + 4 = 5x^4$

$x^8 - 5x^4 + 4 = 0$

$y = x^4$

$y^2 - 5y + 4 = 0$

$(y-4)(y-1) = 0$

$y-4 = 0, y-1 = 0$

$y = 4, y = 1$

$x^4 = 4, x^4 = 1$

$x = \pm 4^{\frac{1}{4}} = \pm(2^2)^{\frac{1}{4}}, x = \pm 1$

$x = \pm\sqrt{2}, x = \pm 1$

43. $A = \dfrac{2B^2C}{3H}$

$3AH = 2B^2C$

$B^2 = \dfrac{3AH}{2C}$

$B = \pm\sqrt{\dfrac{3AH}{2C}}$

44. $2H = 3g(a^2 + b^2)$

$$b^2 = \frac{2H}{3g} - a^2$$

$$b = \pm\sqrt{\frac{2H}{3g} - a^2}$$

45. $20d^2 - xd - x^2 = 0$

$(4d - x)(5d + x) = 0$

$4d - x = 0, \ 5d + x = 0$

$$d = \frac{x}{4}, \ d = -\frac{x}{5}$$

46. $yx^2 - 3x - 7 = 0$

$a = y, \ b = -3, \ c = -7$

$$x = \frac{-(-3) \pm \sqrt{(-3)^2 - 4(y)(-7)}}{2y}$$

$$x = \frac{3 \pm \sqrt{9 + 28y}}{2y}$$

47. $3y^2 - 4ay + 2a = 0$

$$y = \frac{-(-4a) \pm \sqrt{(-4a)^2 - 4(3)(2a)}}{2(3)}$$

$$y = \frac{4a \pm \sqrt{16a^2 - 24a}}{6}$$

$$y = \frac{4a \pm \sqrt{4(4a^2 - 6a)}}{6}$$

$$y = \frac{4a \pm 2\sqrt{4a^2 - 6a}}{6}$$

$$y = \frac{2a \pm \sqrt{4a^2 - 6a}}{3}$$

48. $PV = 5x^2 + 3y^2 + 2x$

$5x^2 + 2x + 3y^2 - PV = 0$

(continued)

48. (continued)

$a = 5, \ b = 2, \ c = 3y^2 - PV$

$$x = \frac{-2 \pm \sqrt{2^2 - 4(5)(3y^2 - PV)}}{2(5)}$$

$$x = \frac{-2 \pm \sqrt{4(1 - 15y^2 + 5PV)}}{10}$$

$$x = \frac{-2 \pm 2\sqrt{1 - 15y^2 + 5PV}}{10}$$

$$x = \frac{-1 \pm \sqrt{1 - 15y^2 + 5PV}}{5}$$

49. $c^2 = a^2 + b^2$

$16^2 = a^2 + 4^2$

$a^2 = 256 - 16 = 240 = 16(15)$

$a = 4\sqrt{15}$

50. $c^2 = a^2 + b^2 = (3\sqrt{2})^2 + 2^2$

$c^2 = 18 + 4 = 22$

$c = \sqrt{22}$

51. $c^2 = a^2 + b^2$

$6^2 = 5^2 + b^2$

$b^2 = 36 - 25 = 11$

$b = \sqrt{11} \approx 3.3$

The car is approximately 3.3 miles from the observer.

52. $A = LW = 203$

$(4W + 1)W = 203$

$4W^2 + W - 203 = 0$

$(W - 7)(4W + 29) = 0$

$W - 7 = 0, \ 4W + 29 = 0$

(continued)

238

52. (continued)

$$W = 7, \ W = -\frac{29}{4} \text{ reject, } W > 0$$

$$L = 4W + 1 = 29$$

The width is 7 m and the length is 29 m.

53. $A = \frac{1}{2}bh = 70$

$$b(2b + 6) = 140$$

$$2b^2 + 6b - 140 = 0$$

$$b^2 + 3b - 70 = 0$$

$$(b - 7)(b + 10) = 0$$

$$b - 7 = 0, \ b + 10 = 0$$

$$b = 7, \ b = -10 \text{ reject, } b > 0$$

$$h = 2b + 6 = 20$$

The base is 7 cm and the altitude is 20 cm.

54. $v = $ speed with no rain

$$200 = vt_1 \Rightarrow t_1 = \frac{200}{v}$$

$$90 = (v - 5)t_2 \Rightarrow t_2 = \frac{90}{v - 5}$$

$$t_1 + t_2 = \frac{200}{v} + \frac{90}{v - 5} = 6$$

$$200(v - 5) + 90v = 6v(v - 5)$$

$$200v - 1000 + 90v = 6v^2 - 30v$$

$$6v^2 - 320v + 1000 = 0$$

$$(v - 50)(6v - 20) = 0$$

$$v - 50 = 0, \ 6v - 20 = 0$$

$$v = 50, \ v = \frac{10}{3} \text{ reject, } v > 5$$

$$v - 5 = 45$$

The speed before the rain was 50 mph and 45 mph in the rain.

55. $v = $ cruising speed

$$60 = vt_1 \Rightarrow t_1 = \frac{60}{v}$$

$$5 = (v - 15)t_2 \Rightarrow t_2 = \frac{5}{v - 15}$$

$$t_1 + t_2 = \frac{60}{v} + \frac{5}{v - 15} = 4$$

$$60(v - 15) + 5v = 4v(v - 15)$$

$$60v - 900 + 5v = 4v^2 - 60v$$

$$4v^2 - 125v + 900 = 0$$

$$(v - 20)(4v - 45) = 0$$

$$v - 20 = 0, \ 4v - 45 = 0$$

$$v = 20, \ v = \frac{45}{4} = 11.25 \text{ reject, } v > 15$$

$$v - 15 = 5$$

The cruising speed is 20 mph and the trolling speed is 5 mph.

56. $(10 + 2x)(6 + 2x) - 10(6) = 100$

$$60 + 32x + 4x^2 - 60 = 100$$

$$4x^2 + 32x - 100 = 0$$

$$x^2 + 8x - 25 = 0$$

$$x = \frac{-8 + \sqrt{8^2 - 4(1)(-25)}}{2} \approx 2.4$$

The walkway should be approximately 2.4 feet wide.

57. $(40 + 2x)(30 + 2x) - 40(3) = 296$

$$1200 + 140x + 4x^2 - 1200 = 296$$

$$4x^2 + 140x - 296 = 0$$

$$x^2 + 35x - 74 = 0$$

$$(x - 2)(x + 37) = 0$$

$$x = 2, \ x = -37 \text{ reject, } x > 0$$

The walkway should be 2 feet wide.

58. $g(x) = -x^2 + 6x - 11$

$\dfrac{-b}{2a} = \dfrac{-6}{2(-1)} = 3$

$g(3) = -3^2 + 6(3) - 11 = -2$

$V(3, -2)$

$g(0) = -0^2 + 6(0) - 11 = -11$

y-intercept: $(0, -11)$

$b^2 - 4ac = 6^2 - 4(-1)(-11) = -8 < 0$

x-intercepts: none

59. $f(x) = x^2 + 10x + 25$

$\dfrac{-b}{2a} = \dfrac{-10}{2(1)} = -5$

$f(-5) = (-5)^2 + 10(-5) + 25 = 0$

$V(-5, 0)$

$f(0) = 25$

y-intercept: $(0, 25)$

$x^2 + 10x + 25 = 0$

$(x + 5)^2 = 0$

$x = -5$

x-intercept: $(-5, 0)$

60. $f(x) = x^2 + 4x + 3$

$\dfrac{-b}{2a} = \dfrac{-4}{2(1)} = -2$

$f(-2) = (-2)^2 + 4(-2) + 3 = -1$

$V(-2, -1)$

$f(0) = 3$

y-intercept: $(0, 3)$

$x^2 + 4x + 3 = 0$

$(x + 3)(x + 1) = 0$

$x = -3, \ x = -1$

x-intercepts: $(-3, 0), \ (-1, 0)$

(continued)

60. (continued)

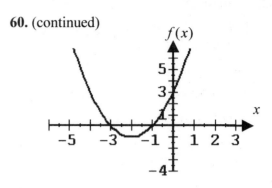

61. $f(x) = x^2 + 6x + 5$

$\dfrac{-b}{2a} = \dfrac{-6}{2(1)} = -3$

$f(-3) = (-3)^2 + 6(-3) + 5 = -4$

$V(-3, -4)$

$f(0) = 5$

y-intercept: $(0, 5)$

$x^2 + 6x + 5 = 0$

$(x + 5)(x + 1) = 0$

$x = -5, \ x = -1$

x-intercepts: $(-5, 0), \ (-1, 0)$

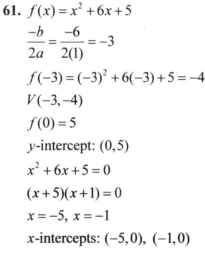

62. $f(x) = -x^2 + 6x - 5$

$\dfrac{-b}{2a} = \dfrac{-6}{2(-1)} = 3$

$f(3) = -3^2 + 6(3) - 5 = 4$

$V(3, 4)$

(continued)

62. (continued)

$f(0) = -5$

y-intercept: $(0.-5)$

$-x^2 + 6x - 5 = 0$

$x^2 - 6x + 5 = 0$

$(x-1)(x-5) = 0$

$x = 1, \ x = 5$

x-intercepts: $(1,0), \ (5,0)$

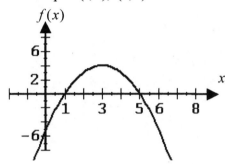

63. $h(t) = -16t^2 + 400t + 40$

$\dfrac{-b}{2a} = \dfrac{-400}{2(-16)} = 12.5$

$h(1.25) = -16(1.25)^2 + 400(12.5) + 40$

$h(12.5) = 2540$

$-16t^2 + 400t + 40 = 0$

$t = \dfrac{-400 - \sqrt{400^2 - 4(-16)(40)}}{2(-16)}$

$t = 25.09960317...$

The maximum height is 2540 feet. The amount of time for the complete flight is 25.1 seconds.

64. $R(x) = x(1200 - x) = -x^2 + 1200x$

$\dfrac{-b}{2a} = \dfrac{-1200}{2(-1)} = 600$

The maximum revenue will occur if the price is \$600 for each unit.

65. $x^2 + 4x - 18 < 0$

$x^2 + 7x - 18 = 0$

$(x+9)(x-2) = 0$

$x = -9, \ x = 2$

Critical points: $-9, \ 2$

Test: $x^2 + 7x - 18$

Region	Test	Results
I	-10	$12 > 0$
II	0	$-18 < 0$
III	3	$12 > 0$

$-9 < x < 2$

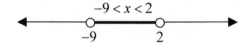

66. $x^2 + 4x - 21 < 0$

$x^2 + 4x - 21 = 0$

$(x+7)(x-3) = 0$

$x = -7, \ x = 3$

Critical points: $-7, \ 3$

Test: $x^2 + 4x - 21$

Region	Test	Results
I	-8	$11 > 0$
II	0	$-21 < 0$
III	4	$11 > 0$

$-7 < x < 3$

67. $x^2 - 9x + 20 > 0$

$x^2 - 9x + 20 = 0$

$(x-5)(x-4) = 0$

$x = 5, \ x = 4$

Critical points: 4, 5

Test: $x^2 - 9x + 20$

(continued)

241

67. (continued)

Region	Test	Results
I	0	$20 > 0$
II	4.5	$-0.25 < 0$
III	6	$2 > 0$

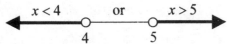

68. $x^2 - 11x + 28 > 0$

$x^2 - 11x + 28 = 0$

$(x - 7)(x - 4) = 0$

$x = 7,\ x = 4$

Critical points: 7, 4

Test: $x^2 - 11x + 28$

Region	Test	Results
I	0	$28 > 0$
II	6	$-2 < 0$
III	8	$4 > 0$

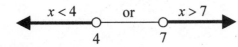

69. $2x^2 - 5x - 3 \le 0$

$2x^2 - 5x - 3 = 0$

$(2x + 1)(x - 3) = 0$

$x = -\dfrac{1}{2},\ x = 3$

Critical points: $-\dfrac{1}{2}$, 3

Test: $2x^2 - 5x - 3$

Region	Test	Results
I	-1	$4 > 0$
II	0	$-3 < 0$
III	4	$9 > 0$

(continued)

69. (continued)

$-\dfrac{1}{2} \le x \le 3$

70. $3x^2 - 5x - 2 \le 0$

$3x^2 - 5x - 2 = 0$

$(3x + 1)(x - 2) = 0$

$x = -\dfrac{1}{3},\ x = 2$

Critical points: $-\dfrac{1}{3}$, 2

Test: $3x^2 - 5x - 2$

Region	Test	Results
I	-1	$6 > 0$
II	0	$-2 < 0$
III	3	$10 > 0$

$-\dfrac{1}{3} \le x \le 2$

71. $16x^2 - 25 > 0$

$16x^2 - 25 = 0$

$x^2 = \dfrac{25}{16}$

$x = \pm\dfrac{5}{4}$

Critical points: $-\dfrac{5}{4},\ \dfrac{5}{4}$

Test: $16x^2 - 25$

Region	Test	Results
I	-2	$39 > 0$
II	0	$-25 < 0$
III	2	$39 > 0$

$x < -\dfrac{5}{4}$ or $x > \dfrac{5}{4}$

242

72. $9x^2 - 4 > 0$

$9x^2 - 4 = 0$

$x^2 = \dfrac{4}{9}$

$x = \pm\dfrac{2}{3}$

Critical points: $-\dfrac{2}{3}, \dfrac{2}{3}$

Test: $9x^2 - 4$

Region	Test	Results
I	−2	$32 > 0$
II	0	$-4 < 0$
III	2	$32 > 0$

$x < -\dfrac{2}{3}$ or $x > \dfrac{2}{3}$

73. $x^2 - 9x > 4 - 7x$

$x^2 - 2x - 4 = 0$

$x = \dfrac{-(-2) \pm \sqrt{(-2)^2 - 4(1)(-4)}}{2}$

$x = 1 \pm \sqrt{5}$

Critical points: $1 - \sqrt{5}, \ 1 + \sqrt{5}$

Test: $x^2 - 2x - 4$

Region	Test	Results
I	−2	$4 > 0$
II	0	$-4 < 0$
III	4	$4 > 0$

$x < 1 - \sqrt{5} \approx -1.2$ or $x > 1 + \sqrt{5} \approx 3.2$

74. $4x^2 - 8x \le 12 + 5x^2$

$-x^2 - 8x - 12 \le 0$

$x^2 + 8x + 12 \ge 0$

$x^2 + 8x + 12 = 0$

(continued)

74. (continued)

$(x + 6)(x + 2) = 0$

$x = -6, \ x = -2$

Critical points: $-6, \ -2$

Test: $x^2 + 8x + 12$

Region	Test	Results
I	−8	$12 > 0$
II	−4	$-4 < 0$
III	0	$12 > 0$

$x \le -6$ or $x \ge -2$

75. $x^2 + 13x > 16 + 7x$

$x^2 + 6x - 16 > 0$

$x^2 + 6x - 16 = 0$

$(x + 8)(x - 2) = 0$

$x + 8 = 0, \ x - 2 = 0$

$x = -8, \ x = 2$

Critical points: $-8, \ 2$

Test: $x^2 + 6x - 16$

Region	Test	Results
I	−10	$24 > 0$
II	0	$-16 < 0$
III	4	$24 > 0$

$x < -8$ or $x > 2$

76. $3x^2 - 12x > -11$

$3x^2 - 12x + 11 > 0$

$3x^2 - 12x + 11 = 0$

$x = \dfrac{12 \pm \sqrt{(-12)^2 - 4(3)(11)}}{2(3)} = 1.4, \ 2.6$

Critical points: $1.4, \ 2.6$

Test: $3x^2 - 12x + 11$

(continued)

76. (continued)

Region	Test	Results
I	0	$11 > 0$
II	2	$-1 < 0$
III	3	$2 > 0$

$x < 1.4$ or $x > 2.6$

77. $-2x^2 + 7x + 12 \le -3x^2 + x$

$x^2 + 6x + 12 \le 0$

$a = 1,\ b = 6,\ c = 12$

$b^2 - 4ac = 6^2 - 4(1)(12) = -12 < 0$

No real solution.

78. $4x^2 + 12x + 9 < 0$

$(2x + 3)^2 < 0$

No real solution.

79. $(x + 4)(x - 2)(3 - x) > 0$

$(x + 4)(x - 2)(3 - x) = 0$

$x = -4,\ x = 2,\ x = 3$

Critical points: $-4,\ 2,\ 3$

Test: $(x + 4)(x - 2)(3 - x)$

Region	Test	Result
I	-5	$56 > 0$
II	0	$-24 < 0$
III	2.5	$\dfrac{13}{8} > 0$
IV	4	$-16 < 0$

$x < -4$ or $2 < x < 3$

80. $(x + 1)(x + 4)(2 - x) < 0$

$(x + 1)(x + 4)(2 - x) = 0$

$x = -1,\ x = -4,\ x = 2$

Critical points: $-1,\ -4,\ 2$

Test: $(x + 1)(x + 4)(2 - x)$

(continued)

80. (continued)

Region	Test	Result
I	-5	$28 > 0$
II	-2	$-8 < 0$
III	0	$8 > 0$
IV	3	$-28 < 0$

$-4 < x < -1$ or $x > 2$

Chapter 8 Test

1. $8x^2 + 9x = 0$

$x(8x + 9) = 0$

$x = 0,\ 8x + 9 = 0$

$x = 0,\ x = -\dfrac{9}{8}$

2. $8x^2 + 10x = 3$

$8x^2 + 10x - 3 = 0$

$(4x - 1)(2x + 3) = 0$

$4x - 1 = 0,\ 2x + 3 = 0$

$x = \dfrac{1}{4},\ x = -\dfrac{3}{2}$

3. $\dfrac{3x}{2} - \dfrac{8}{3} = \dfrac{2}{3x}$

$9x^2 - 16x - 4 = 0$

$(9x + 2)(x - 2) = 0$

$9x + 2 = 0,\ x - 2 = 0$

$x = -\dfrac{2}{9},\ x = 2$

4. $x(x - 3) - 30 = 5(x - 2)$

$x^2 - 3x - 30 = 5x - 10$

$x^2 - 8x - 20 = 0$

(continued)

4. (continued)

$$(x-10)(x+2)=0$$
$$x-10=0,\ x+2=0$$
$$x=10,\ x=-2$$

5. $7x^2-4=52$

$$7x^2=56$$
$$x^2=8$$
$$x=\pm\sqrt{8}=\pm\sqrt{4\cdot2}$$
$$x=\pm2\sqrt{2}$$

6.

$$\frac{2x}{2x+1}-\frac{6}{4x^2-1}=\frac{x+1}{2x-1}$$
$$\frac{2x}{2x+1}-\frac{6}{(2x+1)(2x-1)}=\frac{x+1}{2x-1}$$
$$2x(2x-1)-6=(x+1)(2x+1)$$
$$4x^2-2x-6=2x^2+3x+1$$
$$2x^2-5x-7=0$$
$$(2x-7)(x+1)=0$$
$$2x-7=0,\ x+1=0$$
$$x=\frac{7}{2},\ x=-1$$

7. $2x^2-6x+5=0$

$$a=2,\ b=-6,\ c=5$$
$$x=\frac{-(-6)\pm\sqrt{(-6)^2-4(2)(5)}}{2(2)}$$
$$x=\frac{6\pm\sqrt{-4}}{4}$$
$$x=\frac{6\pm\sqrt{4}\sqrt{-1}}{4}$$
$$x=\frac{6\pm2i}{4}$$
$$x=\frac{3\pm i}{2}$$

8. $2x(x-3)=-3$

$$2x^2-6x+3=0$$
$$a=2,\ b=-6,\ c=3$$
$$x=\frac{-(-6)\pm\sqrt{(-6)^2-4(2)(3)}}{2(2)}$$
$$x=\frac{6\pm\sqrt{12}}{4}=\frac{6\pm\sqrt{4\cdot3}}{4}$$
$$x=\frac{3\pm\sqrt{3}}{2}$$

9. $x^4-9x^2+14=0$

$$y=x^2$$
$$y^2-9y+14=0$$
$$(y-7)(y-2)=0$$
$$y-7=0,\ y-2=0$$
$$y=7,\ y=2$$
$$x^2=7,\ x^2=2$$
$$x=\pm\sqrt{4},\ x=\pm\sqrt{2}$$

10. $3x^{-2}-11x^{-1}-20=0$

$$3-11x-20x^2=0$$
$$20x^2+11x-3=0$$
$$(5x-1)(4x+3)=0$$
$$5x-1=0,\ 4x+3=0$$
$$x=\frac{1}{5},\ x=-\frac{3}{4}$$

11. $x^{\frac{2}{3}}-2x^{\frac{1}{3}}-12=0$

$$y=x^{\frac{1}{3}}$$
$$y^2-2y-12=0$$
$$y=\frac{2\pm\sqrt{(-2)^2-4(1)(-12)}}{2}=\frac{2\pm\sqrt{52}}{2}$$

(continued)

11. (continued)

$$y = \frac{2 \pm \sqrt{4 \cdot 13}}{2} = \frac{2 \pm 2\sqrt{13}}{2} = 1 \pm \sqrt{13}$$

$$x^{\frac{1}{3}} = 1 \pm \sqrt{13}$$

$$x = (1 \pm \sqrt{13})^3$$

12. $B = \dfrac{xyw}{z^2}$

$$z^2 = \frac{xyw}{B}$$

$$z = \pm \sqrt{\frac{xyw}{B}}$$

13. $5y^2 + 2by + 6w = 0$

$$y = \frac{-2b \pm \sqrt{(2b)^2 - 4(5)(6w)}}{2(5)}$$

$$y = \frac{-2b \pm \sqrt{4(b^2 - 30w)}}{10}$$

$$y = \frac{-b \pm \sqrt{b^2 - 30w}}{5}$$

14. $A = LW = (3W + 1)W = 80$

$$3W^2 + W - 80 = 0$$

$$(W - 5)(3W + 16) = 0$$

$$W = 5, \ W = -\frac{16}{3} \text{ reject, } W > 0$$

$L = 3W + 1 = 16$
The length is 16 miles and the width is 5 miles.

15. $c^2 = a^2 + b^2$

$$c^2 = 6^2 + (2\sqrt{3})^2$$

$$c^2 = 36 + 12 = 48$$

$$c = \sqrt{48} = \sqrt{16 \cdot 3}$$

$$c = 4\sqrt{3}$$

16. $6 = vt_1 \Rightarrow t_1 = \dfrac{6}{v}$

$$3 = (v + 1)t_2 \Rightarrow t_2 = \frac{3}{v + 1}$$

$$t_1 + t_2 = 4$$

$$\frac{6}{v} + \frac{3}{v + 1} = 4$$

$$6(v + 1) + 3v = 4v(v + 1)$$

$$6v + 6 + 3v = 4v^2 + 4v$$

$$4v^2 - 5v - 6 = 0$$

$$(v - 2)(4v + 3) = 0$$

$$v - 2 = 0, \ 4v + 3 = 0$$

$$v = 2, \ v = -\frac{3}{4} \text{ reject, } v > 0$$

$$v + 1 = 3$$

They paddled 2 mph for the first part and 3 mph after lunch.

17. $f(x) = -x^2 - 6x - 5$

$$\frac{-b}{2a} = \frac{-(-6)}{2(-1)} = -3$$

$$f(-3) = -(-3)^2 - 6(-3) - 5$$

$$f(-3 = 4$$

$$V(-3, 4)$$

$$f(0) = -5$$

y-intercept: $(0, -5)$

$$x^2 + 6x + 5 = 0$$

$$(x + 1)(x + 5) = 0$$

$$x + 1 = 0, \ x + 5 = 0$$

$$x = -1, \ x = -5$$

x-intercepts: $(-1, 0), \ (-5, 0)$

(continued)

17. (continued)

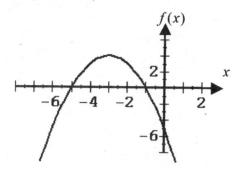

18. $2x^2 + 3x \geq 27$

$2x^2 + 3x - 27 \geq 0$

$2x^2 + 3x - 27 = 0$

$(2x + 9)(x - 3) = 0$

$x = -\dfrac{9}{2}, \ x = 3$

Critical Points: $-\dfrac{9}{2}, \ 3$

Test: $2x^2 + 3x - 27$

Region	Test	Result
I	-5	$8 > 0$
II	0	$-27 < 0$
III	4	$17 > 0$

$x \leq -\dfrac{9}{2} \text{ or } x \geq 3$

19. $-3x^2 + 10x + 8 \geq 0$

$-3x^2 + 10x + 8 \geq 0$

$3x^2 - 10x - 8 \leq 0$

$3x^2 - 10x - 8 = 0$

$(3x + 2)(x - 4) = 0$

$3x + 2 = 0, \ x - 4 = 0$

$x = -\dfrac{2}{3}, \ x = 4$

(continued)

19. (continued)

Critical points: $-\dfrac{2}{3}, \ 4$

Test: $3x^2 - 10x - 8$

Region	Test	Result
I	-1	$5 > 0$
II	0	$-8 < 0$
III	5	$17 > 0$

$-\dfrac{2}{3} \leq x \leq 4$

20. $x^2 + 3x - 7 > 0$

$x = \dfrac{-3 \pm \sqrt{3^2 - 4(1)(-7)}}{2}$

$x = -4.5, \ x = 1.5$

Critical points: $-4.5, \ 1.5$

Test: $x^2 + 3x - 7$

Region	Test	Result
I	-5	$3 > 0$
II	0	$-7 < 0$
III	2	$3 > 0$

$x < -4.5 \text{ or } x > 1.5$

Cumulative Test for Chapters 1-8

1. $(-3x^{-2}y^3)^4 = (-3)^4 x^{-2 \cdot 4} y^{3 \cdot 4}$

$\qquad = \dfrac{81y^{12}}{x^8}$

2. $\dfrac{1}{2}a^3 - 2a^2 + 3a - \dfrac{1}{4}a^3 - 6a + a^2$

$\qquad = \dfrac{1}{4}a^3 - a^2 - 3a$

3. $a(2y + b) = 3ay - 4$

(continued)

3. (continued)

$$2ay + ab = 3ay - 4$$

$$ay = ab + 4$$

$$y = \frac{ab + 4}{a}$$

4. $6x - 3y = -12$

x	y
0	4
-2	0

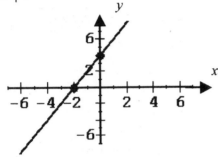

5. $2y + x = 8$

$$y = -\frac{1}{2}x + 4 \Rightarrow m = -\frac{1}{2} \Rightarrow m_{\parallel} = -\frac{1}{2}$$

$$y - y_1 = m_{\parallel}(x - x_1)$$

$$y - (-1) = -\frac{1}{2}(x - 6)$$

$$2y + 2 = -x + 6$$

$$x + 2y = 4$$

6. $V = \frac{4}{3}\pi r^3 = \frac{4}{3}\pi (2)^3 = \frac{32\pi}{3} \text{ in}^3$

7. $125x^3 - 27y^2$

$$= (5x - 3y)(25x^2 + 15xy + 9y^2)$$

8. $\sqrt{72x^3 y^6} = \sqrt{36x^2 y^6 \cdot 2x}$

$$= 6xy^3 \sqrt{2x}$$

9. $(3 + \sqrt{2})(\sqrt{6} + \sqrt{3})$

$$= 3\sqrt{6} + 3\sqrt{3} + \sqrt{12} + \sqrt{6}$$

$$= 4\sqrt{6} + 3\sqrt{3} + \sqrt{4 \cdot 3}$$

$$= 4\sqrt{6} + 3\sqrt{3} + 2\sqrt{3}$$

$$= 4\sqrt{6} + 5\sqrt{3}$$

10. $\dfrac{3}{\sqrt{11}} \cdot \dfrac{\sqrt{11}}{\sqrt{11}} = \dfrac{3\sqrt{11}}{11}$

11. $3x^2 + 12x = 26x$

$$3x^2 - 14x = 0$$

$$x(3x - 14) = 0$$

$$x = 0, \; x = \frac{14}{3}$$

12. $\qquad 12x^2 = 11x - 2$

$$12x^2 - 11x + 2 = 0$$

$$(4x - 1)(3x - 2) = 0$$

$$4x - 1 = 0, \; 3x - 2 = 0$$

$$x = \frac{1}{4}, \; x = \frac{2}{3}$$

13. $\qquad 44 = 3(2x - 3)^2 + 8$

$$3(2x - 3)^2 = 36$$

$$(2x - 3)^2 = 12$$

$$2x - 3 = \pm\sqrt{12}$$

$$2x - 3 = \pm\sqrt{4 \cdot 3}$$

$$2x - 3 = \pm 2\sqrt{3}$$

$$2x = 3 \pm 2\sqrt{3}$$

$$x = \frac{3 \pm 2\sqrt{3}}{2}$$

14. $3 - \dfrac{4}{x} + \dfrac{5}{x^2} = 0$

$3x^2 - 4x + 5 = 0$

$x = \dfrac{-(-4) \pm \sqrt{(-4)^2 - 4(3)(5)}}{2(3)}$

$x = \dfrac{4 \pm \sqrt{-44}}{6}$

$x = \dfrac{4 \pm \sqrt{4 \cdot 11}\sqrt{-1}}{6}$

$x = \dfrac{4 \pm 2i\sqrt{11}}{6}$

$x = \dfrac{2 \pm i\sqrt{11}}{3}$

15. $\sqrt{x-12} = \sqrt{x} - 2$

$x - 12 = x - 4\sqrt{x} + 4$

$4\sqrt{x} = 16$

$\sqrt{x} = 4$

$x = 16$

check:

$\sqrt{16-12} \overset{?}{=} \sqrt{16} - 2$

$2 = 2$

$x = 16$

16. $x^{\frac{2}{3}} + 9x^{\frac{1}{3}} + 18 = 0$

$y = x^{\frac{1}{3}}$

$y^2 + 9y + 18 = 0$

$(y+6)(y+3) = 0$

$y = -6, \ y = -3$

$x^{\frac{1}{3}} = -6, \ x^{\frac{1}{3}} = -3$

$x = -216, \ x = -27$

(continued)

16. (continued)

check:

$(-216)^{\frac{2}{3}} + 9(-216)^{\frac{1}{3}} + 18 \overset{?}{=} 0$

$0 = 0$

$(-27)^{\frac{2}{3}} + 9(-27)^{\frac{1}{3}} + 18 \overset{?}{=} 0$

$0 = 0$

$x = -216, \ x = -27$

17. $2y^2 + 5wy - 7z = 0$

$y = \dfrac{-5w \pm \sqrt{(5w)^2 - 4(2)(7z)}}{2(2)}$

$y = \dfrac{-5w \pm \sqrt{25w^2 - 56z}}{4}$

18. $3y^2 + 16z^2 = 5w$

$3y^2 = 5w - 16z^2$

$y^2 = \dfrac{5w - 16z^2}{3}$

$y = \pm\sqrt{\dfrac{5w - 16z^2}{3}}$

19. $c^2 = a^2 + b^2$

$\left(\sqrt{31}\right)^2 = 4^2 + b^2$

$31 = 16 + b^2$

$b^2 = 15$

$b = \sqrt{15}$

20. $A = \dfrac{1}{2}bh$

$A = \dfrac{1}{2}b(3b+3)$

$45 = \dfrac{1}{2}b(3b+3)$

(continued)

20. (continued)

$$90 = 3b^2 + 3b$$
$$b^2 + b - 30 = 0$$
$$(b-5)(b+6) = 0$$
$$b - 5 = 0, \ b + 6 = 0$$
$$b = 5, \ b = -6 \text{ reject, } b > 0$$
$$h = 3b + 3$$
$$h = 18$$

The base of the triangle is 5 meters and the altitude of the triangle is 18 meters.

21. $f(x) = -x^2 + 8x - 12$
$a = -1, \ b = 8, \ c = -12$

$$x_{\text{vertex}} = \frac{-b}{2a}$$
$$x_{\text{vertex}} = \frac{-8}{2(-1)}$$
$$x_{\text{vertex}} = 4$$
$$y_{\text{vertex}} = f(x_{\text{vertex}})$$
$$y_{\text{vertex}} = f(4) = -4^2 + 8(4) - 12$$
$$y_{\text{vertex}} = 4$$
$$V(4,4)$$
$$f(0) = -12$$
y-intercept: $(0, -12)$
$$-x^2 + 8x - 12 = 0$$
$$x^2 - 8x + 12 = 0$$
$$(x-6)(x-2) = 0$$
$$x - 6 = 0, \ x - 2 = 0$$
$$x = 6, \ x = 2$$
x-intercepts: $(6, 0), \ (2,)$

22.

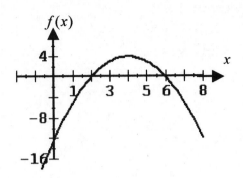

23. $6x^2 - x \le 2$
$$6x^2 - x - 2 \le 0$$
$$6x^2 - x - 2 = 0$$
$$(3x-2)(2x+1) = 0$$
$$3x - 2 = 0, \ 2x + 1 = 0$$
$$x = \frac{2}{3}, \ x = -\frac{1}{2}$$

Critical points: $\frac{2}{3}, \ -\frac{1}{2}$

Test: $6x^2 - x - 2$

Region	Test	Results
I	−1	$5 > 0$
II	0	$-2 < 0$
III	1	$3 > 0$

$$-\frac{1}{2} \le x \le \frac{2}{3}$$

24. $x^2 > -2x + 15$
$$x^2 + 2x - 15 > 0$$
$$x^2 + 2x - 15 = 0$$
$$(x+5)(x-3) = 0$$
$$x + 5 = 0, \ x - 3 = 0$$
$$x = -5, \ x = 3$$

(continued)

24. (continued)

Critical points: -5, 3

Test: $x^2 + 2x - 15$

Region	Test	Results
I	-6	$9 > 0$
II	0	$-15 < 0$
III	5	$20 > 0$

$x < -5$ or $x > 3$

Chapter 9

Pretest Chapter 9

1. $(x-h)^2 + (y-k)^2 = r^2$

$(x-8)^2 + (y-(-2))^2 = \sqrt{7}^2$

$(x-8)^2 + (y+2)^2 = 7$

2. $(x_1, y_1) = (-6, -2),\ (x_2, y_2) = (-3, 4)$

$d = \sqrt{(x_2 - x_1)^2 + (y_2 - y_1)^2}$

$d = \sqrt{(-3-(-6))^2 + (4-(-2))^2}$

$d = \sqrt{45} = \sqrt{9(5)}$

$d = 3\sqrt{5}$

3. $x^2 + y^2 - 2x - 4y + 1 = 0$

$x^2 - 2x + 1 + y^2 - 4y + 4 = -1 + 1 + 4$

$(x-1)^2 + (y-2)^2 = 4 = 2^2$

Center $= (1, 2)$

radius $= 2$

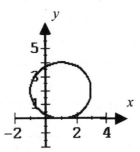

4. $x = (y+1)^2 + 2$

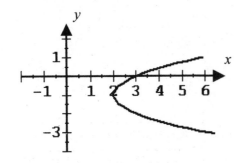

5. $x^2 = y - 4x - 1$

$y = x^2 + 4x + 1$

$y = x^2 + 4x + 4 - 3$

$y = (x+2)^2 - 2$

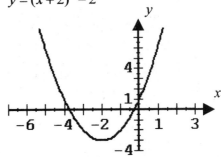

6. $4x^2 + y^2 - 36 = 0$

$4x^2 + y^2 = 36$

$\dfrac{x^2}{9} + \dfrac{y^2}{36} = 1$

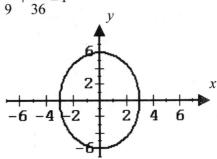

7. $\dfrac{(x+3)^2}{25} + \dfrac{(y-1)^2}{36} = 1$

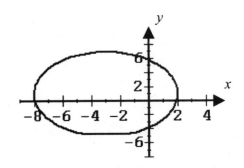

8. $25y^2 - 9x^2 = 225$

$$\frac{y^2}{9} - \frac{x^2}{25} = 1$$

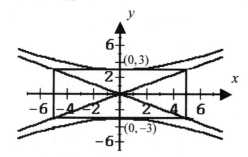

9. $\dfrac{(x-2)^2}{4} - \dfrac{(y+1)^2}{9} = 1$

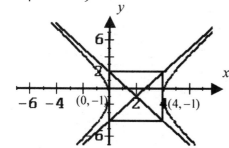

10. $x^2 + y^2 - 25$

$3x + 4y = 0 \Rightarrow y = \dfrac{-3x}{4}$, substitute into

first equation.

$$x^2 + \left(\frac{-3x}{4}\right)^2 = 25$$

$$x^2 + \frac{9x^2}{16} = 25$$

$$16x^2 + 9x^2 = 400$$

$$x^2 = 16$$

$$x = \pm 4$$

$$y = \frac{-3(\pm 4)}{4} = \mp 3$$

$(4, -3)$, $(-4, 3)$ is the solution.

11. $y = x^2 + 1 \Rightarrow x^2 y - 1$, substitute into second equation.

$$4y^2 = 4 - x^2$$

$$4y^2 = 4 - (y - 1)$$

$$4y^2 = 4 - y + 1$$

$$4y^2 + y - 5 = 0$$

$$(y - 1)(4y + 5) = 0$$

$$y - 1 = 0, \quad 4y + 5 = 0$$

$$y = 1, \quad y = -\frac{5}{4}$$

$$x^2 = y - 1 = \begin{cases} 1 - 1 = 0 \\ -\dfrac{5}{4} - 1 = -\dfrac{9}{4} \end{cases}$$

$$x^2 = 0$$

$$x = 0$$

$(0, 1)$ is the solution.

9.1 Exercises

1. Subtract the value of the point and use the absolute value: $|-2 - 4| = |-6| = 6$.

3. The center is $(1, -2)$ and the radius is 3.

5. $d = \sqrt{(x_2 - x_1)^2 + (y_2 - y_1)^2}$

$d = \sqrt{(1 - 2)^2 + (6 - 4)^2} = \sqrt{5}$

7. $d = \sqrt{(x_2 - x_1)^2 + (y_2 - y_1)^2}$

$d = \sqrt{\left(\dfrac{1}{2} - \dfrac{3}{4}\right)^2 + \left(\dfrac{5}{2} - \dfrac{3}{2}\right)^2} = \dfrac{\sqrt{17}}{4}$

9. $d = \sqrt{(x_2 - x_1)^2 + (y_2 - y_1)^2}$

$d = \sqrt{(3 - (-2))^2 + (9 - (-3))^2} = 13$

11. $d = \sqrt{(x_2 - x_1)^2 + (y_2 - y_1)^2}$

$d = \sqrt{(4-0)^2 + (1-(-3))^2} = \sqrt{32}$

$d = 4\sqrt{2}$

13. $d = \sqrt{(x_2 - x_1)^2 + (y_2 - y_1)^2}$

$d = \sqrt{\left(\frac{7}{3} - \frac{1}{3}\right)^2 + \left(\frac{1}{5} - \frac{3}{5}\right)^2}$

$d = \sqrt{\frac{104}{25}}$

$d = \frac{2\sqrt{26}}{5}$

15. $d = \sqrt{(x_2 - x_1)^2 + (y_2 - y_1)^2}$

$d = \sqrt{(1.3 - (-5.7))^2 + (2.6 - 1.6)^2}$

$d = \sqrt{50}$

$d = 5\sqrt{2}$

17. $d = \sqrt{(x_2 - x_1)^2 + (y_2 - y_1)^2}$

$10 = \sqrt{(y-6)^2 + (1-7)^2}$

$100 = y^2 - 4y + 40$

$y^2 - 4y - 60 = 0$

$(y-10)(y+6) = 0$

$y - 10 = 0, \ y + 6 = 0$

$y = 10, \ y = -6$

19. $d = \sqrt{(x_2 - x_1)^2 + (y_2 - y_1)^2}$

$2.5 = \sqrt{(0-1.5)^2 + (y-2)^2}$

$6.25 = 2.25 + y^2 - 4y + 4$

$y^2 - 4y = 0$

$y(y-4) = 0$

$y = 0, \ y - 4 = 0$

$y = 0, \ y = 4$

21. $d = \sqrt{(x_2 - x_1)^2 + (y_2 - y_1)^2}$

$\sqrt{10} = \sqrt{(x-7)^2 + (6-3)^2}$

$10 = x^2 - 14x + 49 + 9$

$x^2 - 14x + 48 = 0$

$(x-6)(x-8) = 0$

$x - 6 = 0, \ x - 8 = 0$

$x = 6, \ x = 8$

23. $d = \sqrt{(x_2 - x_1)^2 + (y_2 - y_1)^2}$

$4 = \sqrt{(4-6)^2 + (y-6)^2}$

$16 = 4 + y^2 - 12y + 36$

$y^2 - 12y + 24 = 0$

$y = \frac{12 \pm \sqrt{12^2 - 4(1)(24)}}{2(1)}$

$y = \frac{12 \pm \sqrt{48}}{2} \approx \begin{cases} 9.5 \\ 2.5 \end{cases}$

$y \approx 9.5$ miles

25. $(x-h)^2 + (y-k)^2 = r^2$

$(x-(-1))^2 + (y-(-7))^2 = \sqrt{5}^2$

$(x+1)^2 + (y+7)^2 = 5$

27. $(x-h)^2 + (y-k)^2 = r^2$

$(x-(-3.5))^2 + (y-0)^2 = 6^2$

$(x+3.5)^2 + y^2 = 36$

29. $(x-h)^2 + (y-k)^2 = r^2$

$\left(x - \frac{7}{4}\right)^2 + (y-0)^2 = \left(\frac{1}{3}\right)^2$

$\left(x - \frac{7}{4}\right)^2 + y^2 = \frac{1}{9}$

31. $x^2 + y^2 = 25$
$C(0,0)$, $r = 5$

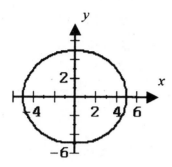

33. $(x-3)^2 + (y-2)^2 = 4 = 2^2$
$C(3,2)$, $r = 2$

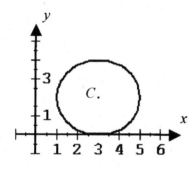

35. $(x+2)^2 + (y-3)^2 = 25 = 5^2$
$C(-2,3)$, $r = 5$

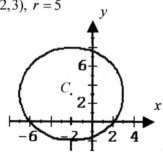

37. $x^2 + y^2 + 6x - 4y - 3 = 0$
$x^2 + 6x + 9 + y^2 - 4y + 4 = 3 + 9 + 4$
$(x+3)^2 + (y-2)^2 = 16 = 4^2$
$C(-3,2)$, $r = 4$

39. $x^2 + y^2 - 12x + 2y - 12 = 0$
$x^2 - 12x + 36 + y^2 + 2y + 1 = 12 + 36 + 1$
$(x-6)^2 + (y+1)^2 = 49 = 7^2$
$C(6,-1)$, $r = 7$

41. $x^2 + y^2 + 3x - 2 = 0$
$x^2 + 3x + y^2 = 2$
$x^2 + 3x + \dfrac{9}{4} + y^2 = 2 + \dfrac{9}{4} = \dfrac{17}{4}$
$\left(x + \dfrac{3}{2}\right)^2 + y^2 = \left(\dfrac{\sqrt{17}}{2}\right)^2$
$C\left(-\dfrac{3}{2}, 0\right)$, $r = \dfrac{\sqrt{17}}{2}$

43. $(x - 42.7)^2 + (y - 29.7)^2 = 25.1^2 = 630.01$

45. $(x - 5.32)^2 + (y + 6.54)^2 = 47.28$

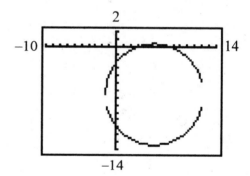

Cumulative Review Problems

47. $9 + \dfrac{3}{x} = \dfrac{2}{x^2}$
$9x^2 + 3x - 2 = 0$
$(3x + 2)(3x - 1) = 0$
$3x + 2 = 0, \ 3x - 1 = 0$
$x = -\dfrac{2}{3}, \ x = \dfrac{1}{3}$

49. $5x^2 - 6x - 7 = 0$

$$x = \frac{-(-6) \pm \sqrt{(-6)^2 - 4(5)(-7)}}{2(5)}$$

$$x = \frac{6 \pm \sqrt{176}}{10} = \frac{6 \pm \sqrt{16(11)}}{10} = \frac{6 \pm 4\sqrt{11}}{10}$$

$$x = \frac{3 \pm 2\sqrt{11}}{5}$$

51. $V = Ah$

$$V = 20 \text{ mi}^2 (150 \text{ ft}) \left(\frac{5280^2 \text{ ft}^2}{\text{mi}^2} \right)$$

$$V = 8.364 \times 10^{10} \text{ ft}^3$$

9.2 Exercises

1. The graph of $y = x^2$ is symmetric about the <u>y-axis</u>. The graph of $x = y^2$ is symmetric about the <u>x-axis</u>.

3. Since $y = 2(x - 3) + 4$ is in standard form, $y = a(x - h)^2 + k$, the vertex is $(h, k) = (3, 4)$.

5. $y = -4x^2$

$V(0, 0)$

y-intercept $= (0, 0)$

x	-2	-1	0	1	2
y	-16	-4	0	-4	-16

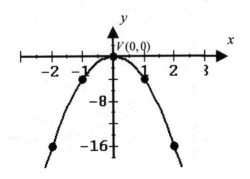

7. $y = x^2 - 6$

$V(0, -6)$

y-intercept: $(0, -6)$

x	-3	-2	0	2	3
y	3	-2	-6	-2	6

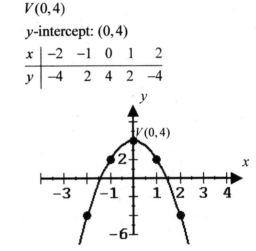

9. $y = -2x^2 + 4$

$V(0, 4)$

y-intercept: $(0, 4)$

x	-2	-1	0	1	2
y	-4	2	4	2	-4

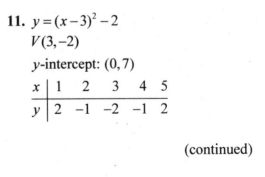

11. $y = (x - 3)^2 - 2$

$V(3, -2)$

y-intercept: $(0, 7)$

x	1	2	3	4	5
y	2	-1	-2	-1	2

(continued)

11. (continued)

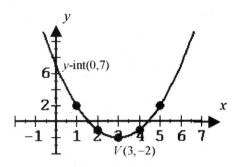

13. $y = 2(x-1)^2 + \dfrac{3}{2}$

$V\left(1, \dfrac{3}{2}\right)$

y-intercept: $\left(0, \dfrac{7}{2}\right)$

x	-1	0	1	2	3
y	$\dfrac{19}{2}$	$\dfrac{7}{2}$	$\dfrac{3}{2}$	$\dfrac{7}{2}$	$\dfrac{19}{2}$

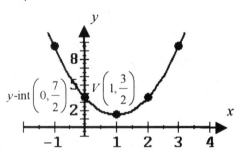

15. $y = -4\left(x + \dfrac{3}{2}\right)^2 + 5$

$V\left(-\dfrac{3}{2}, 5\right)$

y-intercept: $(0, -4)$

x	-3	-2	$-\dfrac{3}{2}$	-1	0
y	-4	4	5	4	-4

(continued)

15. (continued)

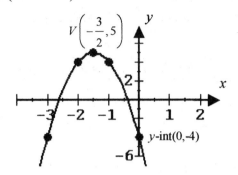

17. $x = \dfrac{1}{4}y^2 - 2$

$V(-2, 0)$

x-intercept: $(-2, 0)$

x	-2	0	0	2	2
y	0	-2	2	-4	4

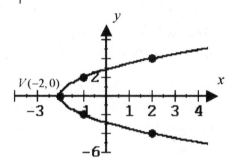

19. $x = (y-2)^2 + 3$

$V(3, 2)$

x-intercept: $(7, 0)$

x	3	4	4	7	7
y	2	1	3	4	0

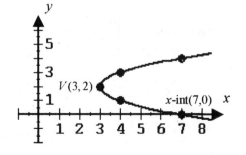

21. $x = -3(y+1) - 2$

$V(-2, -1)$

x-intercept: $(-5, 0)$

x	-2	-5	-5
y	-1	0	-2

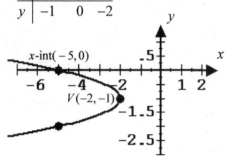

23. $x = \dfrac{1}{2}(y+1)^2 + 2$

$V(2, -1)$

x-intercept: $(2.5, 0)$

x	2	2.5	4	4
y	-1	0	1	-3

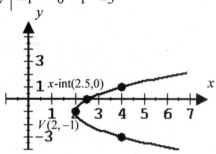

25. $x = -(y-2)^2 + \dfrac{1}{2}$

$V\left(\dfrac{1}{2}, -2\right)$

x-intercept: $\left(-\dfrac{7}{2}, 0\right)$

x	$-\dfrac{7}{2}$	$\dfrac{1}{2}$
y	0	2

(continued)

25. (continued)

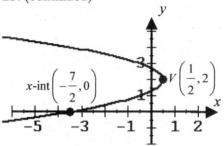

27. $y = x^2 + 12x + 25$

$y = x^2 + 12x + 36 - 9$

$y = (x+6)^2 - 9$

(a) Vertical

(b) Opens upward

(c) $V(-6, -9)$

29. $y = -2x^2 + 4x - 3$

$y = -2(x^2 - 2x + 1) - 3 + 2$

$y = -2(x-1)^2 - 1$

(a) Vertical

(b) Downward

(c) $V(1, -1)$

31. $x = y^2 + 8y + 9$

$x = y^2 + 8y + 16 - 7$

$x = (y+4)^2 - 7$

(a) Horizontal

(b) Opens right

(c) $V(-7, -4)$

33. $y = ax^2$, $8 = a(16)^2$, $a = \dfrac{1}{32}$

$y = \dfrac{1}{32}x^2$

35. $a = \dfrac{1}{4p} = \dfrac{1}{32} \Rightarrow p = 8$

The distance from (0,0) to the focus point is **8** inches.

37. $y = 2x^2 + 6.48x - 0.1312$

$y = 2(x^2 + 3.24x + 2.6244) - 0.1312 - 5.2488$

$y = 2(x + 1.62)^2 - 5.38$

$V(-1.62, -5.38)$

y-intercept: -0.1312

$$\dfrac{-6.48 \pm \sqrt{6.48^2 - 4(2)(-0.1312)}}{2(2)}$$

$= \begin{cases} 0.020121947 \\ -3.260121947 \end{cases} = x\text{-intercepts}$

39. $P = -2x^2 + 200x + 47,000$

$P = -2(x^2 - 100x + 2500) + 47,000 + 5000$

$P = -2(x - 50)^2 + 52,000$

The maximum profit is \$52,000 and the number of items needed is 50.

41. $S = 650d - 2d^2$

$S = -2(d^2 - 325d + 26,406.25) + 52,812.5$

$S = -2(d - 162.5)^2 + 52,812.5$

The maximum sensitivity is 52,812.5 and the dosage is 162.5 milligrams.

Cumulative Review Problems

43. $\sqrt{50x^3} = \sqrt{25x^2 \cdot 2x}$

$\qquad = 5x\sqrt{2x}$ for $x \ge 0$

45. $\sqrt{98x} + x\sqrt{8} - 3\sqrt{50x}$

$= \sqrt{49 \cdot 2x} + x\sqrt{4 \cdot 2} - 3\sqrt{25 \cdot 2x}$

$= 7\sqrt{2x} + 2x\sqrt{2} - 15\sqrt{2x}$

$= 2x\sqrt{2} - 8\sqrt{2x}$

47. $d = rt$

$d = 40t = \dfrac{40}{60}(56 - 15) = 27\dfrac{1}{3}$

Matthew lives $27\dfrac{1}{3}$ miles from work.

49. $8(1050)(0.88) = 7392$

Sir George can expect 7392 blooms if there is heavy rainfall this year.

9.3 Exercises

1. $\dfrac{(x+2)^2}{4} + \dfrac{(y-3)}{9} = 1$ is in the form

$\dfrac{(x-h)^2}{a^2} + \dfrac{(y-k)^2}{b^2} = 1$ where (h,k) is the

center of the ellipse. Therefore, the center of the ellipse is $(-2, 3)$.

3. $\dfrac{x^2}{36} + \dfrac{y^2}{4} = 1 \Rightarrow \dfrac{x^2}{6^2} + \dfrac{y^2}{2^2} = 1$

$a = 6, \ b = 2$

Intercepts: $(\pm 6, 0), \ (0, \pm 2)$

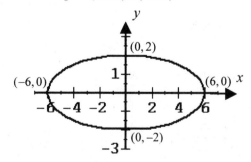

5. $\dfrac{x^2}{81}+\dfrac{y^2}{100}=1 \Rightarrow \dfrac{x^2}{9^2}+\dfrac{y^2}{10^2}=1$

$a=9,\ b=10$

Intercepts: $(\pm 9,0),\ (0,\pm 10)$

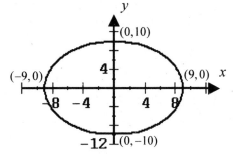

7. $4x^2+y^2-36=0$

$\dfrac{x^2}{9}+\dfrac{y^2}{36}=1 \Rightarrow \dfrac{x^2}{3^2}+\dfrac{y^2}{6^2}=1$

$a=3,\ b=6$

Intercepts: $(\pm 3,0),\ (0,\pm 6)$

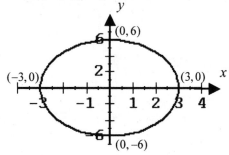

9. $x^2+9y^2=81 \Rightarrow \dfrac{x^2}{81}+\dfrac{y^2}{9}=1$

$\dfrac{x^2}{9^2}+\dfrac{y^2}{3^2}=1$

$a=9,\ b=3$

Intercepts: $(\pm 9,0),\ (0,\pm 3)$

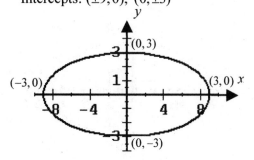

11. $x^2+12y^2=36$

$\dfrac{x^2}{36}+\dfrac{y^2}{3}=1 \Rightarrow \dfrac{x^2}{6^2}+\dfrac{y^2}{\sqrt{3}^2}=1$

$a=6,\ b=\sqrt{3}$

Intercepts: $(\pm 6,0),\ (0,\pm\sqrt{3})$

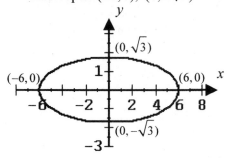

13. $\dfrac{x^2}{9}+\dfrac{y^2}{25}=1 \Rightarrow \dfrac{x^2}{\left(\frac{3}{2}\right)^2}+\dfrac{y^2}{\left(\frac{5}{2}\right)^2}=1$

$a=\dfrac{3}{2},\ b=\dfrac{5}{2}$

Intercepts: $\left(\pm\dfrac{3}{2},0\right),\ \left(0,\pm\dfrac{5}{2}\right)$

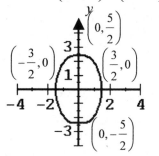

15. $121(x+1)^2+64(y-2)^2=7744$

$\dfrac{(x+1)^2}{64}+\dfrac{(y-2)^2}{121}=1$

$\dfrac{(x+1)^2}{8^2}+\dfrac{(y-2)^2}{11^2}=1$

$a=8,\ b=11,\ C(-1,2)$ (continued)

Intercepts: $(-9,2),(7,2),$

$(-1,12),(-1,-9)$

15. (continued)

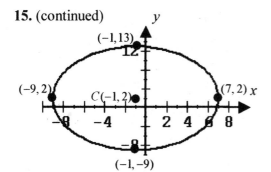

17. $\dfrac{(x-h)^2}{a^2} + \dfrac{(y-k)^2}{b^2} = 1$

$C(0,0) \Rightarrow (h,k) = (0,0)$

$x\text{-int}(9,0) \Rightarrow a = 9$

$y\text{-int}(0,-2) \Rightarrow b = 2$

$\dfrac{x^2}{9^2} + \dfrac{y^2}{2^2} = 1$

$\dfrac{x^2}{81} + \dfrac{y^2}{4} = 1$

19. $\dfrac{(x-h)^2}{a^2} + \dfrac{(y-k)^2}{b^2} = 1$

$C(0,0) \Rightarrow (h,k) = (0,0)$

$x\text{-int}(9,0) \Rightarrow a = 9$

$y\text{-int}(0,3\sqrt{2}) \Rightarrow b = 3\sqrt{2}$

$\dfrac{x^2}{9^2} + \dfrac{y^2}{(3\sqrt{2})^2} = 1$

$\dfrac{x^2}{81} + \dfrac{y^2}{18} = 1$

21. $\dfrac{(x-h)^2}{a^2} + \dfrac{(y-k)^2}{b^2} = 1$

$C(0,0) \Rightarrow (h,k) = (0,0)$

$x\text{-int}(30,0) \Rightarrow a = 30$

$y\text{-int}(0,18) \Rightarrow b = 18$

$\dfrac{x^2}{30^2} + \dfrac{y^2}{18^2} = 1$

$\dfrac{x^2}{900} + \dfrac{y^2}{324} = 1$

23. $\dfrac{(x-h)^2}{a^2} + \dfrac{(y-k)^2}{b^2} = 1$

$\dfrac{(x-5)^2}{9} + \dfrac{(y-2)^2}{1} = 1$

$\dfrac{(x-5)^2}{3^2} + \dfrac{(y-2)^2}{1^2} = 1$

$a = 3, \; b = 1$

$C(5,2)$

Vertices: $(2,2), \; (5,1)$

$\qquad\qquad (8,2), \; (5,3)$

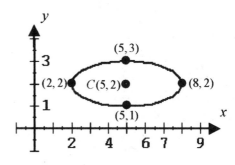

25. $\dfrac{(x-h)^2}{a^2} + \dfrac{(y-k)^2}{b^2} = 1$

$\dfrac{(x+2)^2}{49} + \dfrac{y^2}{25} = 1$

$\dfrac{(x+2)^2}{7^2} + \dfrac{y^2}{5^2} = 1$

$a = 7, \; b = 5$

$C(-2,0)$

Vertices: $(-9,0), \; (-2,5),$

$\qquad\qquad (-2,-5), \; (5,0)$

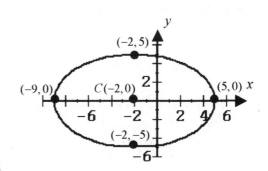

261

27. $\dfrac{(x-h)^2}{a^2}+\dfrac{(y-k)^2}{b^2}=1$

$\dfrac{(x+1)^2}{36}+\dfrac{(y+4)^2}{16}=1$

$\dfrac{(x+1)^2}{6^2}+\dfrac{(y+4)^2}{4^2}=1$

$a=6,\ 2=5$

$C(-1,-4)$

Vertices: $(-7,-4),\ (-1,0),$

$\qquad\qquad (5,-4),\ (-1,-8)$

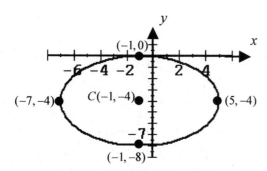

29. $\left(\dfrac{2+6}{2},\dfrac{3+3}{2}\right)=(4,3)$

$a=|6-4|=2$

$b=|7-3|=4$

$\dfrac{(x-4)^2}{2^2}+\dfrac{(y-3)^2}{4^2}=1$

$\dfrac{(x-4)^2}{4}+\dfrac{(y-3)^2}{16}=1$

31. $\dfrac{(-4+5)^2}{4}+\dfrac{(4+a)^2}{9}=1$

$9(1)^2+4(4+a)^2=36$

$4(4+a)^2=27$

$(4+a)^2=\dfrac{27}{4}$

$4+a=\pm\dfrac{\sqrt{27}}{4}$ (continued)

31. (continued)

$a=\pm\dfrac{\sqrt{9\cdot 3}}{4}=\pm\dfrac{3\sqrt{3}}{4}$

$a=\dfrac{-8\pm 3\sqrt{3}}{4}$

33. $\dfrac{(0-3.6)^2}{14.98}+\dfrac{(y-5.3)^2}{28.98}=1$

$(y-5.3)^2=28.98\left(1-\dfrac{12.96}{14.98}\right)$

$y=\pm\sqrt{28.98\left(1-\dfrac{12.967}{14.98}\right)}+5.3$

$y=\begin{cases}7.2768\\3.3232\end{cases}$

$\dfrac{(x-3.6)^2}{14.98}+\dfrac{(0-5.3)^2}{28.98}=1$

$(x-3.6)^2=14.98\left(1-\dfrac{28.09}{28.98}\right)$

$x=\pm\sqrt{14.98\left(1-\dfrac{28.09}{28.98}\right)}+3.6$

$x=\begin{cases}4.2783\\2.9217\end{cases}$

x-intercepts: $(4.2783,0),\ (2.9217,0)$

y-intercepts: $(0,7.2768),\ (0,3.3232)$

35. $A=\pi ab$

$A=3.1416\left(\dfrac{185}{2}\right)\left(\dfrac{154}{2}\right)$

$A=22{,}376.0$ square meters

Cumulative Review Problems

37. $\dfrac{5}{\sqrt{2x}-\sqrt{y}}\cdot\dfrac{\sqrt{2x}+\sqrt{y}}{\sqrt{2x}+\sqrt{y}}=\dfrac{5(\sqrt{2x}+\sqrt{y})}{2x-y}$

39. $\dfrac{1224 \text{ feet}}{1850 \text{ steps}} \cdot \dfrac{12 \text{ inches}}{\text{foot}} = 7.9 \dfrac{\text{inches}}{\text{step}}$

9.4 Exercises

1. $\dfrac{x^2}{a^2} - \dfrac{y^2}{b^2} = 1$

3. This is a horizontal hyperbola with vertices at $(4,0)$ and $(-4,0)$. Draw a fundamental rectangle with corners at $(4,2)$, $(4,-2)$, $(-4,2)$, $(-4,-2)$. Extend the diagonals through the rectangle as asymptotes of the hyperbola. Construct each branch of the hyperbola passing through the vertex and approaching the asymptotes.

5. $\dfrac{x^2}{4} - \dfrac{y^2}{25} = 1 \Rightarrow \dfrac{x^2}{2^2} - \dfrac{y^2}{5^2} = 1$
$a = 2,\ b = 5$

$V(\pm 2, 0),\ y_{\text{asymptote}} = \pm\dfrac{5}{2}x$

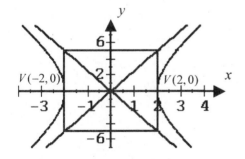

7. $\dfrac{y^2}{36} - \dfrac{x^2}{49} = 1 \Rightarrow \dfrac{y^2}{6^2} - \dfrac{x^2}{7^2} = 1$
$a = 7,\ b = 6$
$V(0, \pm 6)$

$y_{\text{asymptote}} = \pm\dfrac{6}{7}x$

(continued)

7. (continued)

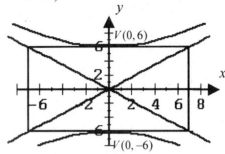

9. $4x^2 - y^2 = 64,\ \dfrac{x^2}{16} - \dfrac{y^2}{64} = 1$

$\dfrac{x^2}{4^2} - \dfrac{y^2}{8^2} = 1,\ a = 4,\ b = 8$

$V(\pm 4, 0),\ y_{\text{asymptote}} = \pm\dfrac{8}{4}x$

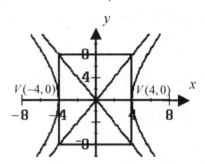

11. $8x^2 - y^2 = 16 \Rightarrow \dfrac{x^2}{2} - \dfrac{y^2}{16} = 1$

$\dfrac{x^2}{\sqrt{2}^2} - \dfrac{y^2}{4^2} = 1,\ a = \sqrt{2},\ b = 4$

$V(\pm\sqrt{2}, 0),\ y_{\text{asymptote}} = \pm\dfrac{4}{\sqrt{2}}x$

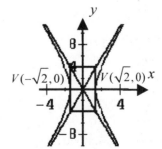

13. $2y^2 - 3x^2 = 54 \Rightarrow \dfrac{y^2}{27} - \dfrac{x^2}{18} = 1$

$\dfrac{y^2}{(3\sqrt{3})^2} - \dfrac{x^2}{(3\sqrt{2})^2} = 1$

$a = 3\sqrt{2}, \; b = 3\sqrt{3}$

$V(0, \pm 3\sqrt{3}), \; y_{\text{asymptote}} = \pm \dfrac{\sqrt{3}}{\sqrt{2}} x$

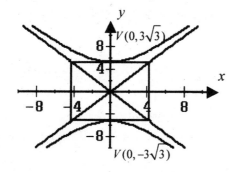

15. $V(\pm 3, 0) \Rightarrow a = 3, \; a^2 = 9$

$y_{\text{asymptote}} = \pm \dfrac{4}{3} x \Rightarrow b = 4, \; b^2 = 16$

$\dfrac{x^2}{9} - \dfrac{y^2}{16} = 1$

17. $V(0, \pm 7) \Rightarrow b = 7, \; b^2 = 49$

$y_{\text{asymptote}} = \pm \dfrac{7}{3} x \Rightarrow a = 3, \; a^2 = 9$

$\dfrac{y^2}{49} - \dfrac{x^2}{9} = 1$

19. $\dfrac{x^2}{a^2} - \dfrac{y^2}{b^2} = 1$, from graph, $a = 120$.

$y_{\text{asymptote}} = \dfrac{3}{1} x \Rightarrow \dfrac{b}{a} = \dfrac{3}{1}$

$\dfrac{b}{120} = \dfrac{3}{1} \Rightarrow b = 360$

$\dfrac{x^2}{120^2} - \dfrac{y^2}{360^2} = 1$

$\dfrac{x^2}{14,400} - \dfrac{y^2}{129,600} = 1$

21. $\dfrac{(x-6)^2}{25} - \dfrac{(y-4)^2}{49} = 1$

$\dfrac{(x-6)^2}{5^2} - \dfrac{(y-4)^2}{7^2} = 1$

$C(6, 4)$

$a = 5, \; b = 7$

$V(6 \pm 5, 4) = (11, 4), \; (1, 4)$

$y_{\text{asymptote}} = \pm \dfrac{7}{5}(x - 6) + 4$

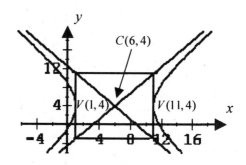

23. $\dfrac{(y+2)^2}{36} - \dfrac{(x+1)^2}{81} = 1$

$\dfrac{(y+2)^2}{6^2} - \dfrac{(x+1)^2}{9^2} = 1$

$C(-1, -2)$

$a = 6, \; b = 9$

$V(-1, -2 \pm 6) = (-1, 4), \; (-1, -8)$

$y_{\text{asymptote}} = \pm \dfrac{6}{9}(x - (-1)) + (-2)$

$y_{\text{asymptote}} = \pm \dfrac{2}{3}(x + 1) - 2$

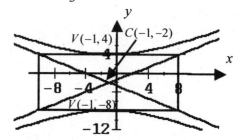

25. $\dfrac{(x+6)^2}{7} - \dfrac{y^2}{3} = 1$

$\dfrac{(x-(-6))^2}{\sqrt{7}^2} - \dfrac{(y-0)^2}{\sqrt{3}^2} = 1$

$C(-6, 0)$

$V(-6 \pm \sqrt{7}, 0)$

27. $C\left(5, \dfrac{0+14}{2}\right) = (5, 7)$

$y_{\text{asymptote}} = \dfrac{7}{5}x = \dfrac{b}{a}x \Rightarrow a = 5, \; b = 7$

$\dfrac{(y-7)^2}{7^2} - \dfrac{(x-5)^2}{5^2} = 1$

$\dfrac{(y-7)^2}{49} - \dfrac{(x-5)^2}{25} = 1$

29. $8x^2 - y^2 = 16\big|_{x=3.5}$

$8(3.5)^2 - y^2 = 16$

$y^2 = 8(3.5)^2 - 16$

$y = \pm\sqrt{8(3.5)^2 - 16} = \pm 9.055385138$

Cumulative Review Problems

31. $12x^2 + x - 6 = (4x+3)(3x-2)$

33. $\dfrac{3}{x^2 - 5x + 6} + \dfrac{2}{x^2 - 4}$

$= \dfrac{3(x+2)}{(x-3)(x-2)(x+2)} + \dfrac{2(x-3)}{(x-2)(x+2)(x-3)}$

$= \dfrac{3x+6+2x-6}{(x-3)(x-2)(x+2)}$

$= \dfrac{5x}{(x-3)(x-2)(x+2)}$

35. (a) $\dfrac{104{,}755 \text{ songs}}{365 \text{ days}} = 286 \dfrac{\text{songs}}{\text{day}}$

35. (b) $\dfrac{287 \text{ songs}}{\text{day}} \cdot \dfrac{4 \text{ min}}{\text{song}} = \dfrac{1148 \text{ min}}{\text{day}}$

$\dfrac{24 \text{ hr}}{\text{day}} \cdot \dfrac{60 \text{ min}}{\text{hr}} - \dfrac{1148 \text{ min}}{\text{day}} = \dfrac{292 \text{ min}}{\text{day}}$

(c) $r \cdot \dfrac{24 \text{ hr}}{\text{day}} \cdot \dfrac{60 \text{ min}}{\text{hr}} = \dfrac{287 \text{ songs}}{\text{day}} \cdot \dfrac{4 \text{ min}}{\text{song}}$

$r = 0.797\overline{2}$

Approximately 79.7% of airtime is music.

37. $\dfrac{(2.1 \times 10^9 + I) \text{ pencils}}{274 \times 10^6 \text{ people}} = \dfrac{10 \text{ pencils}}{\text{person}}$

$I = 10(274 \times 10^6) - 2.1 \times 10^9$

$I = 640{,}000{,}000$ pencils were imported

9.5 Exercises

1. $y^2 = 4x$

$y = x + 1$, substitute into first equation

$(x+1)^2 = 4x$

$x^2 - 2x + 1 = 0$

$(x-1)^2 = 0$

$x - 1 = 0$

$x = 1$

$y = x + 1 = 2$

$(1, 2)$ is the solution.

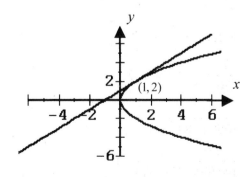

3. $y - 4x = 0 \Rightarrow y = 4x$

$4x^2 + y^2 = 20$

$4x^2 + (4x)^2 = 20$

$4x^2 + 16x^2 = 20x^2 = 20$

$x^2 = 1, \ x = \pm 1$

$y = 4x = 4(\pm 1) = \pm 4$

$(1, 4), \ (-1, -4)$ is the solution

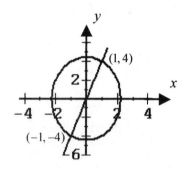

5. $\dfrac{x^2}{1} - \dfrac{y^2}{3} = 1$

$x + y = 1 \Rightarrow y = 1 - x$

$3x^2 - (1 - x)^2 = 3$

$3x^2 - 1 + 2x - x^2 = 3$

$2x^2 + 2x - 4 = 0$

$x^2 + x - 2 = 0$

$(x + 2)(x - 1) = 0$

$x + 2 = 0, \ x - 1 = 0$

$x = -2, \ x = 1$

$y = 1 - x = \begin{cases} 1 - (-2) = 3 \\ 1 - 1 = 0 \end{cases}$

$(-2, 3), \ (1, 0)$ is the solution

7. $x^2 + y^2 - 25 = 0$

$3y = x + 5 \Rightarrow x = 3y - 5$

$(3y - 5)^2 + y^2 - 25 = 0$

$9y^2 - 30y + 25 + y^2 - 25 = 0$

(continued)

7. (continued)

$10y^2 - 30y = 0$

$10y(y - 3) = 0$

$10y = 0, \ y - 3 = 0$

$y = 0, \ y = 3$

$x = 3y - 5 = \begin{cases} 3(0) - 5 = -5 \\ 3(3) - 5 = 4 \end{cases}$

$(-5, 0), \ (4, 3)$ is the solution

9. $x^2 + 2y^2 = 4$

$y = -x + 2$

$x^2 + 2(-x + 2)^2 = 4$

$x^2 + 2x^2 - 8x + 8 = 4$

$3x^2 - 8x + 4 = 0$

$(3x - 2)(x - 2) = 0$

$3x - 2 = 0, \ x - 2 = 2$

$x = \dfrac{2}{3}, \ x = 2$

$y = -x + 2 = \begin{cases} -\dfrac{2}{3} + 2 = \dfrac{4}{3} \\ -2 + 2 = 0 \end{cases}$

$\left(\dfrac{2}{3}, \dfrac{4}{3} \right), \ (2, 0)$ is the solution

11. $\dfrac{x^2}{4} - \dfrac{y^2}{4} = 1$

$x + y - 4 = 0 \Rightarrow x = 4 - y$

$(4 - y)^2 - y^2 = 4$

$16 - 8y + y^2 - y^2 = 4$

$8y = 12$

$y = \dfrac{3}{2}$

(continued)

11. (continued)

$$x = 4 - y = 4 - \frac{3}{2} = \frac{5}{2}$$

$\left(\frac{5}{2}, \frac{3}{2} \right)$ is the solution

13. $2x^2 - 5y^2 = -2 \xrightarrow{\times 2} 4x^2 - 10y^2 = -4$

$\quad 3x^2 + 2y^2 = 35 \xrightarrow{\times 5} 15x^2 + 10y^2 = 175$

$$\underline{\; 19x^2 = 171}$$

$x = \pm 3$

$$y^2 = \frac{35 - 3x^2}{2}$$

$$y = \pm \sqrt{\frac{35 - 3x^2}{2}} \Bigg|_{x = \pm 3} = \pm \sqrt{\frac{35 - 3(\pm 3)^2}{2}}$$

$y = \pm 2$

$(3, \pm 2), \; (-3, \pm 2)$ is the solution

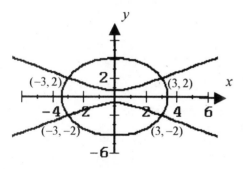

15. $2x^2 + 5y^2 = 42 \xrightarrow{\times -3} -6x^2 - 15y^2 = -126$

$\quad 3x^2 + 4y^2 = 35 \xrightarrow{\times 2} \quad 6x^2 + 8y^2 = 70$

$$\underline{ -7y^2 = -56}$$

$y^2 = 8$

$y = \pm 2\sqrt{x}$

$$x^2 = \frac{42 - 5y^2}{2} = \frac{42 - 5(8)}{2} = 1$$

(continued)

15. (continued)

$\quad x = \pm 1$

$\quad (1, \pm 2\sqrt{2}), (-1, \pm 2\sqrt{2})$ is the solution

17. $x^2 + 2y^2 = 8 \rightarrow x^2 + 2y^2 = 8$

$\quad x^2 - y^2 = 1 \xrightarrow{\times -1} -x^2 + y^2 = -1$

$$\underline{\; 3y^2 = 7}$$

$$y = \pm \frac{\sqrt{21}}{3}$$

$$x^2 = 1 + y^2 = 1 + \frac{7}{3} = \frac{10}{3}$$

$$x = \pm \frac{\sqrt{30}}{3}$$

$$\left(\frac{\sqrt{30}}{3}, \pm \frac{\sqrt{21}}{3} \right), \left(-\frac{\sqrt{30}}{3}, \pm \frac{\sqrt{21}}{3} \right)$$

is the solution

19. $x^2 + y^2 = 7 \rightarrow \quad x^2 + y^2 = 7$

$\quad \dfrac{x^2}{3} - \dfrac{y^2}{9} = 1 \rightarrow 3x^2 - y^2 = 9$

$$\underline{\phantom{\dfrac{x^2}{3} - \dfrac{y^2}{9} = 1 \rightarrow}\; 4x^2 = 16}$$

$x^2 = 4$

$x = \pm 2$

$y^2 = 7 - x^2 = 7 - 4 = 3$

$y = \pm \sqrt{3}$

$(2, \pm \sqrt{3}), \; (-2, \pm \sqrt{3})$ is the solution

21. $xy = 3 \Rightarrow x = \dfrac{3}{y}$

$3y = 3x + 6 = 3 \cdot \dfrac{3}{y} + 6$

$3y^2 - 6y - 9 = 0 \Rightarrow y^2 - 2y - 3 = 0$

(continued)

21. (continued)

$$(y-3)(y+1) = 0$$

$$y - 3 = 0, \ y + 1 = 0$$

$$y = 3, \ y = -1$$

$$x = \frac{3}{y} = \begin{cases} \dfrac{3}{3} = 1 \\ \dfrac{3}{-1} = -3 \end{cases}$$

$(1,3), \ (-3,-1)$ is the solution

23. $xy = 8$

$y = x + 2,$ substitute into first equation

$$x(x+2) = 8$$

$$x^2 + 2x - 8 = 0$$

$$(x+4)(x-2) = 0$$

$$x + 4 = 0, \ x - 2 = 0$$

$$x = -4, \ x = 2$$

$$y = x + 2 = \begin{cases} -4 + 2 = -2 \\ 2 + 2 = 4 \end{cases}$$

$(-4,-2), \ (2,4)$ is the solution

25. $x + y = 5 \Rightarrow x = 5 - y$

$$x^2 + y^2 = 4 \Rightarrow (5-y)^2 + y^2 = 4$$

$$25 - 10y + y^2 + y^2 = 4$$

$$2y^2 - 10y + 21 = 0$$

$$y = \frac{-(-10) \pm \sqrt{(-10)^2 - 4(2)(21)}}{2(2)}$$

$$y = \frac{10 \pm \sqrt{-68}}{4}$$

No real solution

27. $540 = xy \Rightarrow y = \dfrac{540}{x}$

$$39^2 = x^2 + y^2 = x^2 + \left(\frac{540}{x}\right)$$

$$1521 = x^2 + \frac{291,600}{x^2}$$

$$x^4 - 1521x^2 + 291,600 = 0$$

$$(x^2 - 1296)(x^2 - 225) = 0$$

$$x^2 = 1296, \ x^2 = 225$$

$$x = \sqrt{1296} = 36, \ x = \sqrt{225} = 15$$

$$y = \frac{540}{x} = \begin{cases} \dfrac{540}{36} = 15 \\ \dfrac{540}{15} = 36 \end{cases}$$

$(36,15), \ (15,36)$ is the solution

The dimensions of the rectangle are 15 meters by 36 meters.

29. $x^2 + y^2 = 16,000,000 \Rightarrow$

$$y^2 = 16,000,000 - x^2$$

$$25,000,000x^2 - 9,000,000y^2$$

$$= 2.25 \times 10^{14}$$

$$25x^2 - 9y^2 = 2.25 \times 10^8$$

$$25x^2 - 9(16,000,000 - x^2) = 2.25 \times 10^8$$

$$34x^2 = 369,000,000$$

$$x^2 = 10,852,941.18$$

$$x = 3294.380242$$

$$y^2 = 16,000,000 - 10,852,941.18$$

$$y^2 = 5,147,058.824$$

$$y = 2268.713032$$

The hyperbola intersects the circle when $(x, y) \approx (3290, 2270)$

Cumulative Review Problems

31. $\dfrac{6x^4 - 24x^3 - 30x^2}{3x^3 - 21x^2 + 30x} = \dfrac{6x^2(x^2 - 4x - 5)}{3x(x^2 - 7x + 10)}$

$= \dfrac{2x(x-5)(x+1)}{(x-5)(x-2)}$

$= \dfrac{2x(x+1)}{x-2}$

33. $d = 55(5) = v(11)$

$v = 25$ mph

Putting Your Skills to Work

1. Let station A be at $(0,0)$ in an xy coordinate system. The coordinates of station B are $(50,30)$. The epicenter is located on the intersection of the circles $x^2 + y^2 = 40^2$ and

$(x-50)^2 + (y-30)^2 = 20^2$.

Graphing the two circles shows two points of intersection in the first quadrant, $x, y \geq 0$. Solve as follows

$x^2 - 100x + 50^2 + y^2 - 60y + 30^2 = 20^2$

$40^2 - 100x + 50^2 - 60y + 30^2 = 20^2$

$-100x + 4600 - 60y = 0$

$-100x + 4600 - 60\sqrt{1600 - x^2} = 0$

$3\sqrt{1600 - x^2} = -5x + 230$

$9(1600 - x^2) = 25x^2 - 2300x + 52,900$

$34x^2 - 2300x + 38,500 = 0$, solving with the quadratic formula gives

$x = 30.40619117,\ x = 37.24086766$

$y = \sqrt{1600 - x^2} = \begin{cases} 25.98968139 \\ 14.5985539 \end{cases}$

The epicenter is approximately 30.4 miles east and 26.0 miles north of

1. (continued)
station A or approximately 37.2 miles east and 14.6 miles north of station A.

2. The epicenter lies on the intersection of the two circles, $x^2 + y^2 = 50^2$ and

$(x-60)^2 + (y-20)^2 = 30^2$. The points of intersection may be found algebraically or using a graphing calculator, which gives

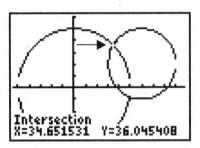

and

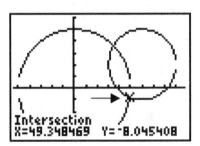

The epicenter is approximately 34.7 miles east and 36.0 miles north of station A or approximately 49.3 miles east and 8.0 miles south of station A.

3. The epicenter lies on the intersection of the three circles

$x^2 + y^2 = 30^2$

$(x-40)^2 + (y-30)^2 = 25^2$

$(x-30)^2 + (y+20)^2 = 27.55^2$.

From a graphing calculator,

(continued)

3. (continued)

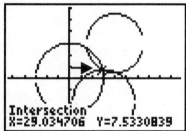

The epicenter is 29.0 miles east and 7.5 miles north of station A.

4. The epicenter lies on the intersection of the three circles

$$x^2 + y^2 = 50^2$$
$$(x-40)^2 + (y-30)^2 = 40^2$$
$$(x-10)^2 + (y-15)^2 = 45.92^2.$$

From a graphing calculator,

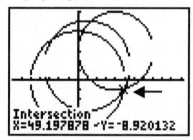

The epicenter is 49.2 miles east and 8.9 miles south of station A.

Chapter 9 Review Problems

1. $d = \sqrt{(7.5-10.5)^2 + (-4-(-6))^2} = \sqrt{13}$

2. $d = \sqrt{(-2-(-7))^2 + (-1-3)^2} = \sqrt{41}$

3. $(x-h)^2 + (y-k)^2 = r^2$
$(x-(-6))^2 + (y-3)^2 = \sqrt{15}^2$
$(x+6)^2 + (y-3)^2 = 15$

4. $(x-h)^2 + (y-k)^2 = r^2$
$(x-0)^2 + (y-(-7))^2 = 5^2$
$x^2 + (y+7)^2 = 25$

5. $x^2 + y^2 - 6x - 8y + 3 = 0$
$x^2 - 6x + 9 + y^2 - 8y + 16 = -3 + 9 + 16$
$(x-3)^2 + (y-4)^2 = 22 = \sqrt{22}^2$
$C(3,4), \ r = \sqrt{22}$

6. $x^2 + y^2 - 10x + 12y + 52 = 0$
$x^2 - 10x + 25 + y^2 + 12y + 36 = -52 + 25 + 36$
$(x-5)^2 + (y+6)^2 = 9 = 3^2, \ C(5,-6), \ r = 3$

7. $x = \dfrac{1}{3}y$

x	y
0	0
3	3
3	-3

$V(0,0)$
y-intercept: $(0,0)$

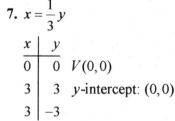

8. $x = \dfrac{1}{2}(y-2)^2 + 4$

x	y
4	2
6	0
6	4

$V(4,2)$
y-intercept: $(6,0)$

(continued)

8. (continued)

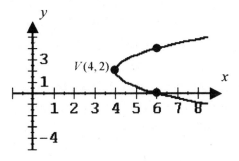

9. $y = -2(x+1)^2$

x	y	
-1	0	$V(-1,0)$
0	-2	y-intercept: $(0,-2)$
-2	-2	

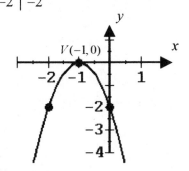

10. $x + 8y = y^2 + 10$

$x = y^2 - 8y + 16 - 6$

$x = (y-4)^2 - 6$

$V(-6,4)$

Opens to right.

11. $x^2 + 6x = y - 4$

$y = x^2 + 6x + 9 - 9 + 4$

$y = (x+3)^2 - 5$

$V(-3,-5)$

Opens upward.

12. $\dfrac{x^2}{\frac{1}{4}} + \dfrac{y^2}{1} = 1 \Rightarrow \dfrac{x^2}{\left(\frac{1}{2}\right)^2} + \dfrac{y^2}{1^2} = 1$

$a = \dfrac{1}{2}, \ b = 1$

$C(0,0)$

Vertices: $(0,1), \ (0,-1),$

$$\left(-\frac{1}{2}, 0\right), \ \left(\frac{1}{2}, 0\right)$$

13. $16x^2 + y^2 - 32 = 0 \Rightarrow \dfrac{x^2}{2} + \dfrac{y^2}{32} = 1$

$\dfrac{x^2}{\sqrt{2}^2} + \dfrac{y^2}{(4\sqrt{2})^2} = 1$

$a = \sqrt{2}, \ b = 4\sqrt{2}$

$C(0,0)$

Vertices: $(0, 4\sqrt{2}), \ (0, -4\sqrt{2})$

$(-\sqrt{2}, 0), \ (\sqrt{2}, 0)$

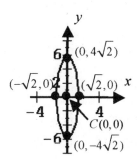

14. $\dfrac{(x+5)^2}{4} + \dfrac{(y+3)^2}{25} = 1$

$\dfrac{(x-(-5))^2}{2^2} + \dfrac{(y-(-3))^2}{5^2} = 1$

$C(-5,-3)$

$a = 2, \ b = 5$

Vertices: $(-3,-3), \ (-7,-3)$

$\qquad\qquad (-5,2), \ (-5,-8)$

15. $\dfrac{(x+1)^2}{9} + \dfrac{(y-2)^2}{16} = 1$

$\dfrac{(x-(-1))^2}{3^2} + \dfrac{(y-2)^2}{4^2} = 1$

$C(-1,2)$

$a = 3, \ b = 4$

Vertices: $(2,2), \ (-4,2)$

$\qquad\qquad (-1,6), \ (-1,-2)$

16. $x^2 - 4y^2 - 16 = 0 \Rightarrow \dfrac{x^2}{16} - \dfrac{y^2}{4} = 1$

$\dfrac{x^2}{4^2} - \dfrac{y^2}{2^2} = 1$

$C(0,0)$

$a = 4, \ b = 2$

Vertices: $(-4,0), \ (4,0)$

$y_{\text{asymptote}} = \pm\dfrac{2}{4}x = \pm\dfrac{1}{2}x$

17. $9y^2 - 25x^2 = 225 \Rightarrow \dfrac{y^2}{25} - \dfrac{x^2}{9} = 1$

$\dfrac{y^2}{5^2} - \dfrac{x^2}{3^2} = 1$

$a = 5, \ b = 3$

$C(0,0)$

Vertices: $(0,5), \ (0,-5)$

$y_{\text{asymptote}} = \pm\dfrac{5}{3}x$

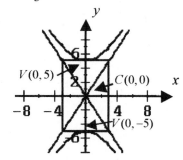

18. $\dfrac{(x-2)^2}{4} - \dfrac{(y+3)^2}{25} = 1$

$\dfrac{(x-2)^2}{2^2} - \dfrac{(y-(-3))^2}{5^2} = 1$

$C(2,-3)$

$a = 2, \ b = 5$

Vertices: $(0,-3), \ (4,-3)$

19. $9(y-2)^2 - (x+5)^2 - 9 = 0$

$\dfrac{(y-2)^2}{1^2} - \dfrac{(x-(-5))^2}{3^3} = 0$

$C(-5,2)$

$a = 3, \ b = 1$

Vertices: $(-5,3), \ (-5,1)$

20. $x^2 + y = 9 \rightarrow x^2 + y = 9$

$y - x = 3 \overset{x-1}{\rightarrow} \quad \underline{x - y = -3}$

$x^2 - x = 6$

(continued)

20. (continued)

$$x^2 - x - 6 = 0$$
$$(x+3)(x-2) = 0$$
$$x+3 = 0, \; x-2 = 0$$
$$x = -3, \; x = 2$$
$$y = x + 3 = \begin{cases} -3+3 = 0 \\ 2+3 = 5 \end{cases}$$

$(-3,0), \; (2,5)$ is the solution.

21. $y^2 + x^2 = 3$

$$x - 2y = 1 \Rightarrow x = 2y + 1$$
$$y^2 + (2y+1)^2 = 3$$
$$y^2 + 4y^2 + 4y + 1 = 3$$
$$5y^2 + 4y - 2 = 0$$

Solving with quadratic formula gives

$$y = \frac{-2 \pm \sqrt{14}}{5}$$

$$x = 2y+1 = \begin{cases} 2\left(\dfrac{-2+\sqrt{14}}{5} \right) + 1 \\[3mm] 2\left(\dfrac{-2-\sqrt{14}}{5} \right) + 1 \end{cases}$$

$$x = \begin{cases} \dfrac{1+2\sqrt{14}}{5} \\[3mm] \dfrac{1-2\sqrt{14}}{5} \end{cases}$$

$\left(\dfrac{1-2\sqrt{14}}{5}, \dfrac{-2-\sqrt{14}}{5} \right)$ and

$\left(\dfrac{1+2\sqrt{14}}{5}, \dfrac{-2+\sqrt{14}}{5} \right)$ is the solution.

22. $2x^2 + y^2 = 17 \Rightarrow y^2 = 17 - 2x^2$

$$x^2 + 2y^2 = 22$$
$$x^2 + 2(17 - 2x^2) = 22$$
$$x^2 + 34 - 4x^2 = 22$$
$$3x^2 = 12$$
$$x^2 = 4$$
$$x = \pm 2$$
$$y = \pm\sqrt{17 - 2x^2}$$
$$y = \pm\sqrt{17 - 2(4)}$$
$$y = \pm 3$$

$(2, \pm 3), \; (-2, \pm 3)$ is the solution.

23. $xy = -2 \Rightarrow y = \dfrac{-2}{x}$

$$x^2 + y^2 = 5 \Rightarrow x^2 + \left(\frac{-2}{x} \right)^2 = 5$$
$$x^4 + 4 = 5x^2$$
$$x^4 - 5x^2 + 4 = 0$$
$$(x^2 - 4)(x^2 - 1) = 0$$
$$x^2 - 4 = 0, \; x^2 - 1 = 0$$
$$x^2 = 4, \; x^2 = 1$$
$$x = \pm 2, \; x = \pm 1$$

$$y = \frac{-2}{x} = \begin{cases} \dfrac{-2}{2} = -1 \\[2mm] \dfrac{-2}{-2} = 1 \\[2mm] \dfrac{-2}{1} = -2 \\[2mm] \dfrac{-2}{-1} = 2 \end{cases}$$

$(2,-1), \; (-2,1), \; (1,-2), \; (-1,2)$ is the solution.

24. $3x^2 - 4y^2 = 12$

$7x^2 - y^2 = 8 \Rightarrow y^2 = 7x^2 - 8$

$3x^2 - 4(7x^2 - 8) = 12$

$3x^2 - 28x^2 + 32 = 12$

$25x^2 = 20 \Rightarrow x^2 = \dfrac{20}{25}$

$y^2 = 7x^2 - 8 = 7 \cdot \dfrac{20}{25} - 8 = -\dfrac{12}{5}$

No real solution.

25. $y = x^2 + 1 \Rightarrow x^2 = y - 1$

$x^2 + y^2 - 8y + 7 = 0$

$y - 1 + y^2 - 8y + 7 = 0$

$y^2 - 7y + 6 = 0$

$(y-1)(y-6) = 0$

$y - 1 = 0, \; y - 6 = 0$

$y = 1, \; y = 6$

$x^2 = y - 1 = \begin{cases} 1 - 1 = 0 \\ 6 - 1 = 5 \end{cases}$

$x = \begin{cases} 0 \\ \pm\sqrt{5} \end{cases}$

$(0,1), \; (\sqrt{5}, 6), \; (-\sqrt{5}, 6)$ is the solution.

26. $2x^2 + y^2 = 18$

$xy = 4 \Rightarrow y = \dfrac{4}{x}$

$2x^2 + \left(\dfrac{4}{x}\right)^2 = 18$

$2x^4 + 16 = 18x^2$

$x^4 - 9x^2 + 8 = 0$

$(x^2 - 8)(x^2 - 1) = 0$

$x^2 - 8 = 0, \; x^2 - 1 = 0$

(continued)

26. (continued)

$x^2 = 8, \; x^2 = 1$

$x = \pm 2\sqrt{2}, \; x = \pm 1$

$y = \dfrac{4}{x} = \begin{cases} \dfrac{4}{2\sqrt{2}} = \sqrt{2} \\[6pt] \dfrac{4}{-2\sqrt{2}} = -\sqrt{2} \\[6pt] \dfrac{4}{1} = 4 \\[6pt] \dfrac{4}{-1} = -4 \end{cases}$

$(2\sqrt{2}, \sqrt{2}), \; (-2\sqrt{2}, -\sqrt{2}),$

$(1,4), \; (-1,-4)$ is the solution.

27. $y^2 - 2x^2 = 2 \xrightarrow{x-2} -2y^2 + 4x^2 = -4$

$2y^2 - 3x^2 = 5 \rightarrow \underline{\quad 2y^2 - 3x^2 = 5 \quad}$

$x^2 = 1$

$x = \pm 1$

$y^2 = 2x^2 + 2 = 2(1) + 2 = 4$

$y = \pm 2$

$(1, \pm 2), \; (-1, \pm 2)$ is the solution.

28. $y^2 = \dfrac{1}{2}x$

$y = x - 1 \Rightarrow x = y + 1$

$y^2 = \dfrac{1}{2}(y + 1)$

$2y^2 - y - 1 = 0$

$(2y + 1)(y - 1) = 0$

$2y + 1 = 0, \; y - 1 = 0$

$y = -\dfrac{1}{2}, \; y = 1$

(continued)

28. (continued)

$$x = y + 1 = \begin{cases} -\dfrac{1}{2} + 1 = \dfrac{1}{2} \\ 1 + 1 = 2 \end{cases}$$

$\left(\dfrac{1}{2}, -\dfrac{1}{2} \right)$, $(2,1)$ is the solution.

29. $y^2 = 2x$

$y = \dfrac{1}{2}x + 1 \Rightarrow x = 2y - 2$

$y^2 = 2(2y - 2) = 4y - 4$

$y^2 - 4y + 4 = 0$

$(y - 2)^2 = 2$

$y = 2$

$x = 2y - 2 = 2(2) - 2 = 4 - 2 = 2$

$x = 2$

$(2, 2)$ is the solution.

30. $y^2 = 4px$

$5^2 = 4p(4)$

$p = \dfrac{25}{16} = 1.5625$

The receiver should be placed 1.56 feet from the center of the dish.

31. $y^2 = 4px$

$y^2 = 4(2)x = 8x$

$y^2 = 8x$

$2.5^2 = 8x$

$x = 0.78125$

The searchlight should be 0.78 feet deep.

Chapter 9 Test

1. $d = \sqrt{(-2 - (-6))^2 + (5 - (-8))^2}$

$d = \sqrt{185}$

2. $y^2 - 6y - x + 13 = 0$

$x - 13 + 9 = y^2 - 6y + 9$

$x = (y - 3)^2 + 4$

Parabola: $V(4, 3)$

x-int: $(13, 0)$

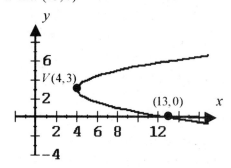

3. $x^2 + y^2 + 6x - 4y + 9 = 0$

$x^2 + 6x + 9 + y^2 - 4y + 4 = -9 + 9 + 4$

$(x + 3)^2 + 9(y - 2)^2 = 4 = 2^2$

Circle: $C(-3, 0)$, $r = 2$

x-int: $(-3, 0)$

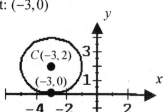

4. $\dfrac{x^2}{25} + \dfrac{y^2}{1} = 1 \Rightarrow \dfrac{x^2}{5^2} + \dfrac{y^2}{1^2} = 1$

Ellipse: $C(0, 0)$

$a = 5$, $b = 1$

Vertices: $(0, 5), (-5, 0), (0, 1), (0, -1)$

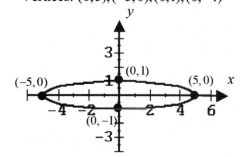

5. $\dfrac{x^2}{10} - \dfrac{y^2}{9} = 1$, hyperbola

$C(0,0)$, $V(\pm\sqrt{10}, 0)$

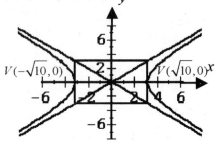

6. $y = -2(x+3)^2 + 4$

Parabola: $C(-3,4)$

y-intercept: $(0,-14)$

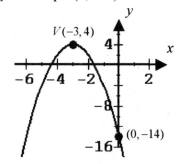

7. $\dfrac{(x+2)^2}{16} + \dfrac{(y-5)^2}{4} = 1$

$\dfrac{(x-(-2))^2}{4^2} + \dfrac{(y-5)^2}{2^2} = 1$

Ellipse: $C(-2,5)$

$a = 2, b = 2$

Vertices: $(-2,7), (-2,3), (-6,5), (2,5)$

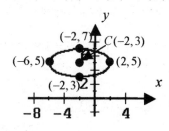

8. $7y^2 - 7x^2 = 28 \Rightarrow \dfrac{y^2}{2^2} - \dfrac{x^2}{2^2} = 1$

Hyperbola: $C(0,0)$, Vertices: $(0,2)$, $(0,-2)$

$a = 2$, $b = 2$, $y_{\text{asymptote}} = \pm\dfrac{2}{2}x = \pm x$

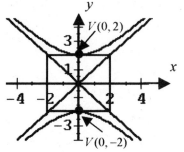

9. $(x-h)^2 + (y-k)^2 = r^2$

$(x-3)^2 + (y-(-5))^2 = \sqrt{8}^2$

$(x-3)^2 + (y+5)^2 = 8$

10. $a = 1$, $b = 3$, $(h,k) = (-4,-2)$

$\dfrac{(x-h)^2}{a^2} + \dfrac{(y-k)^2}{b^2} = 1$

$\dfrac{(x-(-4))^2}{1^2} + \dfrac{(y-(-2))^2}{3^2} = 1$

$\dfrac{(x+4)^2}{1} + \dfrac{(y+2)^2}{9} = 1$

11. $x = (y-k)^2 + h$, $(h,k) = (-7,3)$

$x = (y-3)^2 + (-7)$

$x = (y-3)^2 - 7$

12. $(h,k) = (6,7)$, $a = 3$, $b = 14 - 7 = 7$

$\dfrac{(y-k)^2}{b^2} - \dfrac{(x-h)^2}{a^2} = 1$

$\dfrac{(y-7)^2}{7^2} - \dfrac{(x-6)^2}{3^2} = 1$

$\dfrac{(y-7)^2}{49} - \dfrac{(x-6)^2}{9} = 1$

13. $-2x + y = 5 \Rightarrow y = 2x + 5$

$x^2 + y^2 - 25 = 0$

$x^2 + (2x + 5)^2 - 25 = 0$

$x^2 + 4x^2 + 20x + 25 - 25 = 0$

$5x^2 + 20x = 0$

$5x(x + 4) = 0$

$5x = 0, \; x + 4 = 0$

$x = 0, \; x = -4$

$y = 2x + 5 = 2(0) + 5 = 5$

$y = 2(-4) + 5 = -3$

$(0, 5), \; (-4, -3)$ is the solution.

14. $x^2 + y^2 = 9$

$y = x - 3$

$x^2 + (x - 3)^2 = 9$

$x^2 + x^2 - 6x + 9 = 9$

$x^2 - 3x = 0$

$x(x - 3) = 0$

$x = 0, \; x = 3$

$y = x - 3 = 0 - 3 = -3$

$y = x - 3 = 3 - 3 = 0$

$(0, -3), \; (3, 0)$ is the solution.

15. $4x^2 + y^2 - 4 = 0 \Rightarrow y^2 = 4 - 4x^2$

$9x^2 - 4y^2 - 9 = 0$

$9x^2 - 4(4 - 4x^2) - 9 = 0$

$9x^2 - 16 + 16x^2 - 9 = 0$

$25x^2 = 25$

$x^2 = 1$

$x = \pm 1$

$y = 4 - 4x^2 = 4 - 4(1) = 0$

$(1, 0), \; (-1, 0)$ is the solution.

16. $2x^2 + y^2 = 9$

$xy = 3 \Rightarrow y = -\dfrac{3}{x}$

$2x^2 + \left(-\dfrac{3}{x}\right)^2 = 9$

$2x^4 + 9 = 9x^2$

$2x^4 - 9x^2 + 9 = 0$

$(2x^2 - 3)(x^2 - 3) = 0$

$2x^2 - 3 = 0, \; x^2 - 3 = 0$

$x^2 = \dfrac{3}{2}, \; x^2 = 3$

$x = \pm\sqrt{\dfrac{3}{2} \cdot \dfrac{2}{2}} = \pm\dfrac{\sqrt{6}}{2}, \; x = \pm\sqrt{3}$

$y = -\dfrac{3}{x} = -\dfrac{3}{\frac{\sqrt{6}}{2}} = -\dfrac{6}{\sqrt{6}} \cdot \dfrac{\sqrt{6}}{\sqrt{6}} = -\sqrt{6}$

$y = -\dfrac{3}{x} = -\dfrac{3}{-\frac{\sqrt{6}}{2}} = \sqrt{6}$

$y = -\dfrac{3}{x} = -\dfrac{3}{\sqrt{3}} \cdot \dfrac{\sqrt{3}}{\sqrt{3}} = -\sqrt{3}$

$y = -\dfrac{3}{x} = -\dfrac{3}{-\sqrt{3}} \cdot \dfrac{\sqrt{3}}{\sqrt{3}} = \sqrt{3}$

$(\sqrt{3}, -\sqrt{3}), \; (-\sqrt{3}, \sqrt{3}),$

$\left(\dfrac{\sqrt{6}}{2}, -\sqrt{6}\right), \; \left(-\dfrac{\sqrt{6}}{2}, \sqrt{6}\right)$ is the solution.

Cumulative Test for Chapters 1-9

1. $5(-3) = -3(5)$ illustrates the commutative property of multiplication.

2. $2\{x - 3[x - 2(x + 1)]\}$

(continued)

2. (continued)

$$= 2\{x - 3[x - 2x - 2]\}$$

$$= 2\{x - 3[-x - 2]\}$$

$$= 2\{x + 3x + 6\}$$

$$= 2\{4x + 6\}$$

$$= 8x + 12$$

3. $3(4 - 6)^3 + \sqrt{25} = 3(-2)^3 + \sqrt{25}$

$$= 3(-8) + 5$$

$$= -24 + 5$$

$$= -19$$

4.　$A = 3bt + prt$

$$prt = A - 3bt$$

$$p = \frac{A - 3bt}{rt}$$

5. $x^3 + 125 = x^3 + 5^3 = (x + 5)(x^2 - 5x + 25)$

6. $\dfrac{3}{x - 4} + \dfrac{6}{x^2 - 16}$

$$= \frac{3(x + 4)}{(x - 4)(x + 4)} + \frac{6}{(x - 4)(x + 4)}$$

$$= \frac{3x + 12 + 6}{(x - 4)(x + 4)}$$

$$= \frac{3x + 18}{(x - 4)(x + 4)}$$

7.　$\dfrac{3}{2x + 3} = \dfrac{1}{2x - 3} + \dfrac{2}{4x^2 - 9}$

$$\frac{3}{(2x + 3)} = \frac{1}{2x - 3} + \frac{2}{(2x + 3)(2x - 3)}$$

$$3(2x - 3) = 2x + 3 + 2$$

$$6x - 9 = 2x + 5$$

$$4x = 14$$

$$x = \frac{7}{2}$$

8. $3x - 2y - 9z = 9$

$$x - y + z = 8$$

$$2x + 3y - z = -2$$

Switch the first and second equation.

$$x - y + z = 8$$

$$3x - 2y - 9z = 9$$

$$2x + 3y - z = -2$$

Multiply the first equation by -3 and add to the second equation and multiply the first equation by -2 and add to the third equation

$$x - y + z = 8$$

$$y - 12z = -15$$

$$5y - 3z = -18$$

Multiply the second equation by -5 and add to the third equation

$$x - y + z = 8$$

$$y - 12z = -15$$

$$57z = 57$$

$$z = 1$$

$$y = 12z - 15 = 12(1) - 15$$

$$y = -3$$

$$x - y + z = 8$$

$$x - (-3) + 1 = 8$$

$$x = 4$$

$(4, -3, 1)$ is the solution.

9. $(\sqrt{2} + \sqrt{3})(2\sqrt{6} - \sqrt{3})$

$$= 2\sqrt{12} - \sqrt{6} + 2\sqrt{18} - 3$$

$$= 2\sqrt{4 \cdot 3} + 2\sqrt{9 \cdot 2} - \sqrt{6} - 3$$

$$= 4\sqrt{3} + 6\sqrt{2} - \sqrt{6} - 3$$

10. $\sqrt{8x} + 3x\sqrt{50} - 4x\sqrt{32}$

$$= \sqrt{4 \cdot 2x} + 3x\sqrt{25 \cdot 2} - 4x\sqrt{16 \cdot 2}$$

$$= 2\sqrt{2x} + 15x\sqrt{2} - 16x\sqrt{2}$$

$$= 2\sqrt{2x} - x\sqrt{2}$$

11. $2x + (4x - 1) > 6 - x$

$2x + 4x - 1 > 6 - x$

$6x - 1 > 6 - x$

$7x > 7$

$x > 1$

12. $\dfrac{6(x-4)}{5} \geq \dfrac{3(x+2)}{4}$

$24(x - 4) \geq 15(x + 2)$

$24x - 96 \geq 15x + 30$

$9x \geq 126$

$x \geq 14$

13. $d = \sqrt{(x_2 - x_1)^2 + (y_2 - y_1)^2}$

$d = \sqrt{(-3 - 6)^2 + (-4 - (-1))^2} = \sqrt{90}$

$d = 3\sqrt{10}$

14. $y = -\dfrac{1}{2}(x + 2)^2 - 3$

Parabola: $C(-2, 3)$

y-intercept: $(0, -5)$

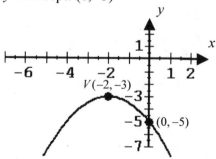

15. $25x^2 + 25y^2 = 125$

$x^2 + y^2 = 5 = \sqrt{5}^2$

Circle: $C(0,0)$, $r = \sqrt{5}$

15. (continued)

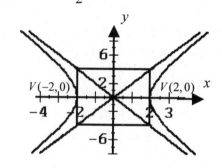

16. $16x^2 - 4y^2 = 64 \Rightarrow \dfrac{x^2}{4} - \dfrac{y^2}{16} = 1$

$\dfrac{x^2}{2^2} - \dfrac{y^2}{4^2} = 1$

Hyperbola: $C(0,0)$

$a = 2,\ b = 4$

Vertices: $(-2, 0),\ (2, 0)$

$y_{\text{asymptote}} = \pm\dfrac{4}{2}x = \pm 2x$

17. $\dfrac{(x-2)^2}{25} + \dfrac{(y-3)^2}{16} = 1$

$\dfrac{(x-2)^2}{5^2} + \dfrac{(y-3)^2}{4^2} = 1$

Ellipse: $C(2,3)$

$a = 5,\ b = 3$

Vertices: $(2,7), (-3,3), (2,-1), (7,3)$

(continued)

(continued)

17. (continued)

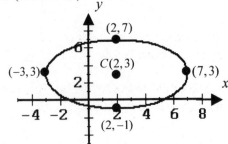

18.
$$y = 2x^2$$
$$y = 2x + 4$$
$$2x^2 = 2x + 4$$
$$x^2 - x - 2 = 0$$
$$(x-2)(x+1) = 0$$
$$x - 2 = 0, \; x + 1 = 0$$
$$x = 2, \; x = -1$$
$$y = 2x^2 = 2(2)^2 = 8$$
$$y = 2(-1)^2 = 2$$
$(2,8), \; (-1,2)$ is the solution.

19. $x^2 + 2y^2 = 16 \Rightarrow x^2 = 16 - 2y^2$
$$4x^2 - y^2 = 24$$
$$4(16 - 2y^2) - y^2 = 24$$
$$64 - 8y^2 - y^2 = 24$$
$$9y^2 = 40$$
$$y = \pm \frac{2\sqrt{10}}{3}$$
$$x^2 = 16 - 2y^2 = 16 - 2\left(\frac{40}{9}\right) = \frac{64}{9}$$
$$x = \pm \frac{8}{3}$$
$\left(\dfrac{8}{3}, \pm\dfrac{2\sqrt{10}}{3}\right), \; \left(-\dfrac{8}{3}, \pm\dfrac{2\sqrt{10}}{3}\right)$ is the solution.

20. $x^2 + y^2 = 25$
$$x - 2y = -5 \Rightarrow x = 2y - 5$$
$$(2y - 5)^2 + y^2 = 25$$
$$4y^2 - 20y + 25 + y^2 = 25$$
$$5y^2 - 20y = 0$$
$$5y(y - 4) = 0$$
$$5y = 0, \; y - 4 = 0$$
$$y = 0, \; y = 4$$
$$x = 2y - 5 = 2(0) - 5 = -5$$
$$x - 2y - 5 = 2(4) - 5 = 3$$
$(3,4), \; (-5,0)$ is the solution

21. $xy = -15 \Rightarrow y = -\dfrac{15}{x}$
$$4x + 3y = 3$$
$$4x + 3\left(-\frac{15}{x}\right) = 3$$
$$4x^2 - 45 = 3x$$
$$4x^2 - 3x - 45 = 0$$
$$(x + 3)(4x - 15) = 0$$
$$x + 3 = 0, \; 4x - 15 = 0$$
$$x = -3, \; x = \frac{15}{4}$$
$$y = -\frac{15}{x} = -\frac{15}{-3} = 5$$
$$y = -\frac{15}{x} = -\frac{15}{\frac{15}{4}} = -4$$
$(-3,5), \; \left(\dfrac{15}{4}, -4\right)$ is the solution

Chapter 10

1. $f(x) = 2x - 6$

(a) $f(-3) = 2(-3) - 6 = -12$

(b) $f(a) = 2a - 6$

(c) $f(2a) = 2(2a) - 6 = 4a - 6$

(d) $f(a + 2) = 2(a + 2) - 6 = 2a - 2$

2. $f(x) = 5x^2 + 2x - 3$

(a) $f(-2) = 5(-2)^2 + 2(-2) - 3 = 13$

(b) $f(a) = 5a^2 + 2a - 3$

(c) $f(a + 1) = 5(a + 1)^2 + 2(a + 1) - 3$

$$= 5(a^2 + 2a + 1) + 2a + 2 - 3$$

$$= 5a^2 + 10a + 5 + 2a - 1$$

$$= 5a^2 + 12a + 4$$

3. $f(x) = \dfrac{3x}{x + 2}$

(a) $f(a) + f(a - 2)$

$$= \frac{3a}{a + 2} + \frac{3(a - 2)}{a - 2 + 2} = \frac{3a^2 + 3(a + 2)(a - 2)}{a(a + 2)}$$

$$= \frac{3a^2 + 3a^2 - 12}{a(a + 2)} = \frac{6a^2 - 12}{a(a + 2)}$$

$$= \frac{6(a^2 - 2)}{a(a + 2)}$$

(b) $f(3a) - f(3) = \dfrac{3(3a)}{3a + 2} - \dfrac{3(3)}{3 + 2}$

$$= \frac{9a}{3a + 2} - \frac{9}{5}$$

$$= \frac{45a - 9(3a + 2)}{5(3a + 2)}$$

$$= \frac{45a - 27a - 18}{5(3a + 2)}$$

$$= \frac{18(a - 1)}{5(3a + 2)}$$

4. Graph passes vertical line test and therefore represents a function.

5. Graph does not pass vertical line test and hence does not represent a function.

6. $f(x) = |x|, \ s(x) = |x - 3|$

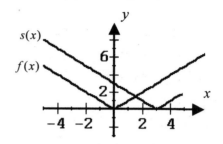

7. $f(x) = x^2, \ h(x) = (x + 2)^2 + 3$

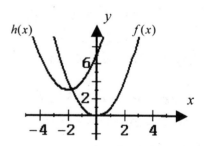

8. $f(x) = \dfrac{2}{x + 6}, \ g(x) = -3x + 1$

(a) $(fg)(x) = f(x)g(x)$

$$= \frac{2}{x + 6}(-3x + 1)$$

$$= \frac{-6x + 2}{x + 6}$$

(b) $(fg)(-4) = \dfrac{-6(-4) + 2}{-4 + 6} = 13$

(c) $f[g(x)] = f(-3x + 1)$

$$= \frac{2}{-3x + 1 + 6} = \frac{2}{-3x + 7}$$

9. $f(x) = 3x - 4$, $g(x) = -2x^2 - 6x + 3$

(a) $(f + g)(x) = f(x) + g(x)$

$$= 3x - 4 - 2x^3 - 6x + 3$$

$$= -2x^3 - 3x - 1$$

(b) $(f + g)(2) = -2(2)^3 - 3(2) - 1 = -23$

(c) $f[g(x)] = f(-2x^3 - 6x + 3)$

$$= 3(-2x^3 - 6x + 3) - 4$$

$$= -6x^3 - 18x + 9 - 4$$

$$= -6x^3 - 18x + 5$$

10. $f(x) = 6x^2 - 5x - 4$, $g(x) = 3x - 4$

(a) $\left(\dfrac{f}{g}\right)(x) = \dfrac{f(x)}{g(x)} = \dfrac{6x^2 - 5x - 4}{3x - 4}$

$$= \dfrac{(3x - 4)(2x + 1)}{(3x - 4)}$$

$$= 2x + 1, \ x \neq \dfrac{4}{3}$$

(b) $\left(\dfrac{f}{g}\right)(-1) = 2(-1) + 1 = -1$

(c) $(f \circ g)(x) = f[g(x)] = f(3x - 4)$

$$= 6(3x - 4)^2 - 5(3x - 4) - 4$$

$$= 6(9x^2 - 24x + 16) - 15x + 20 - 4$$

$$= 54x^2 - 144x + 96 + 16$$

$$= 54x^2 - 159x + 112$$

(d) $(g \circ f)(x) = g[f(x)]$

$$= g(6x^2 - 5x - 4)$$

$$= 3(6x^2 - 5x - 4) - 4$$

$$= 18x^2 - 15x - 12 - 4$$

$$= 18x^2 - 15x - 16$$

11. Graph passes vertical line test and horizontal line test. Graph represents a one-to-one function.

12. Graph passes vertical line test but does not pass horizontal line test. Graph represents a function but not a one-to-one function.

13. Since not two ordered pairs have the same second coordinate, A is a one-to-one function.

14. $F = \{(7,1), (6,3), (2,-1), (-1,5)\}$

      $F^{-1} = \{(1,7), (3,6), (-1,2), (5,-1)\}$

15. $g(x) = 3 - 5x$

   $y = 3 - 5x$

   $x = 3 - 5y$

   $y = \dfrac{3 - x}{5}$

   $g^{-1}(x) = \dfrac{3 - x}{5}$

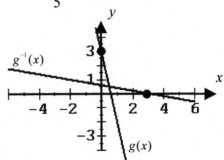

10.1 Exercises

1. $f(x) = 3x - 5$

$$f\left(-\dfrac{2}{3}\right) = 3\left(-\dfrac{2}{3}\right) - 5 = -7$$

3. $f(x) = 3x - 5$

$$f(a - 4) = 3(a - 4) - 5 = 3a - 12 - 5$$

$$= 3a - 17$$

5.

$$g(x) = \frac{1}{2}x - 3$$

$$g(4) + g(a) = \frac{1}{2}(4) - 3 + \frac{1}{2}a - 3$$

$$= \frac{1}{2}a - 4$$

7. $g(x) = \frac{1}{2}x - 3$

$$g(2a) = \frac{1}{2}(2a) - 3 = a - 3$$

9.

$$g(x) = \frac{1}{2}x - 3$$

$$g(2a - 4) = \frac{1}{2}(2a - 4) - 3 = a - 2 - 3$$

$$= a - 5$$

11.

$$g(x) = \frac{1}{2}x - 3$$

$$g(a^2) - g\left(\frac{2}{5}\right) = \frac{1}{2}a^2 - 3 - \left(\frac{1}{2} \cdot \frac{2}{5} - 3\right)$$

$$= \frac{1}{2}a^2 - 3 - \frac{1}{5} + 3$$

$$= \frac{1}{2}a^2 - \frac{1}{5}$$

13. $p(x) = 3x^2 + 4x - 2$

$$p(-2) = 3(-2)^2 + 4(-2) - 2 = 2$$

15. $p(x) = 3x^2 + 4x - 2$

$$p\left(\frac{1}{2}\right) = 3\left(\frac{1}{2}\right)^2 + 4\left(\frac{1}{2}\right) - 2 = \frac{3}{4}$$

17. $p(x) = 3x^2 + 4x - 2$

$$p(a + 1) = 3(a + 1)^2 + 4(a + 1) - 2$$

$$= 3a^2 + 6a + 3 + 4a + 4 - 2$$

$$= 3a^2 + 10a + 5$$

19. $p(x) = 3x^2 + 4x - 2$

$$p\left(-\frac{a^2}{2}\right) = 3\left(-\frac{a^2}{2}\right)^2 + 4\left(-\frac{a^2}{2}\right) - 2$$

$$= \frac{3a^4}{4} - 2a^2 - 2$$

21. $h(x) = \sqrt{x + 5}$

$$h(-1) = \sqrt{-1 + 5} = \sqrt{4} = 2$$

23. $h(x) = \sqrt{x + 5}$

$$h(3) = \sqrt{3 + 5} = \sqrt{8} = 2\sqrt{2}$$

25. $h(x) = \sqrt{x + 5}$

$$h(a^2 - 1) = \sqrt{a^2 - 1 + 5} = \sqrt{a^2 + 4}$$

27. $h(x) = \sqrt{x + 5}$

$$h(3a) = \sqrt{3a + 5}$$

29. $h(x) = \sqrt{x + 5}$

$$h(4a - 1) = \sqrt{4a - 1 + 5} = \sqrt{4a + 4}$$

$$= 2\sqrt{a + 1}$$

31. $h(x) = \sqrt{x + 5}$

$$h(b^2 + b) = \sqrt{b^2 + b + 5}$$

33. $r(x) = \dfrac{7}{x - 3}$

$$r(7) = \frac{7}{7 - 3}$$

$$r(7) = \frac{7}{4}$$

35. $r(x) = \dfrac{7}{x - 3}$

$$r(1.5) = \frac{7}{1.5 - 3} = -4.\overline{6}$$

37. $r(x) = \dfrac{7}{x-3}$

$r(a^2) = \dfrac{7}{a^2 - 3}$

39. $r(x) = \dfrac{7}{x-3}$

$r(a+2) = \dfrac{7}{a+2-3} = \dfrac{7}{a-1}$

41. $r(x) = \dfrac{7}{x-3}$

$r\left(\dfrac{1}{2}\right) + r(8) = \dfrac{7}{\dfrac{1}{2}-3} + \dfrac{7}{8-3}$

$= -\dfrac{7}{5}$

43. $f(x) = 5 - 2x$

$\dfrac{f(x+h) - f(x)}{h}$

$= \dfrac{5 - 2(x+h) - (5-2x)}{h}$

$= \dfrac{5 - 2x - 2h - 5 + 2x}{h}$

$= -2$

45. $f(x) = 2x^2$

$\dfrac{f(x+h) - f(x)}{h}$

$= \dfrac{2(x+h)^2 - 2x^2}{h}$

$= \dfrac{2x^2 + 4xh + 2h^2 - 2x^2}{h}$

$= \dfrac{2h(2x+h)}{h}$

$= 2(2x + 2h)$

$= 4x + 2h$

47. $P = 2.5w^2$

  (a) $P(w) = 2.5w^2$

  (b) $P(20) = 2.5(20)^2 = 1000$ kilowatts

  (c) $P = 2.5(20 + e)^2 = 2.5(400 + 40e + e^2)$

     $P(e) = 2.5e^2 + 100e + 1000$

  (d) $P(2) = 2.5(2)^2 + 100(2) + 1000$

        $= 1210$ kilowatts

49. The function values associated with $p(x) - 13$ would be the function values of $p(x)$ decreased by 13.

$p(3) - 13 \approx 39 - 13 = 26$

51. $f(x) = 3x^2 - 4.6x + 1.23$

$f(3.56a) = 3(3.56a)^2 - 4.6(3.56a) + 1.23$

$= 38.021a^2 - 16.376a + 1.23$

53. $f(x) = 3x^2 - 4.6x + 1.23$

$f(a - 0.152)$

$= 3(a - 0.152)^2 - 4.6(a - 0.152) + 1.23$

$= 3a^2 - 0.912a + 0.069312 - 4.6a + 0.6992 + 1.23$

$= 3a^2 - 5.512a + 1.999$

55. $A(x) = \left(\dfrac{x}{4}\right)^2 + \left(\dfrac{20-x}{4}\right)^2$

$= \dfrac{x^2 + 400 - 40x + x^2}{16}$

$= \dfrac{2x^2 - 40x + 400}{16}$

$A(2) = \dfrac{2(2)^2 - 40(2) + 400}{16} = 20.5$

$A(5) = \dfrac{2(5)^2 - 40(5) + 400}{16} = 15.625$

$A(8) = \dfrac{2(8)^2 - 40(8) + 400}{16} = 13$

Cumulative Review Problems

57. $\dfrac{7}{6} + \dfrac{5}{x} = \dfrac{3}{2x}$

$7x + 30 = 9$

$7x = -21$

$x = -3$

59. $\dfrac{V_{\text{Earth}}}{V_{\text{Mercury}}} = \dfrac{\dfrac{4}{3}\pi\left(\dfrac{7927}{2}\right)^3}{\dfrac{4}{3}\pi\left(\dfrac{3031}{2}\right)^3}$

$V_{\text{Earth}} = 17.88828747...(V_{\text{Mercury}})$

The volume of the Earth is approximately 17.9 times greater than the volume of Mercury.

10.2 Exercises

1. No, $f(x+2)$ means substitute $x+2$ for x in the function, $f(x)$. $f(x)+f(2)$ means evaluate $f(x)$ and $f(2)$ and then add the two results. One example is
$f(x) = 2x + 1$

$f(x+2) = 2(x+2)+1 = 2x+5$

$f(x)+f(2) = 2x+1+2(2)+1 = 2x+6$

3. To obtain the graph of $f(x)+k$, shift the graph of $f(x)$ <u>up</u> k units.

5. Graph fails vertical line test and does not represent a function.

7. Graph passes vertical line test and does represent a function.

9. Graph passes vertical line test and does represent a function.

11. Graph fails vertical line test and does not represent a function.

13. Graph fails vertical line test and does not represent a function.

For Exercises 15, 17, and 19:

x	$f(x) = x^2$
-2	4
-1	1
0	0
1	1
2	4

15. $f(x) = x^2$, $h(x) = x^2 + 4$
Shift $f(x)$ up 4 units.

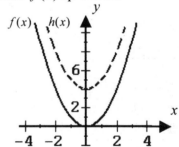

17. $f(x) = x^2$, $p(x) = (x-2)^2 + 1$
Shift $f(x)$ right 2 units and up 1 unit.

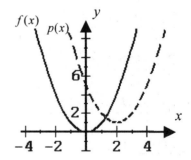

19. $f(x) = x^2$, $g(x) = (x-2)^2 + 1$

Shift $f(x)$ right 2 and up 1 units.

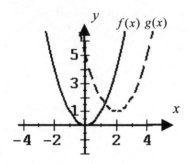

For Exercises 21, 23, and 25:

x	$f(x) = \lvert x \rvert$
–2	2
–1	1
0	0
1	1
2	2

21. $f(x) = \lvert x \rvert$, $r(x) = \lvert x \rvert - 1$

Shift $f(x)$ down 1 unit.

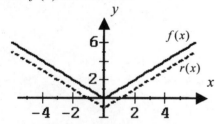

23. $f(x) = \lvert x \rvert$, $s(x) = \lvert x + 4 \rvert$

Shift $f(x)$ left 4 units.

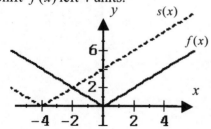

25. $f(x) = \lvert x \rvert$, $t(x) = \lvert x - 3 \rvert - 4$

Shift right 3 and down 4 units.

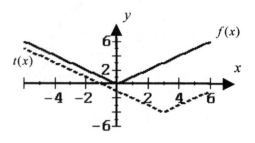

27. $f(x) = x^3$, $j(x) = (x-3)^3 + 3$

Shift $f(x)$ right 3 and up 3 units.

x	$f(x) = x^3$
–2	–8
–1	–1
0	0
1	1
2	8

29. $f(x) = \dfrac{2}{x}$, $g(x) = \dfrac{2}{x} + 3$

Shift $f(x)$ up 3 units.

x	$f(x) = \dfrac{2}{x}$
–2	–1
–1	–2
0	undefined
1	2
2	1

(continued)

29. (continued)

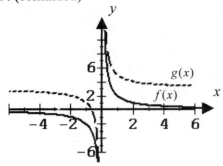

31. $f(x) = x^4, \; f(x) = (x - 3.2)^4 - 2.6$

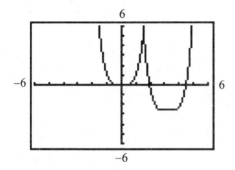

Cumulative Review Problems

33. $\sqrt{12} + 3\sqrt{50} - 4\sqrt{27}$

$= \sqrt{4 \cdot 3} + 3\sqrt{25 \cdot 2} - 4\sqrt{9 \cdot 3}$

$= 2\sqrt{3} + 15\sqrt{2} - 12\sqrt{3}$

$= 15\sqrt{2} - 10\sqrt{3}$

35. $\dfrac{2\sqrt{3} + \sqrt{5}}{\sqrt{3} - 2\sqrt{5}} \cdot \dfrac{\sqrt{3} + 2\sqrt{5}}{\sqrt{3} + 2\sqrt{5}}$

$= \dfrac{6 + 4\sqrt{15} + \sqrt{15} + 10}{3 - 20}$

$= -\dfrac{16 + 5\sqrt{15}}{17}$

37. $\dfrac{\$13,623,120}{28,560} - 250 = \$227$

They overcharged each student $227.

10.3 Exercises

1. $f(x) = -2x + 3, \; g(x) = 2 + 4x$

(a) $(f + g)(x) = f(x) + g(x)$

$= -2x + 3 + 2 + 4x$

$= 2x + 5$

(b) $(f - g)(x) = f(x) - g(x)$

$= -2x + 3 - (2 + 4x)$

$= -2x + 3 - 2 - 4x$

$= -6x + 1$

(c) $(f + g)(2) = 2(2) + 5 = 9$

(d) $(f - g)(-1) = -6(-1) + 1 = 7$

3. $f(x) = x^3 - 4x + 5, \; g(x) = 2x - 1$

(a) $(f + g)(x) = f(x) + g(x)$

$= 2x^2 - 4x + 5 + 2x - 1$

$= 2x^2 - 2x + 4$

(b) $(f - g)(x) = f(x) - g(x)$

$= 2x^2 - 4x + 5 - (2x - 1)$

$= 2x^2 - 6x + 6$

(c) $(f + g)(2) = 2(2)^2 - 2(2) + 4 = 8$

(d) $(f - g)(-1) = 2(-1)^2 - 6(-1) + 6 = 14$

5. $f(x) = x^3 - \dfrac{1}{2}x^2 + x, \; g(x) = x^2 - \dfrac{x}{4} - 5$

(a) $(f + g)(x) = f(x) + g(x)$

$= x^3 - \dfrac{1}{2}x^2 + x + x^2 - \dfrac{x}{4} - 5$

$= x^3 + \dfrac{1}{2}x^2 + \dfrac{3x}{4} - 5$

(b)

(continued)

287

5. (continued)

(b) $(f-g)(x) = f(x) - g(x)$

$$= x^3 - \frac{1}{2}x^2 + x - (x^2 - \frac{x}{4} - 5)$$

$$= x^3 - \frac{1}{2}x^2 + x - x^2 + \frac{x}{4} + 5$$

$$= x^3 - \frac{3}{2}x^2 + \frac{5x}{4} + 5$$

(c) $(f+g)(2) = (2)^3 + \frac{1}{2}(2)^2 + \frac{3(2)}{4} - 5$

$$= \frac{13}{2}$$

(d) $(f-g)(-1) = (-1)^3 - \frac{3}{2}(-1)^2 + \frac{5(-1)}{4} + 5$

$$= \frac{5}{4}$$

7. $f(x) = 3\sqrt{3-x}$, $g(x) = -5\sqrt{3-x}$

(a) $f(x) + g(x) = f(x) + g(x)$

$$= 3\sqrt{3-x} + (-5\sqrt{3-x})$$

$$= -2\sqrt{3-x}$$

(b) $f(x) - g(x) = 3\sqrt{3-x} - (-5\sqrt{3-x})$

$$= 8\sqrt{3-x}$$

(c) $(f+g)(2) = -2\sqrt{3-1} = -2$

(d) $(f-g)(-1) = 8\sqrt{3-(-1)} = 16$

9. $f(x) = x^2 - 3x + 2$, $g(x) = 1 - x$

(a) $(fg)(x) = f(x)g(x)$

$= (x^2 - 3x + 2)(1-x)$

$= x^2 - 3x + 2 - x^3 + 3x^2 - 2x$

$= -x^3 + 4x^2 - 5x + 2$

(b) $(fg)(-3)$

$= -(-3)^3 + 4(-3)^2 - 5(-3) + 2$

$= 80$

11. $f(x) = \frac{2}{x^2}$, $g(x) = x^2 - x$

(a) $(fg)(x) = f(x)g(x) = \frac{2}{x^2}(x^2 - x)$

$$= 2 - \frac{2}{x} = \frac{2(x-1)}{x}$$

(b) $(fg)(-3) = \frac{2(-3-1)}{-3} = \frac{8}{3}$

13. $f(x) = \sqrt{-2x+1}$, $g(x) = -3x$

(a) $(fg)(x) = f(x)g(x) = \sqrt{-2x+1}(-3x)$

$$= -3x\sqrt{-2x+1}$$

(b) $(fg)(-3) = -3(-3)\sqrt{-2(-3)+1} = 9\sqrt{7}$

15. $f(x) = 3x$, $g(x) = 4x - 1$

(a) $\left(\frac{f}{g}\right)(x) = \frac{f(x)}{g(x)} = \frac{3x}{4x-1}$, $x \neq \frac{1}{4}$

(b) $\left(\frac{f}{g}\right)(2) = \frac{3(2)}{4(2)-1} = \frac{6}{7}$

17. $f(x) = x^2 - 1$, $g(x) = x - 1$

(a) $\left(\frac{f}{g}\right)(x) = \frac{f(x)}{g(x)} = \frac{x^2-1}{x-1}$

$$= \frac{(x-1)(x+1)}{x-1}$$

$$= x+1, \ x \neq 1$$

(b) $\left(\frac{f}{g}\right)(2) = 2 + 1 = 3$

19. $f(x) = x^2 + 10x + 25$, $g(x) = x + 5$

(a) $\left(\frac{f}{g}\right)(x) = \frac{f(x)}{g(x)} = \frac{x^2+10x+25}{x+5}$

$$= \frac{(x+5)(x+5)}{(x+5)}$$

$$= x+5, \ x \neq -5$$

19. (b) $\left(\dfrac{f}{g}\right)(2) = 2 + 5 = 7$

21. $f(x) = 4x - 1$, $g(x) = 4x^2 + 7x - 2$

(a) $\left(\dfrac{f}{g}\right)(x) = \dfrac{f(x)}{g(x)} = \dfrac{4x - 1}{4x^2 + 7x - 2}$

$\qquad = \dfrac{(4x - 1)}{(4x - 1)(x + 2)}$

$\qquad = \dfrac{1}{x + 2}, \; x \neq -2, \dfrac{1}{4}$

(b) $\left(\dfrac{f}{g}\right)(2) = \dfrac{1}{2 + 2} = \dfrac{1}{4}$

23. $f(x) = 3x + 2$, $g(x) = x^2 - 2x$
$(f - g)(x) = f(x) - g(x)$
$\qquad = 3x + 2 - (x^2 - 2x)$
$\qquad = -x^2 + 5x + 2$

25. $f(x) = 3x + 2$, $g(x) = x^2 - 2x$
$(fg)(x) = f(x)g(x)$
$\qquad = (3x + 2)(x^2 - 2x)$
$\qquad = 3x^3 - 6x^2 + 2x^2 - 4x$
$\qquad = 3x^3 - 4x^2 - 4x$

27. $(fg)(x) = 3x^3 - 4x^2 - 4x$
$(fg)(-1) = 3(-1)^3 - 4(-1)^2 - 4(-1) = -3$

29. $f(x) = 3x + 2$, $h(x) = \dfrac{x - 2}{3}$

$\left(\dfrac{f}{h}\right)(x) = \dfrac{f(x)}{h(x)}$

$\qquad = \dfrac{3x + 2}{\dfrac{x - 2}{3}}$

$\qquad = \dfrac{9x + 6}{x - 2}, \; x \neq 2$

31. $f(x) = 2 - 3x$, $g(x) = 2x + 5$
$f[g(x)] = f[2x + 5]$
$\qquad = 2 - 3(2x + 5)$
$\qquad = 2 - 6x - 15$
$\qquad = -6x - 13$

33. $f(x) = 3x^2$, $g(x) = x - 4$
$f[g(x)] = f[x - 4]$
$\qquad = 3(x - 4)^2$
$\qquad = 3(x^2 - 8x + 16)$
$\qquad = 3x^2 - 24x + 48$

35. $f(x) = 4 - 3x$, $g(x) = 2x^2 - 1$
$f[g(x)] = f[2x^2 - 1]$
$\qquad = 4 - 3(2x^2 - 1)$
$\qquad = 4 - 6x^2 + 3$
$\qquad = -6x^2 + 7$

37. $f(x) = \dfrac{3}{x + 1}$, $g(x) = 2x - 1$
$f[g(x)] = f(2x - 1)$

$\qquad = \dfrac{3}{2x - 1 + 1}$

$\qquad = \dfrac{3}{2x}, \; x \neq 0$

39. $f(x) = |x + 3|$, $g(x) = 2x - 1$
$f[g(x)] = f[2x - 1]$
$\qquad = |2x - 1 + 3|$
$\qquad = |2x + 2| \text{ or }$
$\qquad = |2(x + 1)|$
$\qquad = |2||x + 1|$
$\qquad = 2|x + 1|$

41. $f(x) = x^2 + 2$, $g(x) = 3x + 5$

$$f[g(x)] = f(3x + 5)$$
$$= (3x + 5)^2 + 2$$
$$= 9x^2 + 30x + 25 + 2$$
$$= 9x^2 + 30x + 27$$

43. $f(x) = x^2 + 2$, $g(x) = 3x + 5$

$$g[f(x)] = g\left[x^2 + 2\right]$$
$$= 3(x^2 + 2) + 5$$
$$= 3x^2 + 6 + 5$$
$$= 3x^2 + 11$$

45. From Exercise 43,

$$g[f(x)] = 3x^2 + 11$$
$$g[f(3)] = 3(3)^2 + 11$$
$$= 38$$

47. $p(x) = \sqrt{x - 1}$, $f(x) = x^2 + 2$

$$(p \circ f)(x) = p[f(x)]$$
$$= p\left[x^2 + 2\right]$$
$$= \sqrt{x^2 + 2 - 1}$$
$$= \sqrt{x^2 + 1}$$

49. $g(x) = 3x + 5$, $h(x) = \dfrac{1}{x}$

$$(g \circ h)(\sqrt{2}) = g\left[h(\sqrt{2})\right]$$
$$= g\left[\frac{1}{\sqrt{2}}\right]$$
$$= 3 \cdot \frac{1}{\sqrt{2}} \cdot \frac{\sqrt{2}}{\sqrt{2}} + 5$$
$$= \frac{3\sqrt{2}}{2} + 5$$

51. $p(x) = \sqrt{x - 1}$, $f(x) = x^2 + 2$

$$(p \circ f)(-5) = p[f(-5)]$$
$$= p\left[(-5)^2 + 2\right]$$
$$= p(27)$$
$$= \sqrt{27 - 1}$$
$$= \sqrt{26}$$

53. $K[C(F)] = K\left[\dfrac{5F - 160}{9}\right]$

$$= \frac{5F - 160}{9} + 273$$
$$= \frac{5F - 160 + 9(273)}{9}$$
$$= \frac{5F + 2297}{9}$$

55. $\quad v[r(h)] = v[3.5h]$

$$= 31.4(3.5h)^2$$
$$= 384.65h^2$$
$$384.65h^2\big|_{h=8} = 24{,}617.6 \text{ ft}^3$$

Cumulative Review Problems

57. $25x^4 - 1 = (5x^2)^2 - 1 = (5x^2 - 1)(5x^2 + 1)$

59. $3x^2 - 7x + 2 = (3x - 1)(x - 2)$

61. $x = $ number of 60 sec commercials

$$60x + 30(20 - x) = 14(60)$$
$$60x + 600 - 30x = 840$$
$$30x = 240$$
$$x = 8$$
$$20 - x = 12$$

She should play eight commercials that are 60 sec long and twelve that are 30 sec long.

10.4 Exercises

1. A one-to-one function is a function in which no ordered pairs <u>have the same second coordinate</u>.

3. The graphs of a function f and its inverse f^{-1} are symmetric about the line <u>$y = x$</u>.

5. $B = \{(0,1),(1,0),(10,0)\}$ is not one-to-one since two ordered pairs, $(1,0)$ and $(10,0)$, have the same second coordinate.

7. $F = \left\{\left(\dfrac{2}{3},2\right),\left(3,-\dfrac{4}{5}\right),\left(-\dfrac{2}{3},-2\right),\left(-3,\dfrac{4}{5}\right)\right\}$

   is a one-to-one function since no two ordered pairs have the same second coordinate.

9. $E = \left\{(1,3),\left(\dfrac{1}{2},-5\right),(-1,-3),\left(-5,\dfrac{1}{2}\right)\right\}$

   is a one-to-one function since no two ordered pairs have the same second coordinate.

11. Graph of function passes horizontal line test and therefore, function is one-to-one.

13. Graph of function does not pass horizontal line test and therefore, function is not one-to-one.

15. Graph of function passes horizontal line test and therefore, function is one-to-one.

17. Yes, it passes the vertical line test. No, it does not pass the horizontal line test.

19. $J = \{(8,2),(1,1),(0,0),(-8,-2)\}$
    $J^{-1} = \{(2,8),(1,1),(0,0),(-2,-8)\}$

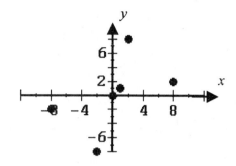

21. $f(x) = 4x - 5,\ f(x) \to y$
    $$y = 4x - 5,\ x \leftrightarrow y$$
    $$x = 4y - 5$$
    $$4y = x + 5$$
    $$y = \frac{x+5}{4}$$
    $$f^{-1}(x) = \frac{x+5}{4}$$

23. $f(x) = x^3 - 2,\ f(x) \to y$
    $$y = x^3 - 2,\ x \leftrightarrow y$$
    $$x = y^3 - 2$$
    $$y^3 = x + 2$$
    $$y = \sqrt[3]{x+2}$$
    $$f^{-1}(x) = \sqrt[3]{x+2}$$

25. $f(x) = -\dfrac{4}{x},\ f(x) \to y$
    $$y = -\frac{4}{x},\ x \leftrightarrow y$$
    $$y = -\frac{4}{x}$$
    $$f^{-1}(x) = -\frac{4}{x}$$

27. $f(x) = -\dfrac{3}{x-2}, \ f(x) \to y$

$y = -\dfrac{3}{x-2}, \ x \leftrightarrow y$

$x = -\dfrac{3}{y-2}$

$y - 2 = -\dfrac{3}{x}$

$y = 2 - \dfrac{3}{x}$

$f^{-1}(x) = 2 - \dfrac{3}{x}$

29. No. $f(x) = 2x^2 + 3$ is vertical parabola and fails the horizontal line test; it is not one-to-one and therefore, does not have an inverse.

31. $g(x) = 2x + 5, \ g(x) \to y$

$y = 2x + 5, \ x \leftrightarrow y$

$x = 2y + 5$

$2y = x - 5$

$y = \dfrac{x-5}{2}$

$g^{-1}(x) = \dfrac{x-5}{2}$

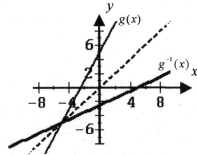

$g(x)$ and $g^{-1}(x)$ are symmetrical about the line $y = x$.

33. $h(x) = \dfrac{1}{2}x - 2, \ h(x) \to y$

$y = \dfrac{1}{2}x - 2, \ x \leftrightarrow y$

$x = \dfrac{1}{2}y - 2$

$2x = y - 4$

$y = 2x + 4$

$f^{-1}(x) = 2x + 4$

$h(x)$ and $h^{-1}(x)$ are symmetrical about the line $y = x$.

35. $k(x) = 3 - 2x, \ k(x) \to y$

$y = 3 - 2x, \ x \leftrightarrow y$

$x = 3 - 2y$

$2y = 3 - x$

$y = \dfrac{3-x}{2}$

$k^{-1}(x) = \dfrac{3-x}{2}$

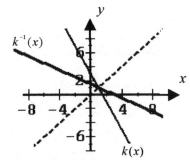

37. $f(x) = 0.0063x - 5, \ f(x) \rightarrow y$

$y = 0.0063x - 5, \ x \leftrightarrow y$

$x = 0.0063y - 5$

$0.0063y = x + 5$

$y = \dfrac{x+5}{0.0063}$

$f^{-1}(x) = \dfrac{x+5}{0.0063}$

The inverse function tells how many Spanish pesetas are given by the bank for x dollars. No, because of the bank fee the inverse function will not work for Manuela's transaction.

39. $f\left[f^{-1}(x)\right] = f\left(\dfrac{1}{2}x - \dfrac{3}{4}\right)$

$= 2\left(\dfrac{1}{2}x - \dfrac{3}{4}\right) + \dfrac{3}{2}$

$= x - \dfrac{3}{2} + \dfrac{3}{2}$

$= x$

$f^{-1}\left[f(x)\right] = f^{-1}\left(2x + \dfrac{3}{2}\right)$

$= \dfrac{1}{2}\left(2x + \dfrac{3}{2}\right) - \dfrac{3}{4}$

$= x + \dfrac{3}{4} - \dfrac{3}{4}$

$= x$

Cumulative Review Problems

41. $x^{2/3} + 7x^{1/3} + 12 = 0$

$(x^{1/3} + 4)(x^{1/3} + 3) = 0$

$x^{1/3} + 4 = 0, \ x^{1/3} + 3 = 0$

$x^{1/3} = -4, \quad x^{1/3} = -3$

$x = -64, \quad x = -27$

43. $\text{ratio} = \dfrac{4.6}{4.2} = \dfrac{23}{21}$

45. $\dfrac{1}{16}(12,800,000) = 800,000$ people

Putting Your Skills to Work

1. $p(x) = x^3 - 80x^2 + 1900x + 2000$

x	0	10	20	30
y	2000	14,000	16,000	14,000

x	40	50	60
y	14,000	22,000	44,000

2. 1960 corresponds to $x = 20$ and 1980 corresponds to $x = 40$. From the graph, the smallest number of cubic feet of lumber produced during this period appears to be sometime in 1975 or 1976.

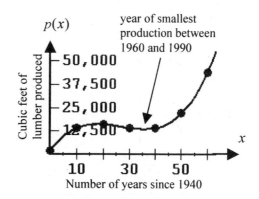

3. $p(x) = -x^4 + 15x^3 - 66x^2 + 80x + 50$

x	0	1	2	3	4	5	6	7	8
$p(x)$	50	78	50	20	18	50	98	120	50

From the table, 78 per hour at 12 noon and 120 per hour at 6 P.M.

4. From the graph, the hourly demand for Big Macs is lowest at $x = 3.5$ which corresponds to 2:30 P.M.

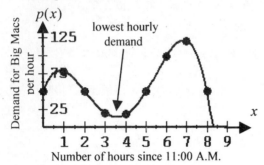

Chapter 10 Review Problems

1. $f(x) = \frac{1}{2}x + 3$

$$f(a-1) = \frac{1}{2}(a-1) + 3 = \frac{1}{2}a + \frac{5}{2}$$

2. $f(x) = \frac{1}{2}x + 3$

$$f(a+2) = \frac{1}{2}(a+2) + 3 = \frac{1}{2}a + 4$$

3. $f(x) = \frac{1}{2}x + 3$

$$f(a-1) - f(a) = \frac{1}{2}(a-1) + 3 - \left(\frac{1}{2}a + 3\right)$$

$$= \frac{1}{2}a - \frac{1}{2} + 3 - \frac{1}{2}a - 3$$

$$= -\frac{1}{2}$$

4. $f(x) = \frac{1}{2}x + 3$

$$f(a+2) - f(a) = \frac{1}{2}(a+2) + 3 - \left(\frac{1}{2}a + 3\right)$$

$$= \frac{1}{2}a + 1 + 3 - \frac{1}{2}a - 3$$

$$= 1$$

5. $f(x) = \frac{1}{2}x + 3$

$$f(2a+3) = \frac{1}{2}(2a+3) + 3$$

$$= a + \frac{3}{2} + 3$$

$$= a + \frac{9}{2}$$

6. $f(x) = \frac{1}{2}x + 3$

$$f(2a-3) = \frac{1}{2}(2a-3) + 3$$

$$= a - \frac{3}{2} + 3$$

$$= a + \frac{3}{2}$$

7. $p(x) = -2x^2 + 3x - 1$

$$p(-3) = -2(-3)^2 + 3(-3) - 1$$

$$= -28$$

8. $p(x) = -2x^2 + 3x - 1$

$$p(4) = -2(4)^2 + 3(4) - 1 = -21$$

9. $p(x) = -2x^2 + 3x - 1$

$p(2a) + p(-2)$

$$= -2(2a)^2 + 3(2a) - 1 + (-2(-2)^2 + 3(-2) - 1)$$

$$= -8a^2 + 6a - 1 + (-8 - 6 - 1)$$

$$= -8a^2 + 6a - 16$$

10. $p(x) = -2x^2 + 3x - 1$

$p(3a) + p(3)$

$$= -2(3a)^2 + 3(3a) - 1 + (-2(3)^2 + 3(3) - 1)$$

$$= -18a^2 + 9a - 1 + (-18 + 9 - 1)$$

$$= -18a^2 + 9a - 11$$

11. $p(x) = -2x^2 + 3x - 1$

$p(a+2) = -2(a+2)^2 + 3(a+2) - 1$

$\qquad = -2(a^2 + 4a + 4) + 3(a+2) - 1$

$\qquad = -2a^2 - 8a - 8 + 3a + 6 - 1$

$\qquad = -2a^2 - 5a - 3$

12. $p(x) = -2x^2 + 3x - 1$

$p(a-3) = -2(a-3)^2 + 3(a-3) - 1$

$\qquad = -2(a^2 - 6a + 9) + 3a - 9 - 1$

$\qquad = -2a^2 + 12a - 18 + 3a - 10$

$\qquad = -2a^2 + 15a - 28$

13. $h(x) = |2x - 1|$

$h(8a) = |2(8a) - 1| = |16a - 1|$

14. $h(x) = |2x - 1|$

$h(7a) = |2(7a) - 1| = |14a - 1|$

15. $h(x) = |2x - 1|$

$h\left(\dfrac{1}{4}a\right) = \left|2\left(\dfrac{1}{4}a\right) - 1\right|$

$\qquad = \left|\dfrac{1}{2}a - 1\right|$

16. $h(x) = |2x - 1|$

$h\left(\dfrac{3}{2}a\right) = \left|2\left(\dfrac{3}{2}a\right) - 1\right| = |3a - 1|$

17. $h(x) = |2x - 1|$

$h(a - 5) = |2(a-5) - 1| = |2a - 11|$

18. $h(x) = |2x - 1|$

$h(a + 4) = |2(a+4) - 1| = |2a + 7|$

19. $r(x) = \dfrac{3x}{x+4}, \ x \neq -4$

$r(5) = \dfrac{3(5)}{5+4} = \dfrac{15}{9} = \dfrac{5}{3}$

20. $r(x) = \dfrac{3x}{x+4}, \ x \neq -4$

$r(-6) = \dfrac{3(-6)}{-6+4}$

$\qquad = \dfrac{-18}{-2} = 9$

21. $r(x) = \dfrac{3x}{x+4}, \ x \neq -4$

$r(a+3) = \dfrac{3(a+3)}{a+3+4}$

$\qquad = \dfrac{3a+9}{a+7}$

22. $r(x) = \dfrac{3x}{x+4}, \ x \neq -4$

$r(a-2) = \dfrac{3(a-2)}{a-2+4}$

$\qquad = \dfrac{3a-6}{a+2}$

23. $r(x) = \dfrac{3x}{x+4}, \ x \neq -4$

$r(3) + r(a) = \dfrac{3(3)}{3+4} + \dfrac{3(a)}{a+4}$

$\qquad = \dfrac{9}{7} + \dfrac{3a}{a+4}$

$\qquad = \dfrac{9(a+4) + 7(3a)}{7(a+4)}$

$\qquad = \dfrac{9a + 36 + 21a}{7a + 28}$

$\qquad = \dfrac{30a + 36}{7a + 28}$

24. $r(x) = \dfrac{3x}{x+4}$, $x \neq -4$

$$r(a) + r(-2) = \frac{3a}{a+4} + \frac{3(-2)}{-2+4}$$

$$= \frac{3a}{a+4} + \frac{-6}{2}$$

$$= \frac{3a}{a+4} - 3$$

$$= \frac{3a - 3(a+4)}{a+4}$$

$$= \frac{-12}{a+4}$$

25. $f(x) = 7x - 4$

$$\frac{f(x+h) - f(x)}{h} = \frac{7(x+h) - 4 - (7x - 4)}{h}$$

$$= \frac{7x + 7h - 4 - 7x + 4}{h}$$

$$= 7$$

26. $f(x) = 6x - 5$

$$\frac{f(x+h) - f(x)}{h} = \frac{6(x+h) - 5 - (6x - 5)}{h}$$

$$= \frac{6x + 6h - 5 - 6x + 5}{h}$$

$$= 6$$

27. $f(x) = 2x^2 - 5x$

$$\frac{f(x+h) - f(x)}{h}$$

$$= \frac{2(x+h)^2 - 5(x+h) - (2x^2 - 5x)}{h}$$

$$= \frac{2x^2 + 4xh + 2h^2 - 5x - 5h - 2x^2 + 5x}{h}$$

$$= \frac{4xh + 2h^2 - 5h}{h}$$

$$= 4x + 2h - 5$$

28. $f(x) = 2x - 3x^2$

$$\frac{f(x+h) - f(x)}{h}$$

$$= \frac{2(x+h) - 3(x+h)^2 - (2x - 3x^2)}{h}$$

$$= \frac{2x + 2h - 3x^2 - 6xh - 3h^2 - 2x + 3x^2}{h}$$

$$= \frac{2h - 6xh - 3h^2}{h}$$

$$= 2 - 6x - 3h$$

$$= -6x - 3h + 2$$

29. (a) Yes, the graph passes the vertical line test and therefore represents a function.
(b) Yes, the graph passes the horizontal line test and therefore represents a one-to-one function.

30. (a) No, the graph fails the vertical line test and therefore does not represent a function.
(b) No, unless the graph represents a function first it cannot represent a one-to-one function.

31. (a) Yes, the graph passes the vertical line test and therefore represents a function.
(b) No, the graph fails the horizontal line test and therefore does not represent a one-to-one function.

32. (a) Yes, the graph passes the vertical line test and therefore represents a function.
(b) No, the graph fails the horizontal line test and therefore does not represent a one-to-one function.

33. (a) No, the graph fails the vertical line test and therefore does not represent a function.

33. (b) No, unless the graph represents a
function first it cannot represent a
one-to-one function.

34. (a) Yes, the graph passes the vertical line
test and therefore represents a function.
(b) Yes, the graph passes the horizontal
line test and therefore represents a
one-to-one function.

35. $f(x) = x^2$

$g(x) = (x+2)^2 + 4$ is $f(x)$ shifted left 2
units and up 4 units.

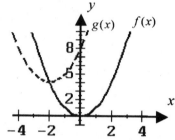

36. $f(x) = |x|$

$g(x) = |x+3|$ is $f(x)$ shifted left 3 units.

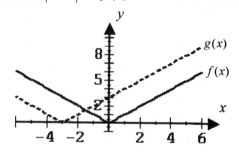

37. $f(x) = |x|$

$g(x) = |x-4|$ is $f(x)$ shifted
right 2 units.

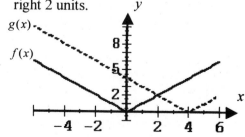

38. $f(x) = |x|$

$h(x) = |x| + 3$ is $f(x)$ shifted up 3 units.

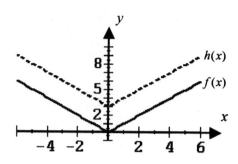

39. $f(x) = |x|$

$h(x) = |x| - 2$ is $f(x)$ shifted down
2 units.

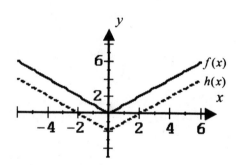

40. $f(x) = x^3$

$r(x) = (x+3)^3 + 1$ is $f(x)$ shifted left
3 units and up 1unit.

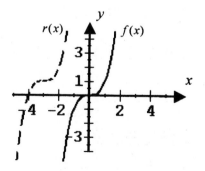

41. $f(x) = x^3$

$r(x) = (x-1)^3 + 5$ is $f(x)$ shifted right 1 unit and up 5 units.

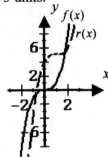

42. $f(x) = \dfrac{2}{x},\ x \neq 0$

$r(x) = \dfrac{2}{x+3} - 2,\ x \neq -3$ is $f(x)$ shifted left 3 units and down 2 units.

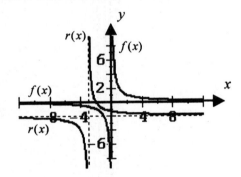

43. $f(x) = \dfrac{4}{x},\ x \neq 0$

$r(x) = \dfrac{4}{x+2},\ x \neq -2$ is $f(x)$ shifted left 2 units.

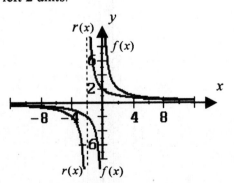

In Exercises 44-63,

$$f(x) = 3x+5;\ g(x) = \frac{2}{x},\ x \neq 0$$

$$s(x) = \sqrt{x-2},\ x \geq 2;\ h(x) = \frac{x+1}{x-4},\ x \neq 4$$

$$p(x) = 2x^2 - 3x + 4;\ t(x) = -\frac{1}{2}x - 3$$

44. $(f+p)(x) = f(x) + p(x)$
$$= 3x + 5 + 2x^2 - 3x + 4$$
$$= 2x^2 + 9$$

45. $(f+t)(x) = f(x) + t(x)$
$$= 3x + 5 + \left(-\frac{1}{2}x - 3\right)$$
$$= \frac{5}{2}x + 2$$

46. $(t-f)(x) = t(x) - f(x)$
$$= -\frac{1}{2}x - 3 - (3x+5)$$
$$= -\frac{7}{2}x - 8$$

47. $(p-f)(x) = p(x) - f(x)$
$$= 2x^2 - 3x + 4 - (3x+5)$$
$$= 2x^2 - 3x + 4 - 3x - 5$$
$$= 2x^2 - 6x - 1$$

48. From Exercise 43,
$$(p-f)(x) = 2x^2 - 6x - 1$$
$$(p-f)(2) = 2(2)^2 - 6(2) - 1$$
$$= 2(4) - 12 - 1$$
$$= 8 - 13$$
$$= -5$$

49. From Exercise 46,

$$(t-f)(x) = -\frac{7}{2}x - 8$$

$$(t-f)(-3) = -\frac{7}{2}(-3) - 8$$

$$= \frac{5}{2}$$

50. $(fg)(x) = f(x)g(x)$

$$= (3x+5)\left(\frac{2}{x}\right)$$

$$= \frac{6x+10}{x}, \ x \neq 0$$

51. $(tp)(x) = t(x)p(x)$

$$= \left(-\frac{1}{2}x - 3\right)(2x^2 - 3x + 4)$$

$$= -x^3 + \frac{3x^2}{2} - 2x - 6x^2 + 9x - 12$$

$$= -x^3 - \frac{9}{2}x^2 + 7x - 12$$

52. $\left(\dfrac{g}{h}\right)(x) = \dfrac{g(x)}{h(x)} = \dfrac{\dfrac{2}{x}}{\dfrac{x+1}{x-4}}$

$$= \frac{2}{x} \cdot \frac{x-4}{x+1}$$

$$= \frac{2x-8}{x^2+1}, \ x \neq 0, 4, -1$$

53. $\left(\dfrac{g}{f}\right)(x) = \dfrac{g(x)}{f(x)} = \dfrac{\dfrac{2}{x}}{3x+5}$

$$= \frac{2}{x} \cdot \frac{1}{3x+5}$$

$$= \frac{2}{3x^2+5x}, \ x \neq 0, -\frac{5}{3}$$

54. From Exercise 52,

$$\left(\frac{g}{h}\right)(x) = \frac{2x-8}{x^2+x}, \ x \neq -1, 0, 4$$

$$\left(\frac{g}{h}\right)(-2) = \frac{2(-2)-8}{(-2)^2+(-2)} = -6$$

55. From Exercise 53,

$$\left(\frac{g}{f}\right)(x) = \frac{2}{3x^2+5x}, \ x \neq 0, -\frac{5}{3}$$

$$\left(\frac{g}{f}\right)(-3) = \frac{2}{3(-3)^2+5(-3)} = \frac{1}{6}$$

56. $f[t(x)] = f\left[-\frac{1}{2}x - 3\right]$

$$= 3\left(-\frac{1}{2}x - 3\right) + 5$$

$$= -\frac{3}{2}x - 4$$

57. $h[f(x)] = h(3x+5)$

$$= \frac{3x+5+1}{3x+5-4}$$

$$= \frac{3x+6}{3x+1}, \ x \neq -\frac{1}{3}$$

58. $s[p(x)] = s\left[2x^2 - 3x + 4\right]$

$$= \sqrt{2x^2 - 3x + 4 - 2}$$

$$= \sqrt{2x^2 - 3x + 2}$$

59. $s[t(x)] = s\left[-\frac{1}{2}x - 3\right]$

$$= \sqrt{-\frac{1}{2}x - 3 - 2}$$

$$= \sqrt{-\frac{1}{2}x - 5}, \ x \leq -10$$

60. From Exercise 58,

$$s[p(x)] = \sqrt{2x^2 - 3x + 2}$$

$$s[p(2)] = \sqrt{2(2)^2 - 3(2) + 2} = 2$$

61. From Exercise 59,

$$s[t(x)] = \sqrt{-\frac{1}{2}x - 5}, \ x \le 10$$

$$s[t(-18)] = \sqrt{-\frac{1}{2}(-18) - 5} = 2$$

62. $f[g(x)] = f\left[\dfrac{2}{x}\right], \ x \ne 0$

$$= 3\left(\frac{2}{x}\right) + 5 = \frac{6}{x} + 5$$

$$= \frac{6 + 5x}{x}$$

$$g[f(x)] = g[3x + 5]$$

$$= \frac{2}{3x + 5}$$

$$f[g(x)] \ne g[f(x)]$$

63. $p[g(x)] = p\left[\dfrac{2}{x}\right]$

$$= 2\left(\frac{2}{x}\right)^2 - 3\left(\frac{2}{x}\right) + 4$$

$$= \frac{8 - 6x + 4x^2}{x^2}$$

$$g[p(x)] = g[2x^2 - 3x + 4]$$

$$= \frac{2}{2x^2 - 3x + 4}$$

$$p[g(x)] \ne g[p(x)]$$

64. $B = \{(3, 7), (7, 3), (0, 8), (0, -8)\}$

(a) $D = \{0, 3, 7\}$

(continued)

64. (continued)

(b) $R = \{-8, 3, 7, 8\}$

(c) No, the set does not define a function since two of the ordered pairs have the same first coordinate.

(d) No, since the set does not define a function it cannot define a one-to-one function.

65. $A = \{(100, 10), (200, 20), (300, 30), (400, 10)\}$

(a) $D = \{100, 200, 300, 400\}$

(b) $R = \{10, 20, 30\}$

(c) Yes, the set defines a function since no two of the ordered pairs have the same first coordinate.

(d) No, the set does not define a one-to-one function since two of the ordered pairs have the same second coordinate.

66. $D = \left\{\left(\dfrac{1}{2}, 2\right), \left(\dfrac{1}{4}, 4\right), \left(-\dfrac{1}{3}, 3\right), \left(4, \dfrac{1}{4}\right)\right\}$

(a) domain $= \left\{\dfrac{1}{2}, \dfrac{1}{4}, -\dfrac{1}{3}, 4\right\}$

(b) $R = \left\{2, 4, -3, \dfrac{1}{4}\right\}$

(c) Yes, the set defines a function since no two of the ordered pairs have the same first coordinate.

(d) Yes, the set defines a one-to-one function since it is a function and no two ordered pairs have the same second coordinate.

67. $C = \{(12, 6), (0, 6), (0, -1), (-6, -12)\}$

(a) $D = \{12, 0, -6\}$

(b) $R = \{-1, -12, 6\}$

(continued)

67. (continued)

(c) No, the set does not define a function since two of the ordered pairs have the same first coordinate.

(d) No, since the set does not define a function it cannot define a one-to-one function.

68. $E = \{(0,1),(1,2),(2,9),(-1,-2)\}$

(a) $D = \{-1,0,1,2\}$

(b) $R = \{-2,1,2,9\}$

(c) Yes, the set defines a function since no two of the ordered pairs have the same first coordinate.

(d) Yes, the set defines a one-to-one function since it is a function and no two ordered pairs have the same second coordinate.

69. $F = \{(3,7),(2,1),(0,-3),(1,1)\}$

(a) $D = \{0,1,2,3\}$

(b) $R = \{-3,1,7\}$

(c) Yes, the set defines a function since no two of the ordered pairs have the same first coordinate.

(d) No, the set does not define a one-to-one function since two of the ordered pairs have the same second coordinate.

70. $A = \left\{\left(3,\dfrac{1}{3}\right),\left(-2,-\dfrac{1}{2}\right),\left(-4,-\dfrac{1}{4}\right),\left(5,\dfrac{1}{5}\right)\right\}$

$A^{-1} = \left\{\left(\dfrac{1}{3},3\right),\left(-\dfrac{1}{2},-2\right),\left(-\dfrac{1}{4},-4\right),\left(\dfrac{1}{5},5\right)\right\}$

71. $B = \{(1,10),(3,7),(0,-3),(1,1)\}$

$B^{-1} = \{(10,1),(7,3),(-3,0),(1,1)\}$

72. $f(x) = -\dfrac{3}{4}x + 2,\ f(x) \to y$

$y = -\dfrac{3}{4}x + 2,\ x \leftrightarrow y$

$x = -\dfrac{3}{4}y + 2$

$y = -\dfrac{4}{3}x + \dfrac{8}{3},\ y \to f^{-1}(x)$

$f^{-1}(x) = -\dfrac{4}{3}x + \dfrac{8}{3}$

73. $g(x) = -8 - 4x,\ g(x) \to y$

$y = -8 - 4x,\ x \leftrightarrow y$

$x = -8 - 4y$

$y = -\dfrac{1}{4}x - 2,\ y \to g^{-1}(x)$

$g^{-1}(x) = -\dfrac{1}{4}x - 2$

74. $h(x) = \dfrac{x+2}{3},\ h(x) \to y$

$y = \dfrac{x+2}{3},\ x \leftrightarrow y$

$x = \dfrac{y+2}{3}$

$y = 3x - 2,\ y \to h^{-1}(x)$

$h^{-1}(x) = 3x - 2$

75. $j(x) = \dfrac{1}{x-3},\ j(x) \to y$

$y = \dfrac{1}{x-3},\ x \leftrightarrow y$

$x = \dfrac{1}{y-3}$

$y = \dfrac{1}{x} + 3,\ y \to j^{-1}(x)$

$j^{-1}(x) = \dfrac{1}{x} + 3$

76. $p(x) = \sqrt[3]{x+1}$, $p(x) \to y$

$y = \sqrt[3]{x+1}$, $x \leftrightarrow y$

$x = \sqrt[3]{y+1}$

$y = x^3 - 1$, $y \to p^{-1}(x)$

$p^{-1}(x) = x^3 - 1$

77. $r(x) = x^3 + 2$, $r(x) \to y$

$y = x^3 + 2$, $x \leftrightarrow y$

$x = y^3 + 2$

$y = \sqrt[3]{x-2}$, $y \to r^{-1}(x)$

$r^{-1}(x) = \sqrt[3]{x-2}$

78. $f(x) = \dfrac{-x-2}{3}$, $f(x) \to y$

$y = \dfrac{-x-2}{3}$, $x \leftrightarrow y$

$x = \dfrac{-y-2}{3}$

$y = -3x - 2$, $y \to f^{-1}(x)$

$f^{-1}(x) = -3x - 2$

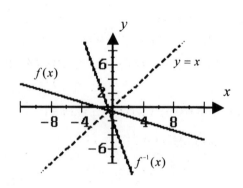

79. $f(x) = -\dfrac{3}{4}x + 1$, $f(x) \to y$

$y = -\dfrac{3}{4}x + 1$, $x \leftrightarrow y$

(continued)

79. (continued)

$x = -\dfrac{3}{4}y + 1$

$y = -\dfrac{4}{3}x + \dfrac{4}{3}$, $y \to f^{-1}(x)$

$f^{-1}(x) = -\dfrac{4}{3}x + \dfrac{4}{3}$

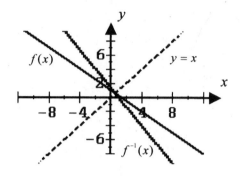

Chapter 10 Test

1. $f(x) = \dfrac{3}{4}x - 2$

$f(-8) = \dfrac{3}{4}(-8) - 2 = -8$

2. $f(x) = \dfrac{3}{4}x - 2$

$f(2a) = \dfrac{3}{4}(2a) - 2 = \dfrac{3}{2}a - 2$

3. $f(x) = \dfrac{3}{4}x - 2$

$f(a) - f(2) = \dfrac{3}{4}a - 2 - \left(\dfrac{3}{4}(2) - 2\right)$

$= \dfrac{3}{4}a - 2 - \dfrac{3}{2} + 2$

$= \dfrac{3}{4}a - \dfrac{3}{2}$

4. $f(x) = 3x^2 - 2x + 4$

$f(-6) = 3(-6)^2 - 2(-6) + 4 = 124$

5. $f(x) = 3x^2 - 2x + 4$

$f(a+1) = 3(a+1)^2 - 2(a+1) + 4$

$= 3a^2 + 6a + 3 - 2a - 2 + 4$

$= 3a^2 + 4a + 5$

6. $f(x) = 3x^2 - 2x + 4$

$f(a) + f(1) = 3a^2 - 2a + 4 + 3(1)^2 - 2(1) + 4$

$= 3a^2 - 2a + 9$

7. $f(x) = 3x^2 - 2x + 4$

$f(-2a) - 2 = 3(-2a)^2 - 2(-2a) + 4 - 2$

$= 12a^2 + 4a + 2$

8. (a) Graph passes vertical line test and therefore represents a function.
(b) Graph fails horizontal line test and does not represent a one-to-one function.

9. (a) Graph passes vertical line test and therefore represents a function.
(b) Graph passes horizontal line test and therefore represents a one-to-one function.

10. $f(x) = x^2$

$g(x) = (x-1)^2 + 3$ is $f(x)$ shifted right 1 unit and up 3 units.

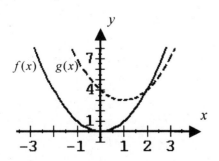

11. $f(x) = |x|$

$g(x) = |x+1| + 2$ is $f(x)$ shifted left 1 unit and up 2 units.

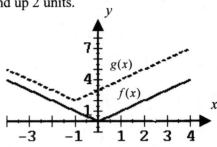

12. $f(x) = 3x^2 - x - 6, \ g(x) = -2x^2 + 5x + 7$

(a)

$(f+g)(x) = f(x) + g(x)$

$= 3x^2 - x - 6 + (-2x^2 + 5x + 7)$

$= 3x^2 - x - 6 - 2x^2 + 5x + 7$

$= x^2 + 4x + 1$

(b)

$(f-g)(x) = f(x) - g(x)$

$= 3x^2 - x - 6 - (-2x^2 + 5x + 7)$

$= 3x^2 - x - 6 + 2x^2 - 5x - 7$

$= 5x^2 - 6x - 13$

(c) From (b)

$(f-g)(x) = 5x^2 - 6x - 13$

$(f-g)(-2) = 5(-2)^2 - 6(-2) - 13 = 19$

13. $f(x) = \dfrac{3}{x}, \ x \neq 0; \ g(x) = 2x - 1$

(a) $(fg)(x) = f(x)g(x)$

$= \dfrac{3}{x}(2x - 1)$

$= \dfrac{6x - 3}{x}, \ x \neq 0$

(continued)

13. (continued)

(b) $\left(\dfrac{f}{g}\right)(x) = \dfrac{f(x)}{g(x)}$

$$= \dfrac{\dfrac{3}{x}}{2x-1}$$

$$= \dfrac{3}{2x^2 - x}, \; x \neq 0, \dfrac{1}{2}$$

(c) $g[f(x)] = g\left[\dfrac{3}{x}\right]$

$$= 2 \cdot \dfrac{3}{x} - 1$$

$$= \dfrac{6}{x} - 1, \; x \neq 0$$

14. $f(x) = \dfrac{1}{2}x - 3, \; g(x) = 4x + 5$

(a) $(f \circ g)(x) = f[g(x)]$

$$= f[4x+5]$$

$$= \dfrac{1}{2}(4x+5) - 3$$

$$= 2x - \dfrac{1}{2}$$

(b) $(g \circ f)(x) = g[f(x)]$

$$= g\left[\dfrac{1}{2}x - 3\right]$$

$$= 4\left(\dfrac{1}{2}x - 3\right) + 5$$

$$= 2x - 7$$

(c) $f[f(x)] = f\left(\dfrac{1}{2}x - 3\right)$

$$= \dfrac{1}{2}\left(\dfrac{1}{2}x - 3\right) - 3$$

$$= \dfrac{1}{4}x - \dfrac{9}{2}$$

15. $B = \{(1,8),(8,1),(9,10),(-10,9)\}$

(a) Yes, the function is one-to-one since no two ordered pairs have the same second coordinate.

(b) $B^{-1} = \{(8,1),(1,8),(10,9),(9,-10)\}$

16. $A = \{(1,5),(2,1),(4,-7),(0,7)\}$

(a) Yes, the function is one-to-one since no two ordered pairs have the same second coordinate.

(b) $A^{-1} = \{(5,1),(1,2),(-7,4),(7,0)\}$

17. $f(x) = \dfrac{1}{2}x - \dfrac{1}{5}, \; f(x) \to y$

$$y = \dfrac{1}{2}x - \dfrac{1}{5}, \; x \leftrightarrow y$$

$$x = \dfrac{1}{2}y - \dfrac{1}{5}$$

$$y = 2x + \dfrac{2}{5}, \; y \to f^{-1}(x)$$

$$f^{-1}(x) = 2x + \dfrac{2}{5}$$

18. $f(x) = -3x - 2, \; f(x) \to y$

$$y = -3x + 2, \; x \leftrightarrow y$$

$$x = -3y + 2$$

$$y = -\dfrac{1}{3}x + \dfrac{2}{3}, \; y \to f^{-1}(x)$$

$$f^{-1}(x) = -\dfrac{1}{3}x + \dfrac{2}{3}$$

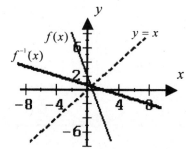

19. $f(x) = \dfrac{3}{7}x + \dfrac{1}{2}$, $f^{-1}(x) = \dfrac{14x - 7}{6}$

$$f^{-1}[f(x)] = f^{-1}\left[\dfrac{3}{7}x + \dfrac{1}{2}\right]$$

$$= \dfrac{14\left[\dfrac{3}{7}x + \dfrac{1}{2}\right] - 7}{6}$$

$$= \dfrac{6x + 7 - 7}{6} = \dfrac{6x}{6}$$

$$= x$$

Cumulative Test for Chapters 1-10

1. $5(-3) = -3(5)$ illustrates the commutative property of multiplication.

2. $3(4-6)^3 + \sqrt{25} = 3(-2)^3 + \sqrt{25}$

$$= 3(-8) + 5$$
$$= -24 + 5$$
$$= -19$$

3. $2\{x - 3[x - 2(x+1)]\}$

$$= 2\{x - 3[x - 2x - 2]\}$$
$$= 2\{x - 3[-x - 2]\}$$
$$= 2\{x + 3x + 6\}$$
$$= 2\{4x + 6\}$$
$$= 8x + 12$$

4. $A = 3bt + prt$

$$prt = A - 3bt$$
$$p = \dfrac{A - 3bt}{rt}$$

5. $x^3 + 125 = x^3 + 5^3$

$$= (x + 5)(x^2 - 5x + 25)$$

6. $\dfrac{3}{x-4} + \dfrac{6}{x^2 - 16}$

$$= \dfrac{3(x+4)}{(x-4)(x+4)} + \dfrac{6}{(x-4)(x+4)}$$

$$= \dfrac{3x + 12 + 6}{(x-4)(x+4)}$$

$$= \dfrac{3x + 18}{(x-4)(x+4)}$$

7. $\dfrac{3}{2x+3} = \dfrac{1}{2x-3} + \dfrac{2}{4x^2 - 9}$

$$\dfrac{3}{(2x+3)} = \dfrac{1}{2x-3} + \dfrac{2}{(2x+3)(2x-3)}$$
$$3(2x-3) = 2x + 3 + 2$$
$$6x - 9 = 2x + 5$$
$$4x = 14$$
$$x = \dfrac{7}{2}$$

8. $3x - 2y - 9z = 9$

$$x - y + z = 8$$
$$2x + 3y - z = -2$$

Switch the first and second equation.

$$x - y + z = 8$$
$$3x - 2y - 9z = 9$$
$$2x + 3y - z = -2$$

Multiply the first equation by -3 and add to the second equation and multiply the first equation by -2 and add to the third equation

$$x - y + z = 8$$
$$y - 12z = -15$$
$$5y - 3z = -18$$

Multiply the second equation by -5 and add to the third equation

(continued)

8. (continued)
$$x - y + z = 8$$
$$y - 12z = -15$$
$$57z = 57$$
$$z = 1$$
$$y = 12z - 15 = 12(1) - 15$$
$$y = -3$$
$$x - y + z = 8$$
$$x - (-3) + 1 = 8$$
$$x = 4$$
$(4, -3, 1)$ is the solution.

9. $\sqrt{8x} + 3x\sqrt{50} - 4x\sqrt{32}$
$$= \sqrt{4 \cdot 2x} + 3x\sqrt{25 \cdot 2} - 4x\sqrt{16 \cdot 2}$$
$$= 2\sqrt{2x} + 15x\sqrt{2} - 16x\sqrt{2}$$
$$= 2\sqrt{2x} - x\sqrt{2}$$

10. $(\sqrt{2} + \sqrt{3})(2\sqrt{6} - \sqrt{3})$
$$= 2\sqrt{12} - \sqrt{6} + 2\sqrt{18} - 3$$
$$= 2\sqrt{4 \cdot 3} + 2\sqrt{9 \cdot 2} - \sqrt{6} - 3$$
$$= 4\sqrt{3} + 6\sqrt{2} - \sqrt{6} - 3$$

11. $d = \sqrt{(x_2 - x_1)^2 + (y_2 - y_1)^2}$
$$d = \sqrt{(-3 - 6)^2 + (-4 - (-1))^2} = \sqrt{90}$$
$$d = 3\sqrt{10}$$

12. $12x^2 - 11x + 2 = (4x - 1)(3x - 2)$

13. $x^4 - 10x^2 + 9 = (x^2 - 9)(x^2 - 1)$
$$= (x + 3)(x - 3)(x + 1)(x - 1)$$

14. $(x - h)^2 + (y - k)^2 = r^2$
$$(x - (-3))^2 + (y - 6)^2 = 14^2$$
$$(x + 3)^2 + (y - 6)^2 = 169$$

15. $f(x) = 3x^2 - 2x + 1$
(a) $f(-2) = 3(-2)^2 - 2(-2) + 1 = 17$
(b) $f(a - 2) = 3(a - 2)^2 - 2(a - 2) + 1$
$$= 3a^2 - 12a + 12 - 2a + 4 + 1$$
$$= 3a^2 - 14a + 17$$
(c) $f(a) + f(-2) = 3a^2 - 2a + 1 + 17$
$$= 3a^2 - 2a + 18$$

16. $f(x) = x^3$, $g(x) = (x + 2)^3 + 4$ is $f(x)$ shifted right 2 units and up 4 units.

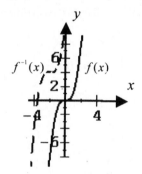

17. $f(x) = 2x^2 - 5x - 6$, $g(x) - 5x + 3$

(a) $(fg)(x) = f(x)g(x)$
$$= (2x^2 - 5x - 6)(5x + 3)$$
$$= 10x^3 + 6x^2 - 25x^2 - 15x - 30x + 18$$
$$= 10x^3 - 19x^2 - 45x + 18$$

(b) $\left(\dfrac{f}{g}\right)(x) = \dfrac{f(x)}{g(x)}$
$$= \frac{2x^2 - 5x - 6}{5x + 3}, \ x \neq -\frac{3}{5}$$

(c) $f[g(x)] = f[5x + 3]$
$$= 2(5x + 3)^2 - 5(5x + 3) - 6$$
$$= 50x^2 + 60x + 18 - 25x - 15 - 6$$
$$= 50x^2 + 35x - 3$$

18. $A = \{(3,6),(1,8),(2,7),(4,4)\}$

(a) Yes, A is a function; no two ordered pairs have the same first coordinate.

(b) Yes, A is a one-to-one function since no two ordered pairs have the same second coordinate.

(c) $A^{-1} = \{(6,3),(8,1),(7,2),(4,4)\}$

19. $f(x) = 7x - 3,\ f(x) \to y$

$y = 7x - 3,\ x \leftrightarrow y$

$x = 7y - 3$

$y = \dfrac{x+3}{7},\ y \to f^{-1}(x)$

$f^{-1}(x) = \dfrac{x+3}{7}$

20. $f(x) = 5x^3 - 3x^2 - 6$

(a) $f(5) = 5(5)^3 - 3(5)^2 - 6 = 544$

(b) $f(-3) = 5(-3)^3 - 3(-3)^2 - 6 = -168$

(c) $f(2a) = 5(2a)^3 - 3(2a)^2 - 6$

$\qquad = 40a^3 - 12a^{-2} - 6$

21. (a) $f(x) = -\dfrac{2}{3}x + 2,\ f(x) \to y$

$y = -\dfrac{2}{3}x + 2,\ x \leftrightarrow y$

$x = -\dfrac{2}{3}y + 2$

$y = -\dfrac{3}{2}x + 3,\ y \to f^{-1}(x)$

$f^{-1}(x) = -\dfrac{3}{2}x + 3$

(continued)

21. (continued)

(b)

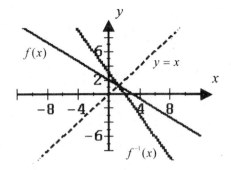

22. From Exercise 21,

$f\left[f^{-1}(x)\right] = f\left[-\dfrac{3}{2}x + 3\right]$

$\qquad = -\dfrac{2}{3}\left[-\dfrac{3}{2}x + 3\right] + 2$

$\qquad = x - 2 + 2$

$\qquad = x$

Chapter 11

Pretest Chapter 11

1. $f(x) = 2^{-x}$

x	$y = f(x) = 2^{-x}$
-2	4
-1	2
0	1
1	$\dfrac{1}{2}$
2	$\dfrac{1}{4}$

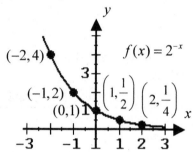

2. $3^{2x-1} = 27$

$3^{2x-1} = 3^3$

$2x - 1 = 3$

$2x = 4$

$x = 2$

3. $A = P(1+r)^t$

$A = 10,000(1+0.12)^4 = 15,735.1936$

In 4 years Nancy will have \$15,735.19.

4. $\dfrac{1}{49} = 7^{-2}$

$\log_7\left(\dfrac{1}{49}\right) = -2$

5. $\log_5 x = 3$

$x = 5^3 = 125$

6. Let $N = \log_{10}(10,000)$

$10^N = 10,000$

$10^N = 10^4$

$N = 4$

7. $\log_5\left(\dfrac{x^2 y^5}{z^3}\right) = \log_5(x^2 y^5) - \log_5 z^3$

$= \log_5 x^2 + \log_5 y^5 - \log_5 z^3$

$= 2\log_5 x + 5\log_5 y - 3\log_5 z$

8. $\dfrac{1}{2}\log_4 x - 3\log_4 w = \log_4 x^{1/2} - \log_4 w^3$

$= \log_4 \sqrt{x} - \log_4 w^3$

$= \log_4\left(\dfrac{\sqrt{4}}{w^3}\right)$

9. $\log_3 x + \log_3 2 = 4$

$\log_3(2x) = 4$

$2x = 3^4$

$2x = 81$

$x = \dfrac{81}{2}$

10. $\log x = 3.9170$

$x = 10^{3.9170} = 8260.3795$

11. $\ln 4.79 = 1.5665$

12. $\log_6 5.02 = \dfrac{\ln 5.02}{\ln 6} = 0.9005$

13. $\log 0.7523 = -0.1236$

14. $\ln x = 22.976$

$e^{\ln x} = e^{22.976}$

$x = 9.5137 \times 10^9$

15. $\log x - \log(x+3) = -1$

$\log \dfrac{x}{x+3} = -1$

$\dfrac{x}{x+3} = 10^{-1}$

$10x = x+3$

$9x = 3$

$x = \dfrac{3}{9} = \dfrac{1}{3}$

check:

$$\log \frac{1}{3} - \log\left(\frac{1}{3}+3\right) \overset{?}{=} -1$$

$$\log\left(\frac{\frac{1}{3}}{\frac{1}{3}+3}\right) \overset{?}{=} -1$$

$$\log \frac{1}{10} = \log 1 - \log 10 = 0 - 1 = -1$$

16. $4^{2x+1} = 9$

$\log 4^{2x+1} = \log 9$

$(2x+1)\log 4 = \log 9$

$2x+1 = \dfrac{\log 9}{\log 4}$

$x = \dfrac{1}{2}\left(\dfrac{\log 9}{\log 4} - 1\right) \approx 0.2925$

17. $A = P(1+r)^t$

$7000 = 2000(1+0.06)^t$

(continued)

17. (continued)

$(1.06)^t = \dfrac{7}{2}$

$\log(1.06)^t = \log \dfrac{7}{2}$

$t \log(1.06) = \log \dfrac{7}{2}$

$t = \dfrac{\log \dfrac{7}{2}}{\log(1.06)} = 21.49968154$

It would take about 21 years for $2000 to grow to $7000 and 6% compounded yearly.

11.1 Exercises

1. The exponential function is an equation of the form

$\underline{f(x) = b^x \text{ where } b > 0, \ b \neq 0, \text{ and } x}$

$\underline{\text{is a real number.}}$

3. $f(x) = 3^x$

x	$y = f(x) = 3^x$
-2	$\dfrac{1}{9}$
-1	$\dfrac{1}{3}$
0	1
1	3
2	9

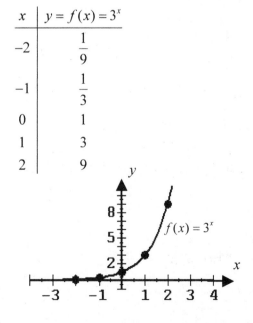

5. $f(x) = 2^{-x}$

x	$y = f(x) = 2^{-x}$
-2	4
-1	2
0	1
1	$\dfrac{1}{2}$
2	$\dfrac{1}{4}$

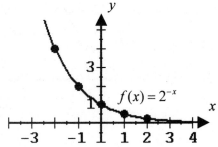

9. $f(x) = 2^{x+5}$

x	$y = f(x) = 2^{x+5}$
-7	$\dfrac{1}{4}$
-6	$\dfrac{1}{2}$
-5	1
-4	2
-3	4

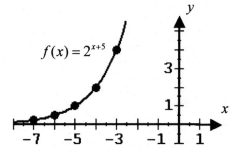

7. $f(x) = 3^{-x}$

x	$y = f(x) = 3^{-x}$
-2	9
-1	3
0	1
1	$\dfrac{1}{3}$
2	$\dfrac{1}{9}$

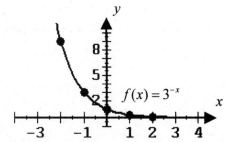

11. $f(x) = 3^{x-4}$

x	$y = f(x) = 3^{x-4}$
6	9
5	3
4	1
3	$\dfrac{1}{3}$
2	$\dfrac{1}{9}$

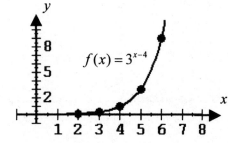

13. $f(x) = 2^{x+2}$

x	$y = f(x) = 2^x + 2$
-2	$\dfrac{9}{4}$
-1	$\dfrac{5}{2}$
0	3
1	4
2	6

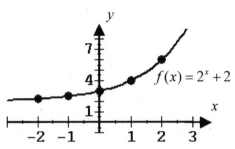

15. $f(x) = e^{x-1}$

x	$y = f(x) = e^{x-1}$
-2	0.05
-1	0.14
0	0.37
1	1
2	2.7

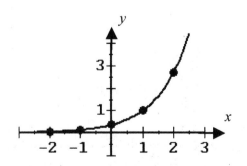

17. $f(x) = 2e^x$

x	$y = f(x) = 2e^x$
-2	0.27
-1	0.74
0	2
1	5.44
2	14.8

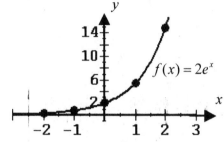

19. $f(x) = e^{1-x}$

x	$y = f(x) = e^{1-x}$
-2	20.1
-1	7.39
0	2.72
1	0.37
2	0.14

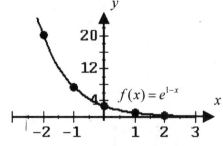

21. $2^x = 4$

$2^x = 2^2$

$x = 2$

23. $2^x = 1$

$\quad 2^x = 2^0$

$\quad\quad x = 0$

25. $2^x = \dfrac{1}{8}$

$\quad 2^x = \dfrac{1}{2^3} = 2^{-3}$

$\quad\quad x = -3$

27. $3^x = 81$

$\quad 3^x = 3^4$

$\quad\quad x = 4$

29. $3^x = 1$

$\quad 3^x = 3^0$

$\quad\quad x = 0$

31. $3^{-x} = \dfrac{1}{9} = \dfrac{1}{3^2}$

$\quad 3^{-x} = 3^{-2}$

$\quad\quad -x = -2$

$\quad\quad\quad x = 2$

33. $4^x = 256$

$\quad 4^x = 4^4$

$\quad\quad x = 4$

35. $7^{x-1} = 49$

$\quad 7^{x-1} = 7^2$

$\quad x - 1 = 2$

$\quad\quad\quad x = 3$

37. $8^{3x-1} = 64 = 8^2$

$\quad 3x - 1 = 2$

$\quad\quad 3x = 3$

$\quad\quad\quad x = 1$

39. $A = P\left(1 + \dfrac{r}{n}\right)^{nt}$

$A = 2000\left(1 + \dfrac{0.063}{1}\right)^{1(3)} = 2402.314094$

Alicia will have \$2402.31 after 3 years.

41. $A = P\left(1 + \dfrac{r}{n}\right)^{nt}$

$A = 2000\left(1 + \dfrac{0.14}{4}\right)^{4(5)} = 3979.577727$

$A = 2000\left(1 + \dfrac{0.14}{12}\right)^{12(5)} = 4011.219586$

She will have \$3979.58 if it is compounded quarterly and \$4011.22 if it is compounded monthly.

43. $B(t) = 4000(2^t)$

$\quad B(3) = 4000(2^3) = 32,000$

$\quad B(9) = 4000(2^9) = 2,048,000$

At the end of 3 hours there will be 32,000 bacteria in the culture and at the end of 9 hours there will be 2,048,000 bacteria in the culture.

45. $f(t) = (1 - 0.08)^t = 0.92^t$

$\quad f(5) = 0.92^5 = 0.65908...$

$\quad f(25) = 0.92^{25} = 0.1243... > 0.10$

In 5 years approximately 65.9% of the homeowners will still be using septic tanks. At the present rate, 12.4% of the homeowners will still be using septic tanks in 25 years so the goal of less than 10% will not be achieved.

47. $A = Ce^{-0.0004297t}$

$A = 6e^{-0.0004297(1000)} \approx 3.91$

There will be approximately 3.91 mg of radium in the container after 1000 years.

49. $P = 14.7e^{-0.21d}$

$P = 14.7e^{-0.21(2)} \approx 9.66$

The man will experience a pressure of approximately 9.66 lb/in^2 .

51. $N = 0.00472e^{0.11596t}$

$N = 0.00472e^{0.11596(80)} \approx 50.4$

$N = 0.00472e^{0.11596(90)} \approx 160.8$

$\dfrac{160.8 - 50.4}{50.4} \approx 2.19$

About 50.4 million stocks were traded in 1980 and about 160.8 million stocks were traded in 1990, an increase of about 219%.

53. From the graph the world's population reached three billion people sometime in 1955.

55. $5.68(1.017)^{10} \approx 6.7$. The world's population would be about 6.7 billion people in 2005.

57. $f(x) = \dfrac{e^x + e^{-x}}{2}$

X	Y1
-1	1.5431
-.5	1.1276
0	1
.5	1.1276
1	1.5431
1.5	2.3524
2	3.7622

Y1🔳cosh(X)

57. (continued)

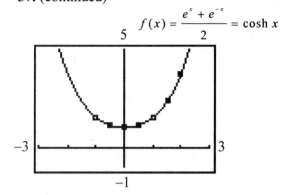

$$f(x) = \dfrac{e^x + e^{-x}}{2} = \cosh x$$

Cumulative Review Problems

59. $5 - 2(3 - x) = 2(2x + 5) + 1$

$5 - 6 + 2x = 4x + 10 + 1$

$2x = -12$

$x = -6$

11.2 Exercises

1. A logarithm is an <u>exponent</u> .

3. In the equation $y = \log_b x$, the domain (the permitted values of x) is $\underline{x > 0}$.

5. $81 = 3^4 \Leftrightarrow \log_3 81 = 4$

7. $36 = 6^2 \Leftrightarrow \log_6 36 = 2$

9. $\dfrac{1}{25} = 5^{-2} \Leftrightarrow \log_5 \dfrac{1}{25} = -2$

11. $\dfrac{1}{32} = 2^{-5} \Leftrightarrow \log_2 \dfrac{1}{32} = -5$

13. $y = e^5 \Leftrightarrow \log_e y = 5$

15. $2 = \log_3 9 \Leftrightarrow 3^2 = 9$

17. $0 = \log_5 1 \Leftrightarrow 5^0 = 1$

19. $\dfrac{1}{2} = \log_{16} 4 \Leftrightarrow 16^{1/2} = 4$

21. $-2 = \log_{10}(0.01) \Leftrightarrow 10^{-2} = 0.01$

23. $-4 = \log_3\left(\dfrac{1}{81}\right) \Leftrightarrow 3^{-4} = \dfrac{1}{81}$

25. $\dfrac{2}{3} = \log_e x \Leftrightarrow e^{2/3} = x$

27. $\log_2 x = 4 \Leftrightarrow 2^4 = x$
$\qquad x = 16$

29. $\log_{10} x = -3 \Leftrightarrow 10^{-3} = x$
$\qquad x = \dfrac{1}{1000}$

31. $\log_4 64 = y \Leftrightarrow 4^y = 64 = 4^3$
$\qquad y = 3$

33. $\log_8\left(\dfrac{1}{64}\right) = y \Leftrightarrow 8^y = \dfrac{1}{64} = \dfrac{1}{8^2} = 8^{-2}$
$\qquad y = -2$

35. $\log_a 144 = 2 \Leftrightarrow a^2 = 144 = 12^2$
$\qquad a = 12$

37. $\log_a 1000 = 3 \Leftrightarrow a^3 = 1000 = 10^3$
$\qquad a = 3$

39. $\log_{25} 5 = w \Leftrightarrow 25^w = 5$
$\qquad (5^2)^w = 5^{2w} = 5^1$
$\qquad 2w = 1$
$\qquad w = \dfrac{1}{2}$

41. $\log_3\left(\dfrac{1}{3}\right) = w$
$\qquad 3^w = \dfrac{1}{3} = \dfrac{1}{3^1} = 3^{-1}$
$\qquad w = -1$

43. $\log_{15} w = 0 \Leftrightarrow 15^0 = w$
$\qquad w = 1$

45. $\log_w 81 = -2 \Leftrightarrow w^{-2} = 81$
$\qquad w^2 = \dfrac{1}{81}$
$\qquad w = \pm\dfrac{1}{9}, \text{ pick } + \text{ since } w > 0$
$\qquad w = \dfrac{1}{9}$

47. $\log_{10}(0.001) = x \Leftrightarrow 10^x = 0.001 = 10^{-3}$
$\qquad x = -3$

49. $\log_2 128 = x \Leftrightarrow 2^x = 128 = 2^7$
$\qquad x = 7$

51. $\log_{23} 1 = x \Leftrightarrow 23^x = 1 = 23^0$
$\qquad x = 0$

53. $\log_6 \sqrt{6} = x \Leftrightarrow 6^x = \sqrt{6} = 6^{1/2}$
$\qquad x = \dfrac{1}{2}$

55. $\log_2 64 = x \Leftrightarrow 2^x = 64 = 2^6$
$\qquad x = 6$

57. $\log_4 x = y$
$\qquad 4^y = x$

(continued)

314

57. (continued)

$x = 4^y$	y
$\dfrac{1}{16}$	-2
$\dfrac{1}{4}$	-1
1	0
4	1
16	2

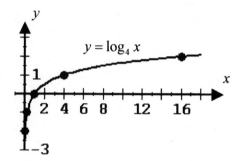

59. $\log_{1/3} x = y \Leftrightarrow (1/3)^y = x$

$x = (1/3)^y$	y
9	-2
3	-1
1	0
$1/3$	1
$1/9$	2

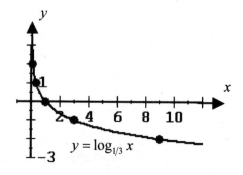

61. $\log_{10} x = y \Leftrightarrow 10^y = x$

$x = 10^y$	y
$1/100$	-2
$1/10$	-1
1	0
10	1
100	2

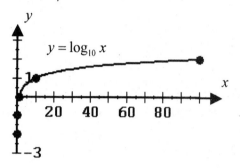

63. $f(x) = \log_3 x,\ f(x) \to y$

$y = \log_3 x \Leftrightarrow 3^y = x$

$x = 3^y$	y
$1/9$	-2
$1/3$	-1
1	0
3	1
9	2

$f^{-1}(x) = 3^x,\ f^{-1}(x) \to y$

$y = 3^x$

x	$y = 3^x$
-2	$1/9$
-1	$1/3$
0	0
1	3
2	9

(continued)

63. (continued)

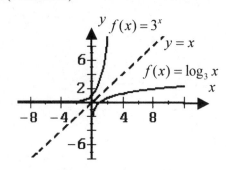

65. $pH = -\log_{10}\left[H^+\right]$

$$pH = -\log_{10}\left[10^{-2.5}\right]$$

$$-pH = \log_{10}\left[10^{-2.5}\right]$$

$$10^{-pH} = 10^{-2.5}$$

$$-pH = -2.5$$

$$pH = 2.5$$

The pH is 2.5.

67. $pH = -\log_{10}\left[H^+\right]$

$$8 = -\log_{10}\left[H^+\right]$$

$$-8 = \log_{10}\left[H^+\right]$$

$$10^{-8} = H^+$$

The concentration of hydrogen ions in the solution is 10^{-8}.

69. $pH = -\log_{10}\left[H^+\right]$

$$9.25 = -\log_{10}\left[H^+\right]$$

$$-9.25 = \log_{10}\left[H^+\right]$$

$$H^+ = 10^{-9.25} = 5.623 \times 10^{-10}$$

The concentration of hydrogen ions in the solution is 5.623×10^{-10}.

71. $$N = 1200 + (2500)(\log_{10} d)$$

$$\log_{10} d = \frac{N - 1200}{2500}$$

$$\log_{10} 100,000 = \frac{N - 1200}{2500}$$

$$10^{\frac{N-1200}{2500}} = 100,000 = 10^5$$

$$\frac{N - 1200}{2500} = 5$$

$$N - 1200 = 12,500$$

$$N = 13,700$$

13,700 sets of software were sold.

73. $$N = 1200 + (2500)(\log_{10} d)$$

$$\log_{10} d = \frac{18,700 - 1200}{2500}$$

$$\log_{10} d = 7$$

$$d = 10^7$$

$$d = 10,000,000$$

They should spend $10,000,000.

75. $5^{\log_5 4} = y \Leftrightarrow \log_5 4 = \log_5 y \rightarrow y = 4$

Cumulative Review Problems

77. $y = -\dfrac{2}{3}x + 5$,

x	y
0	5
3	3

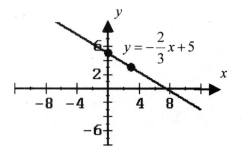

79. $m = \dfrac{y_2 - y_1}{x_2 - x_1} = \dfrac{3-2}{-6-(-1)} = -\dfrac{1}{5}$

81. (a) $C(t) = P(1.04)^t$

$C(5) = 4400(1.04)^5 = \$5353.27$

(b) $C(t) = P(1.04)^t$

$C(10) = 16,500(1.04)^{10} = \$24,424.03$

83. Cost for train:

$98 + 0.75(2)(23) = 132.5$

Cost to drive:

$2.50(23) + 120 + 1(23) = 200.50$

$200.5 - 132.5 = 68$

It is cheaper to take the train. He would save $68 per month by using the train.

11.3 Exercises

1. $\log_7 MN = \log_7 M + \log_7 N$

3. $\log_5(7 \cdot 11) = \log_5 7 + \log_5 11$

5. $\log_b 9f = \log_b 9 + \log_b f$

7. $\log_9 \left(\dfrac{2}{7} \right) = \log_9 2 - \log_9 7$

9. $\log_a \left(\dfrac{G}{7} \right) = \log_a G - \log_a 7$

11. $\log_a \left(\dfrac{E}{F} \right) = \log_a E - \log_a F$

13. $\log_8 a^7 = 7 \log_8 a$

15. $\log_b A^{-2} = -2 \log_b A$

17. $\log_5 \sqrt{w} = \log_5 w^{1/2}$

$\qquad = \dfrac{1}{2} \log_5 w$

19. $\log_5 \sqrt{x} y^3 = \log_5 x^{1/2} + \log_5 y^3$

$\qquad = \dfrac{1}{2} \log_5 x + 3 \log_5 y$

21. $\log_{13} \left(\dfrac{5B}{A^2} \right) = \log_{13}(5B) - \log_{13} A^2$

$\qquad\qquad = \log_{13} 5 + \log_{13} B - 2 \log_{13} A$

23. $\log_2 \left(\dfrac{5xy^4}{\sqrt{z}} \right)$

$= \log_2(5xy^4) - \log_2 \sqrt{z}$

$= \log_2 5 + \log_2 x + \log_2 y^4 - \log_2 z^{1/2}$

$= \log_2 5 + \log_2 x + 4 \log_2 y - \dfrac{1}{2} \log_2 z$

25. $\log_b \sqrt[3]{\dfrac{x}{y^2 z}}$

$= \log_b \left(\dfrac{x}{y^2 z} \right)^{1/3}$

$= \dfrac{1}{3} \log_b \left(\dfrac{x}{y^2 z} \right)$

$= \dfrac{1}{3} \left[\log_b x - \log_b(y^2 z) \right]$

$= \dfrac{1}{3} \left[\log_b x - \log_b y^2 - \log_b z \right]$

$= \dfrac{1}{3} \left[\log_b x - 2 \log_b y - \log_b z \right]$

$= \dfrac{1}{3} \log_b x - \dfrac{2}{3} \log_b y - \dfrac{1}{3} \log_b z$

27. $\log_4 13 + \log_4 y + \log_4 3$
$= \log_4(13 \cdot y \cdot 3) = \log_4(39y)$

29. $5\log_3 x - \log_3 7 = \log_3 x^5 - \log_3 7$
$= \log_3 \dfrac{x^5}{7}$

31. $\dfrac{2}{3}\log_b x + \dfrac{1}{2}\log_b y - 3\log_b z$
$= \log_b x^{2/3} + \log_b y^{1/2} - \log_b z^3$
$= \log_b \sqrt[3]{x^2} + \log_b \sqrt{y} - \log_b z^3$
$= \log_b \dfrac{\sqrt[3]{x^2}\sqrt{y}}{z^3}$

33. $\log_3 3 = 1$

35. $\log_e e = 1$

37. $\log_9 1 = 0$

39. $\log_5 5 + \log_5 1 = 1 + 0 = 1$

41. $\log_8 x = \log_8 7 \Rightarrow x = 7$

43. $\log_5(2x + 7) = \log_5(29)$
$2x + 7 = 29$
$2x = 22$
$x = 11$

45. $\log_3 1 = x \Rightarrow x = 0$

47. $\log_7 7 = x \Rightarrow x = 1$

49. $\log_{10} x + \log_{10} 25 = 2$
$\log_{10}(25x) = 2 \Leftrightarrow 10^2 = 25x$
$25x = 100 \rightarrow x = 4$

51. $\log_2 7 = \log_2 x - \log_2 3$
$\log_2 7 = \log_2 \dfrac{x}{3}$
$\dfrac{x}{3} = 7$
$x = 21$

53. $3\log_5 x = \log_5 8$
$\log_5 x^3 = \log_5 8$
$x^3 = 8$
$x = 2$

55. $\log_e x - \log_e 2 = 2$
$\log_e \dfrac{x}{2} = 2 \Leftrightarrow e^2 = \dfrac{x}{2}$
$x = 2e^2$

57. $\log_6(5x + 21) - \log_6(x + 3) = 1$
$\log_6 \dfrac{5x + 21}{x + 3} = 1 \Rightarrow \dfrac{5x + 21}{x + 3} = 6$
$5x + 21 = 6x + 18$
$x = 3$

59. $5^{\log_5 4} + 3^{\log_3 2} = 4 + 2 = 6$

61. Let $\log_b M = x, \; \log_b N = y \Leftrightarrow$
$b^x = M, \; b^y = N$
$\dfrac{M}{N} = \dfrac{b^x}{b^y}$
$\dfrac{M}{N} = b^{x-y}$
$\log_b \dfrac{M}{N} = \log_b b^{x-y}$
$\log_b \dfrac{M}{N} = x - y$
$\log_b \dfrac{M}{N} = \log_b M - \log_b N$

Cumulative Review Problems

63. $A = \pi r^2 = \pi(4)^2 \approx 50.27 \text{ m}^2$

65. $5x + 3y = 9 \xrightarrow{\times 2} 10x + 6y = 18$

$7x - 2y = 25 \xrightarrow{\times 3} \underline{21x - 6y = 75}$

$\qquad\qquad\qquad 31x \qquad = 93$

$x = 3$

$5(3) + 3y = 9$

$3y = -6$

$y = -2$

$(3, -2)$ is the solution.

67. $\dfrac{9.30 \times 10^8 - 8.01 \times 10^8}{8.01 \times 10^8} = 0.161048\ldots$

$\dfrac{E - 9.3 \times 10^8}{9.3 \times 10^8} = 0.161048\ldots$

$E \approx 1.08 \times 10^9$

Emissions increased 16.1% from 1996 to 2000. Emissions in 2004 will be 1.08×10^9 metric tons.

69. $a = \dfrac{v_2 - v_1}{t} = \dfrac{0 - 38}{3} = -\dfrac{38 \text{ ft}}{3 \text{ sec}}$

$v_2 = v_1 + at = 38 + \dfrac{-38}{3}(2) = \dfrac{38}{3} \dfrac{\text{ft}}{\text{sec}}$

$\dfrac{38 \text{ ft}}{3 \text{ sec}} \cdot \dfrac{\text{mile}}{5280 \text{ ft}} \cdot \dfrac{3600 \text{ sec}}{\text{hour}} \approx 8.64 \text{ mph}$

The speed after 2 seconds was approximately 8.64 mph.

$\dfrac{38 \text{ ft}}{\text{second}} \cdot \dfrac{\text{mile}}{5280 \text{ ft}} \cdot \dfrac{3600 \text{ seconds}}{\text{hour}}$

$\approx 25.9 \text{ mph} < 35 \text{ mph}$

He was not over the speed limit.

11.4 Exercises

1. $\log 5.13 \approx 0.7101173651$

3. $\log 25.6 \approx 1.408239965$

5. $\log 356 \approx 2.551449998$

7. $\log 125,000 \approx 5.096910013$

9. $\log 0.0123 \approx -1.910094889$

11. Error. You cannot take the log of a negative number.

13. $\log x = 2.016$

$x = 10^{2.016} \approx 103.7528416$

15. $\log x = 1.7860$

$x = 10^{1.7860} \approx 61.09420249$

17. $\log x = 3.9304$

$x = 10^{3.9304} \approx 8519.223264$

19. $\log x = 6.4683$

$x = 10^{6.4683} \approx 2939679.609$

21. $\log x = -3.3893$

$x = 10^{-3.3893} \approx 0.000408037$

23. $\log x = 2.0030$

$x = 10^{2.0030} \approx 100.6391669$

25. $\text{antilog}(7.6215) \approx 41831168.87$

27. $\text{antilog}(-1.0826) \approx 0.0826799109$

29. $\ln 5.62 \approx 1.726331664$

31. $\ln 107 \approx 4.672828834$

33. $\ln 136,000 \approx 11.82041016$

35. $\ln 0.00579 \approx -5.151622987$

37. $\ln x = 0.95$
$x = e^{0.95} \approx 2.585709659$

39. $\ln x = 2.4$
$x = e^{2.4} \approx 11.02317638$

41. $\ln x = -0.13$
$x = e^{-0.13} \approx 0.8780954309$

43. $\ln x = -2.7$
$x = e^{-2.7} \approx 0.0672055127$

45. $\text{antilog}_e(6.1582) \approx 472.5766708$

47. $\text{antilog}_e(-2.1298) \approx 0.1188610637$

49. $\log_3 9.2 = \dfrac{\log 9.2}{\log 3} \approx 2.020006063$

51. $\log_7(7.35) = \dfrac{\log 7.35}{\log 7} \approx 1.025073184$

53. $\log_6 0.127 = \dfrac{\log 0.127}{\log 6} \approx -1.151699337$

55. $\log_{11} 128 = \dfrac{\log 128}{\log 11} \approx 2.023453784$

57. $\log_4 0.07733 = \dfrac{\ln 0.07733}{\ln 4} \approx -1.846414$

59. $\log_{18} 98,625 = \dfrac{\ln 98,625}{\ln 18} \approx 3.978408669$

61. $\ln 1537 \approx 7.337587744$

63. $\text{antilog}_e(-1.874) \approx 0.1535083985$

65. $\log x = 8.5634$
$x = 10^{8.5634} \approx 3.65931672 \times 10^8$

67. $\log_4 x = 0.8645 \Leftrightarrow x = 4^{0.8645}$
$x \approx 3.314979618$

69. $y = \log_6 x$

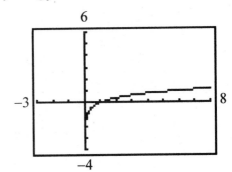

71. $y = \log_{0.2} x$

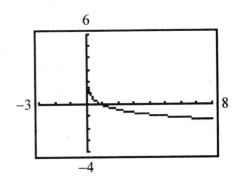

73. $N = 32.82 + 1.0249 \ln x$
$N = 32.82 + 1.0249 \ln 10 \approx 35.18$ in 2000
$N = 32.82 + 1.0249 \ln 20 \approx 35.89$ in 2010
$\dfrac{35.89 - 35.18}{35.18} \approx 0.02$, a 2% increase

75. $R = \log x$
$R = \log 184,000$
$R \approx 5.26$

77. $R = \log x$

$5.4 = \log x \Leftrightarrow x = 10^{5.4} \approx 251188.6432$

The shock wave is about 251,000 times greater than the smallest detectable shock wave.

Cumulative Review Problems

79. $2y^2 + 4y - 3 = 0$

$y = \dfrac{-4 \pm \sqrt{4^2 - 4(2)(-3)}}{2(2)} = \dfrac{-4 \pm \sqrt{40}}{4}$

$y = \dfrac{-4 \pm \sqrt{4 \cdot 10}}{4} = \dfrac{-4 \pm 2\sqrt{10}}{4}$

$y = \dfrac{-2 \pm \sqrt{10}}{2}$

81. Let x, y, z, w, s be the distances between adjacent exits, then:

$x + y + z + w + s = 36$

$x + y \qquad\quad = 12 \rightarrow$ first eqn

$\quad y + z \qquad = 15$

$\qquad z + w \;\; = 12 \rightarrow$ first eqn

$\qquad\quad w + s = 15$

$12 + 12 + s = 36$

$\qquad\qquad s = 12$

$w + s = w + 12 = 15$

$\qquad\quad w = 3$

$z + w = z + 3 = 12$

$\qquad\quad z = 9$

$y + z = y + 9 = 15$

$\qquad\quad y = 6$

$x + y = x + 6 = 12$

$\qquad\quad x = 6$

(continued)

81. (continued)

distance between Exit 1 and Exit 4:
$x + y + z = 6 + 6 + 9 = 21$ miles

distance between Exit 1 and Exit 5:
$x + y + z + w$
$= 6 + 6 + 9 + 3 = 12 + 12 = 24$ miles

11.5 Exercises

1. $\log_8(x+12) + \log_8 4 = 2$

$\log_8(4x + 48) = 2 \Leftrightarrow 4x + 48 = 8^2$

$4x + 48 = 64$

$4x = 16$

$x = 4$

check: $\log_8(4+12) + \log_8 4 \overset{?}{=} 2$

$\log_8 16 + \log_8 4 \overset{?}{=} 2$

$\log_8 64 \overset{?}{=} 2$

$2 = 2$

3. $\log_5(x-2) + \log_5 3 = 2$

$\log_5(3x - 6) = 2 \Leftrightarrow 3x - 6 = 5^2 = 25$

$3x = 31$

$x = \dfrac{31}{3}$

check: $\log_5\left(\dfrac{31}{3} - 2\right) + \log_5 3 \overset{?}{=} 2$

$\log_5\left(3\left(\dfrac{31}{3} - 2\right)\right) \overset{?}{=} 2$

$\log_5 25 \overset{?}{=} 2$

$2 = 2$

5. $\quad \log_3(2x-1) = 2 - \log_3 4$

$\log_3(2x-1) + \log_3 4 = 2$

$\log_3(8x-4) = 2 \Leftrightarrow 8x - 4 = 3^2 = 9$

$8x = 13$

$x = \dfrac{13}{8}$

check: $\log_3\left(2\cdot\dfrac{13}{8}-1\right) \overset{?}{=} 2 - \log_3 4$

$\log_3\left(\dfrac{9}{4}\right) \overset{?}{=} 2 - \log_3 4$

$\log_3 9 - \log_3 4 \overset{?}{=} 2 - \log_3 4$

$2 - \log_3 4 = 2 - \log_3 4$

7. $\quad \log(30x+40) = 2 + \log(x-1)$

$\log(30x+40) - \log(x-1) = 2$

$\log\left(\dfrac{30x+40}{x-1}\right) = 2 \Leftrightarrow \dfrac{30x+40}{x-1} = 10^2$

$30x + 40 = 100x - 100$

$70x = 140$

$x = 2$

check: $\log(30(2)+40) \overset{?}{=} 2 + \log(2-1)$

$\log 100 \overset{?}{=} 2 + \log(1)$

$2 \overset{?}{=} 2 + \log(1) = 2 + 0$

$2 = 2$

9. $\quad 2 + \log_6(x-1) = \log_6(12x)$

$\log_6(x-1) - \log_6(12x) = -2$

$\log_6\left(\dfrac{x-1}{12x}\right) = -2 \Leftrightarrow \dfrac{x-1}{12x} = 6^{-2}$

$36x - 36 = 12x$

$24x = 36, \; x = \dfrac{3}{2}$

(continued)

9. (continued)

check: $2 + \log_6\left(\dfrac{3}{2}-1\right) \overset{?}{=} \log_6\left(12\cdot\dfrac{3}{2}\right)$

$2 + \log_6\dfrac{1}{2} \overset{?}{=} \log_6\dfrac{36}{2}$

$2 + \log_6 1 - \log_6 2 \overset{?}{=} \log_6\dfrac{36}{2}$

$2 + 0 - \log_6 2 \overset{?}{=} \log_6 36 - \log_6 2$

$2 - \log_6 2 = 2 - \log_6 2$

11. $\log(x+20) - \log x = 2$

$\log(x+20) - \log x = 2$

$\log\left(\dfrac{x+20}{x}\right) = 2 \Leftrightarrow \dfrac{x+20}{x} = 10^2$

$100x = x + 20$

$99x = 20$

$x = \dfrac{20}{99}$

check: $\log\left(\dfrac{20}{99}+20\right) - \log\left(\dfrac{20}{99}\right) \overset{?}{=} 2$

$\log\left(\dfrac{2000}{99}\right) - \log\left(\dfrac{20}{99}\right) \overset{?}{=} 2$

$\log\left(\dfrac{100\cdot 20}{99}\right) - \log\left(\dfrac{20}{99}\right) \overset{?}{=} 2$

$\log 100 + \log\left(\dfrac{20}{99}\right) - \log\left(\dfrac{20}{99}\right) \overset{?}{=} 2$

$2 = 2$

13. $2\log_3 4 = \log_3 x - \log_3(x-1)$

$\log_3 4^2 = \log_3\left(\dfrac{x}{x-1}\right)$

$\dfrac{x}{x-1} = 4^2 = 16$

$x = 16x - 16$

(continued)

13. (continued)

$$15x = 16$$

$$x = \frac{16}{15}$$

check: $2\log_3 4 \overset{?}{=} \log_3\left(\frac{16}{15}\right) - \log_3\left(\frac{16}{15} - 1\right)$

$$\log_3 4^2 \overset{?}{=} \log_3\left(\frac{16}{15}\right) - \log_3\left(\frac{1}{15}\right)$$

$$\log_3 4^2 \overset{?}{=} \log_3\left(\frac{\frac{16}{15}}{\left(\frac{1}{15}\right)}\right) = \log_3 16$$

$$\log_3 16 = \log_3 16$$

15. $\qquad 1 + \log(x-2) = \log(6x)$

$$\log(6x) - \log(x-2) = 1$$

$$\log\frac{6x}{x-2} = 1 \Leftrightarrow \frac{6x}{x-2} = 10^1$$

$$6x = 10x - 20$$

$$4x = 20$$

$$x = 5$$

check: $1 + \log(5-2) \overset{?}{=} \log(6 \cdot 5) = \log 30$

$$1 + \log 3 \overset{?}{=} \log(10 \cdot 3)$$

$$1 + \log 3 \overset{?}{=} \log 10 + \log 3$$

$$1 + \log 3 = 1 + \log 3$$

17. $\log_2(x+5) - 2 = \log_2 x$

$$\log_2(x+5) - \log_2 x = 2$$

$$\log_2 \frac{x+5}{x} = 2 \Leftrightarrow \frac{x+5}{x} = 2^2 = 4$$

$$x + 5 = 4x$$

$$3x = 5$$

$$x = \frac{5}{3}$$

(continued)

17. (continued)

check: $\log_2\left(\frac{5}{3} + 5\right) - 2 \overset{?}{=} \log_2 \frac{5}{3}$

$$\log_2 \frac{20}{3} - 2 \overset{?}{=} \log_2 \frac{5}{3}$$

$$\log_2\left(4 \cdot \frac{5}{3}\right) - 2 \overset{?}{=} \log_2 \frac{5}{3}$$

$$\log_2 4 + \log_2 \frac{5}{3} - 2 \overset{?}{=} \log_2 \frac{5}{3}$$

$$2 + \log_2 \frac{5}{3} - 2 \overset{?}{=} \log_2 \frac{5}{3}$$

$$\log_2 \frac{5}{3} = \log_2 \frac{5}{3}$$

19. $\qquad 2\log_7 x = \log_7(x+4) + \log_7 2$

$$\log_7 x^2 - \log_7(x+4) = \log_7 2$$

$$\log_7 \frac{x^2}{(x+4)} = \log_7 2$$

$$\frac{x^2}{(x+4)} = 2$$

$$x^2 - 2x - 8 = 0$$

$$(x-4)(x+2) = 0$$

$$x = 4, \ x = -2, \ \text{reject, gives } \log_7(\text{negative})$$

check: $2\log_7 4 \overset{?}{=} \log_7(4+4) + \log_7 2$

$$\log_7 4^2 \overset{?}{=} \log_7(8) + \log_7 2$$

$$\log_7 16 \overset{?}{=} \log_7(8 \cdot 2)$$

$$\log_7 16 = \log_7 16$$

21. $\ln 10 - \ln x = \ln(x-3)$

$$\ln \frac{10}{x} = \ln(x-3)$$

$$\frac{10}{x} = x - 3 \qquad \text{(continued)}$$

$$x^2 - 3x - 10 = 0$$

21. (continued)

$$(x-5)(x+2)=0$$

$x=5$, $x=-2$, reject, gives ln(negative)

check: $\ln 10 - \ln 5 \overset{?}{=} \ln(5-3)$

$$\ln\frac{10}{5}\overset{?}{=}\ln 2$$

$$\ln 2 = \ln 2$$

23. $8^{x-1}=11$

$$\log 8^{x-1} = \log 11$$

$$(x-1)\log 8 = \log 11$$

$$x-1 = \frac{\log 11}{\log 8}$$

$$x = 1 + \frac{\log 11}{\log 8}$$

$$x = \frac{\log 8}{\log 8} + \frac{\log 11}{\log 8}$$

$$x = \frac{\log 8 + \log 11}{\log 8}$$

25. $2^{3x+4}=17$

$$\log 2^{3x+4} = \log 17$$

$$(3x+4)\log 2 = \log 17$$

$$3x+4 = \frac{\log 17}{\log 2}$$

$$3x = \frac{\log 17}{\log 2} - 4 \cdot \frac{\log 2}{\log 2}$$

$$x = \frac{\log 17}{3\log 2} - 4 \cdot \frac{\log 2}{3\log 2}$$

$$x = \frac{\log 17 - 4\log 2}{3\log 2}$$

27. $6^{4x-1}=225$

$$\log 6^{4x-1} = \log 225$$

$$(4x-1)\log 6 = \log 225$$

(continued)

27. (continued)

$$4x-1 = \frac{\log 225}{\log 6}$$

$$4x = 1 + \frac{\log 225}{\log 6}$$

$$4x = \frac{\log 6 + \log 225}{\log 6}$$

$$x = \frac{\log 6 + \log 225}{4\log 6}$$

$$x \approx 1.006$$

29. $5^x = 4^{x+1}$

$$\log 5^x = \log 4^{x+1}$$

$$x\log 5 = (x+1)\log 4$$

$$x\log 5 = x\log 4 + \log 4$$

$$x(\log 5 - \log 4) = \log 4$$

$$x = \frac{\log 4}{\log 5 - \log 4}$$

$$x \approx 6.213$$

31. $e^{x-1} = 28$

$$\ln e^{x-1} = \ln 28$$

$$(x-1)\ln e = \ln 28$$

$$x = 1 + \ln 28$$

$$x \approx 4.332$$

33. $88 = e^{2x+1}$

$$\ln 88 = \ln e^{2x+1}$$

$$\ln 88 = (2x+1)\ln e$$

$$\ln 88 = (2x+1)(1)$$

$$\ln 88 = 2x+1$$

$$2x = \ln 88 - 1$$

$$x = \frac{\ln 88 - 1}{2}$$

$$x \approx 1.739$$

35.
$$A = P(1+r)^t$$
$$5000 = 1500(1+0.08)^t$$
$$1.08^t = \frac{10}{3}$$
$$\ln 1.08^t = \ln\frac{10}{3}$$
$$t = \frac{\ln\frac{10}{3}}{\ln 1.08} = 15.64392564...$$

It will take approximately 16 years.

37.
$$A = P(1+r)^t$$
$$3P = P(1+0.06)^t$$
$$1.06^t = 3$$
$$\ln 1.06^t = \ln 3$$
$$t\ln 1.06 = \ln 3$$
$$t = \frac{\ln 3}{\ln 1.06}$$
$$t = 18.85417668...$$

It will take approximately 19 years.

39.
$$A = P(1+r)^t$$
$$6500 = 5000(1+r)^6$$
$$1.3 = (1+r)^6$$
$$\ln 1.3 = \ln(1+r)^6 = 6\ln(1+r)$$
$$\ln(1+r) = \frac{\ln 1.3}{6}$$
$$e^{\ln(1+r)} = e^{\frac{\ln 1.3}{6}}$$
$$1+r = e^{\frac{\ln 1.3}{6}}$$
$$r = e^{\frac{\ln 1.3}{6}} - 1$$
$$r = 0.0446975079...$$

The rate is approximately 4.5%.

41.
$$A = A_0 e^{rt}$$
$$9 = 6e^{0.02t}, \; 1.5 = e^{0.02t}$$
$$\ln 1.5 = \ln e^{0.02t}$$
$$\ln 1.5 = 0.02t$$
$$t = \frac{\ln 1.5}{0.02} = 20.27325541...$$

It would take approximately 20 years.

43.
$$A = A_0 e^{rt}$$
$$4A_0 = A_0 e^{0.02t}, \; 4 = e^{0.02t}$$
$$\ln 4 = \ln e^{0.02t}$$
$$\ln 4 = 0.02t$$
$$t = \frac{\ln 4}{0.02} = 69.31471806...$$

It would take approximately 69 years.

45.
$$N = 20,800(1.264)^x$$
$$N = 20,800(1.264)^{13}$$
$$N = 437,295.79$$

In 2003 there will be approximately 437,000 employees.

47.
$$N = 20,800(1.264)^x$$
$$274,000 = 20,800(1.264)^x$$
$$1.264^x = \frac{274,000}{20,800}$$
$$\ln 1.264^x = \ln\frac{274,000}{20,800}$$
$$x = \frac{\ln\frac{274,000}{20,800}}{\ln 1.264} = 11.00461354...$$

The number of employees will reach 274,000 sometime in 2001.

49.

$$A = A_0 e^{rt}$$

$$3.5 = 3e^{0.03t}$$

$$e^{0.03t} = \frac{7}{6}$$

$$\ln e^{0.03t} = \ln \frac{7}{6}$$

$$0.03t = \ln \frac{7}{6}$$

$$t = \frac{\ln \frac{7}{6}}{0.03} = 5.138355994\ldots$$

There will be 3.5 million people in approximately 5 years.

51.

$$A = A_0 e^{rt}$$

$$1800 = 200e^{0.04t}$$

$$e^{0.04t} = 9$$

$$0.04t = \ln 9$$

$$t = \frac{\ln 9}{0.04} = 54.93061443\ldots$$

It will take approximately 55 hours.

53. $A = A_0 e^{rt}$

$$A = 24,500e^{0.05(13)} = 46930.75031\ldots$$

By the end of 2010 approximately 46,931 people will be infected.

55. $R = \log\left(\dfrac{I}{I_0}\right)$

$$6.8 = \log\left(\frac{I_N}{I_0}\right) = \log I_N - \log I_0$$

$$7.2 = \log\left(\frac{I_J}{I_0}\right) = \log I_J - \log I_0$$

(continued)

55. (continued)

Subtracting the two equations gives

$$-0.4 = \log I_N - \log I_J = \log \frac{I_N}{I_J}$$

$$10^{\log \frac{I_N}{I_J}} = 10^{-0.4}$$

$$\frac{I_N}{I_J} = 10^{-0.4}$$

$$I_J = 10^{0.4} I_N \approx 2.5 I_N$$

The Japan earthquake was about 2.5 times as intense as the Northridge earthquake.

57. $R = \log\left(\dfrac{I}{I_0}\right)$

$$8.3 = \log\left(\frac{I_S}{I_0}\right) = \log I_S - \log I_0$$

$$6.8 = \log\left(\frac{I_J}{I_0}\right) = \log I_J - \log I_0$$

Subtracting the two equations gives

$$1.5 = \log I_S - \log I_J = \log \frac{I_S}{I_J}$$

$$10^{\log \frac{I_S}{I_J}} = 10^{1.5}$$

$$\frac{I_S}{I_J} = 10^{1.5}$$

$$I_S = 10^{1.5} I_J \approx 32 I_J$$

The intensity of the San Francisco earthquake was approximately 32 times greater than the intensity of the Japanese earthquake.

59. Graph $y_1 = 300e^{0.12x}$ and $y_2 = 750 + 100x$ on the same screen and use the *intersect* feature to solve.

(continued)

59. (continued)

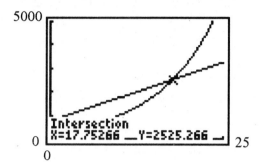

The populations will become equal in approximately 17.8 years.

Cumulative Review Problems

61. $\sqrt{98x^3y^2} = \sqrt{49x^2y^2 \cdot 2x}$
$$= 7xy\sqrt{2x}$$

63. 9 years old: 1 student, given in problem
12 years old: 54 students, also given
13 years old: $54 + 13 = 87$ students
14 or 15 years old: 80 students, given
10 years old + 11 years old:
$248 - 80 - 87 - 54 - 1 = 26$ students

Let $x =$ the number of students 10 years old and $y =$ the number of students 11 years old.

$x + y = 26$

$x + 10 = y \rightarrow x + x + 10 = 26$

$2x = 16,\ x = 8$ students 10 years old

$y = x + 10 = 18$ students 11 years old

Summary:
    9 years old: 1 student
    10 years old: 8 students
    11 years old: 18 students
    12 years old: 54 students
    13 years old: 87 students
    14,15 years old: 80 students

Putting Your Skills to Work

1. $T = C + (T_0 - C)e^{-kt}$
$$140 = 72 + (220 - 72)e^{-k(30)}$$
$$68 = 148e^{-k(30)}$$
$$e^{k(30)} = \frac{148}{68}$$
$$\ln e^{k(30)} = \ln \frac{148}{68}$$
$$k(30) = \ln \frac{148}{68}$$
$$k = \frac{\ln \dfrac{148}{68}}{30} \approx 0.026$$

2. $T = C + (T_0 - C)e^{-kt}$
$$T = 72 + (220 - 72)e^{-0.26t}$$

3. $T = 72 + (220 - 72)e^{-0.26t}$
$$90 = 72 + (220 - 72)e^{-0.026t}$$
$$18 = 148e^{-0.026t}$$
$$e^{0.026t} = \frac{148}{18}$$
$$0.026t = \ln \frac{148}{18}$$
$$t = \frac{\ln \dfrac{148}{18}}{0.026} = 81.03232753\ldots$$

It will take approximately 81 minutes for the pie to cool.

4. $T = C + (T_0 - C)e^{-kt}$
ceramic: $T = 70 + (190 - 70)e^{-0.13722(20)}$
$$T = 77.7°F$$
foam: $T = 70 + (190 - 70)e^{-0.05(20)}$
$$T = 114.1°F \qquad \text{(continued)}$$

4. (continued)

cardboard: $T = 70 + (190 - 70)e^{-0.08(20)}$

$T = 94.2°F$

5. ceramic: $125 = 70 + (190 - 70)e^{-0.13722t}$

$125 = 70 + (190 - 70)e^{-0.13722t}$

$e^{0.13722t} = \dfrac{120}{55}$

$0.13722t = \ln\dfrac{120}{55}$

$t = \dfrac{\ln\dfrac{120}{55}}{0.13722} \approx 5.7 \text{ minutes}$

foam: $125 = 70 + (190 - 70)e^{-0.05t}$

$125 = 70 + (190 - 70)e^{-0.05t}$

$e^{0.05t} = \dfrac{120}{55}$

$0.05t = \ln\dfrac{120}{55}$

$t = \dfrac{\ln\dfrac{120}{55}}{0.05} \approx 15.6 \text{ minutes}$

cardboard: $125 = 70 + (190 - 70)e^{-0.08t}$

$125 = 70 + (190 - 70)e^{-0.08t}$

$e^{0.08t} = \dfrac{120}{55}$

$0.08t = \ln\dfrac{120}{55}$

$t = \dfrac{\ln\dfrac{120}{55}}{0.08} \approx 9.8 \text{ minutes}$

6. Coffee cools the slowest in a foam container and fastest in a ceramic container.

Chapter 11 Review Problems

1. $f(x) = 4^{3+x}$

x	$y = f(x) = 4^{3+x}$
-5	$1/16$
-4	$1/4$
-3	1
-2	4
-1	16

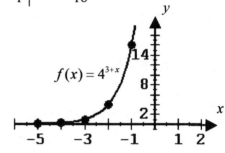

2. $f(x) = e^{x-3}$

x	$y = f(x) = e^{x-3}$
1	0.14
2	0.37
3	1
4	2.72
5	7.39

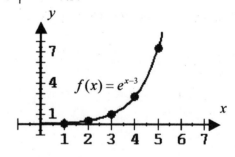

3. $5^{x+2} = 125 = 5^3$

$x + 2 = 3$

$x = 1$

4. $-3 = \log_{10}(0.001) \Leftrightarrow 10^{-3} = 0.001$

5. $\dfrac{1}{32} = 2^{-5} \Leftrightarrow \log_2 \dfrac{1}{32} = -5$

6. $\log_w 16 = 4 \Leftrightarrow w^4 = 16 = 2^4, \ w = 2$

7. $\log_3 x = -2 \Leftrightarrow 3^{-2} = x, \ x = \dfrac{1}{9}$

8. $\log_8 x = 0 \Leftrightarrow 8^0 = x, \ x = 1$

9. $\log_7 w = -1 \Leftrightarrow 7^{-1} = w, \ w = \dfrac{1}{7}$

10. $\log_w 27 = 3 \Leftrightarrow w^3 = 27 = 3^3, \ w = 3$

11. $\log_{10} w = -3 \Leftrightarrow 10^{-3} = w, \ w = 0.001$

12. $\log_{10} 1000 = x \Leftrightarrow 10^x = 1000 = 10^3, \ x = 3$

13. $\log_2 64 = x \Leftrightarrow 2^x = 64 = 2^6, \ x = 6$

14. $\log_2 \dfrac{1}{4} = x \Leftrightarrow 2^x = \dfrac{1}{4} = 2^{-2}, \ x = -2$

15. $\log_5 125 = x \Leftrightarrow 5^x = 125 = 5^3, \ x = 3$

16. $\log_3 x = y \Leftrightarrow 3^y = x$

$x = 3^y$	y
1/9	-2
1/3	-1
1	0
3	1
9	2

(continued)

16. (continued)

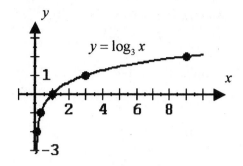

17. $\log_2\left(\dfrac{5x}{\sqrt{w}}\right) = \log_2(5x) - \log_2 \sqrt{w}$

$$= \log_2(5x) - \log_2 w^{1/2}$$

$$= \log_2 5 + \log_2 x - \dfrac{1}{2}\log_2 w$$

18. $\log_2 x^3 \sqrt{y} = \log_2 x^3 + \log_2 \sqrt{y}$

$$= 3\log_2 x + \log_2 y^{1/2}$$

$$= 3\log_2 x + \dfrac{1}{2}\log_2 y$$

19. $\log_3 x + \log_3 w^{1/2} - \log_3 2$

$$= \log_3 x + \log_3 \sqrt{w} - \log_3 2$$

$$= \log_3(x\sqrt{w}) - \log_3 2$$

$$= \log_3 \dfrac{x\sqrt{w}}{2}$$

20. $4\log_8 w - \dfrac{1}{3}\log_8 z = \log_8 w^4 - \log_8 z^{1/3}$

$$= \log_8 w^4 - \log_8 \sqrt[3]{z}$$

$$= \log_8 \dfrac{w^4}{\sqrt[3]{z}}$$

21. $\log_e e^6 = 6\log_e e$

$$= 6(1)$$

$$= 6$$

22. $\log_5 100 - \log_5 x = \log_5 4$

$$\log_5 \frac{100}{x} = \log_5 4$$

$$\frac{100}{x} = 4$$

$$x = 25$$

23. $\log_8 x + \log_8 3 = \log_8 75$

$$\log_8 (3x) = \log_8 75$$

$$3x = 75$$

$$x = 25$$

24. $\log 23.8 = 1.376576957$

25. $\log 0.0817 = -1.087777943$

26. $\ln 3.92 = 1.366091654$

27. $\ln 803 = 6.688354714$

28. $\log n = 1.1367 \Leftrightarrow n = 10^{1.1367}$

$$n = 13.69935122$$

29. $\ln n = 1.7 \Leftrightarrow n = e^{1.7}$

$$n = 5.473947392$$

30. $\log_8 2.81 = \dfrac{\ln 2.81}{\ln 8}$

$$\log_8 2.81 = 0.4968567101$$

31. $\log_7 (x+3) + \log_7 5 = 2$

$$\log_7 (5(x+3)) = 2$$

$$5(x+3) = 7^2$$

$$5x + 15 = 49$$

$$5x = 34$$

$$x = \frac{34}{5}$$

(continued)

31. (continued)

check: $\log_7 \left(\dfrac{34}{5} + 3 \right) + \log_7 5 \overset{?}{=} 2$

$$\log_7 \left(\frac{49}{5} \right) + \log_7 5 \overset{?}{=} 2$$

$$\log_7 49 - \log_7 5 + \log_7 5 \overset{?}{=} 2$$

$$\log_7 7^2 - \log_7 5 + \log_7 5 \overset{?}{=} 2$$

$$2\log_7 7 - \log_7 5 + \log_7 5 \overset{?}{=} 2$$

$$2 = 2$$

32. $\log_3 (2x+3) = \log_3 2 - 3$

$$\log_3 (2x+3) - \log_3 2 = -3$$

$$\log_3 \frac{2x+3}{2} = -3$$

$$\frac{2x+3}{2} = 3^{-3} = \frac{1}{3^3} = \frac{1}{27}$$

$$27(2x+3) = 2$$

$$54x + 81 = 2$$

$$54x = -79$$

$$x = \frac{-79}{54}$$

check: $\log_3 \left(2 \cdot \dfrac{-79}{54} + 3 \right) \overset{?}{=} \log_3 2 - 3$

$$\log_3 \left(\frac{2}{27} \right) \overset{?}{=} \log_3 2 - 3$$

$$\log_3 2 - \log_3 27 \overset{?}{=} \log_3 2 - 3$$

$$\log_3 2 - \log_3 3^3 \overset{?}{=} \log_3 2 - 3$$

$$\log_3 2 - 3\log_3 3 \overset{?}{=} \log_3 2 - 3$$

$$\log_3 2 - 3\log_3 3 \overset{?}{=} \log_3 2 - 3$$

$$\log_3 2 - 3(1) \overset{?}{=} \log_3 2 - 3$$

$$\log_3 2 - 3 = \log_3 2 - 3$$

33. $\log_5(x+1) - \log_5 8 = \log_5 x$

$\log_5(x+1) - \log_5 x = \log_5 8$

$\log_5 \dfrac{x+1}{x} = \log_5 8$

$\dfrac{x+1}{x} = 8$

$x+1 = 8x$

$7x = 1$

$x = \dfrac{1}{7}$

check: $\log_5\left(\dfrac{1}{7}+1\right) - \log_5 8 \overset{?}{=} \log_5\left(\dfrac{1}{7}\right)$

$\log_5\left(\dfrac{8}{7}\right) - \log_5 8 \overset{?}{=} \log_5\left(\dfrac{1}{7}\right)$

$\log_5\left(\dfrac{1}{7}\cdot 8\right) - \log_5 8 \overset{?}{=} \log_5\left(\dfrac{1}{7}\right)$

$\log_5\left(\dfrac{1}{7}\right) + \log_5 8 - \log_5 8 \overset{?}{=} \log_5\left(\dfrac{1}{7}\right)$

$\log_5\left(\dfrac{1}{7}\right) = \log_5\left(\dfrac{1}{7}\right)$

34. $2\log_3(x+3) - \log_3(x+1) = 3\log_3 2$

$\log_3(x+3)^2 - \log_3(x+1) = \log_3 2^3$

$\log_3\left(\dfrac{(x+3)^2}{(x+1)}\right) = \log_3 8$

$\dfrac{(x+3)^2}{(x+1)} = 8$

$(x+3)^2 = 8(x+1)$

$x^2 + 6x + 9 = 8x + 8$

$x^2 - 2x + 1 = 0$

$(x-1)^2 = 0$

$x - 1 = 0$

$x = 1$

(continued)

34. (continued)

check: $2\log_3(1+3) - \log_3(1+1) \overset{?}{=} 3\log_3 2$

$2\log_3(4) - \log_3(2) \overset{?}{=} 3\log_3 2$

$\log_3(4^2) - \log_3(2) \overset{?}{=} 3\log_3 2$

$\log_3(16) - \log_3(2) \overset{?}{=} 3\log_3 2$

$\log_3 \dfrac{16}{2} \overset{?}{=} 3\log_3 2$

$\log_3 8 \overset{?}{=} 3\log_3 2$

$\log_3 2^3 \overset{?}{=} 3\log_3 2$

$3\log_3 2 = 3\log_3 2$

35. $\log_2(x-2) + \log_2(x+5) = 3$

$\log_2\big((x-2)(x+5)\big) = 3$

$(x-2)(x+5) = 2^3 = 8$

$x^2 + 3x - 10 = 8$

$x^2 + 3x - 18 = 0$

$(x+6)(x-3) = 0$

$x = 3,\ x = -6,$ reject -6 since it gives

$\log_2(\text{negative}).$

check: $\log_2(3-2) + \log_2(3+5) \overset{?}{=} 3$

$\log_2(1) + \log_2(8) \overset{?}{=} 3$

$0 + \log_2(2^3) \overset{?}{=} 3$

$3\log_2 2 \overset{?}{=} 3$

$3 = 3$

36. $\log_5(x+1) + \log_5(x-3) = 1$

$\log_5\big((x+1)(x-3)\big) = 1$

$(x+1)(x-3) = 5^1 = 5$

$x^2 - 2x - 3 = 5$

(continued)

36. (continued)

$x^2 - 2x - 8 = 0$

$(x-4)(x+2) = 0$

$x = 4, \ x = -2, \ \text{reject} \ -2 \ \text{since it}$

$\qquad\qquad \text{gives} \ \log_5(\text{negative})$

check: $\log_5(4+1) + \log_5(4-3) \overset{?}{=} 1$

$\qquad\qquad \log_5(5) + \log_5(1) \overset{?}{=} 1$

$\qquad\qquad\qquad 1 + 0 \overset{?}{=} 1$

$\qquad\qquad\qquad\qquad 1 = 1$

37. $\log(2t+1) + \log(4t-1) = 2\log 3$

$\log\big((2t+3)(4t-1)\big) = \log 3^2 = \log 9$

$(2t+3)(4t-1) = 9$

$8t^2 + 10t - 3 = 9$

$8t^2 + 10t - 12 = 0$

$4t^2 + 5t - 6 = 0$

$(4t-3)(t+2) = 0$

$t = \dfrac{3}{4}, \ t = -2, \ \text{reject} \ -2 \ \text{since it gives}$

$\qquad\qquad \log(\text{negative})$

check:

$\log\left(2\cdot\dfrac{3}{4}+3\right) + \log\left(4\cdot\dfrac{3}{4}-1\right) \overset{?}{=} 2\log 3$

$\log\left(\dfrac{9}{2}\right) + \log(2) \overset{?}{=} 2\log 3$

$\log\left(\dfrac{9}{2}\cdot 2\right) \overset{?}{=} 2\log 3$

$\log(9) \overset{?}{=} 2\log 3$

$\log(3^2) \overset{?}{=} 2\log 3$

$2\log 3 = 2\log 3$

38. $\log(2t+4) - \log(3t+1) = \log 6$

$\log\left(\dfrac{2t+4}{3t+1}\right) = \log 6$

$\dfrac{2t+4}{3t+1} = 6$

$2t+4 = 18t + 6$

$16t = -2$

$t = -\dfrac{1}{8}$

check: $\log\left(2\cdot\dfrac{-1}{8}+4\right) - \log\left(3\cdot\dfrac{-1}{8}+1\right) \overset{?}{=} \log 6$

$\log\left(\dfrac{15}{4}\right) - \log\left(\dfrac{5}{8}\right) \overset{?}{=} \log 6$

$\log\left(\dfrac{\dfrac{15}{4}}{\dfrac{5}{8}}\right) \overset{?}{=} \log 6$

$\log\left(\dfrac{\dfrac{15}{4}}{\dfrac{5}{8}}\right) \overset{?}{=} \log 6$

$\log 6 = \log 6$

39. $\qquad 3^x = 14$

$\log 3^x = \log 14$

$x \log 3 = \log 14$

$x = \dfrac{\log 14}{\log 3}$

40. $\qquad 5^x = 4^{x+2}$

$\log 5^x = \log 4^{x+2}$

$x \log 5 = (x+2)\log 4 = x\log 4 + 2\log 4$

$x(\log 5 - \log 4) = 2\log 4$

$x = \dfrac{2\log 4}{\log 5 - \log 4}$

41.
$$16e^{x+1} = 56$$
$$\ln(16e^{x+1}) = \ln 56$$
$$\ln 16 + \ln e^{x+1} = \ln 56$$
$$(x+1)\ln e = \ln 56 - \ln 16 = \ln \frac{56}{16}$$
$$x+1 = \ln 3.5$$
$$x = -1 + \ln 3.5$$

42.
$$e^{2x} = 30.6$$
$$\ln e^{2x} = \ln 30.6$$
$$2x \ln 3 = \ln 30.6$$
$$2x = \ln 30.6$$
$$x = \frac{\ln 30.6}{2}$$

43.
$$2^{3x+1} = 5^x$$
$$\ln 2^{3x+1} = \ln 5^x$$
$$(3x+1)\ln 2 = x \ln 5$$
$$3x \ln 2 + \ln 2 = x \ln 5$$
$$x(3\ln 2 - \ln 5) = -\ln 2$$
$$x = \frac{\ln 2}{\ln 5 - 3\ln 2}$$
$$x \approx -1.4748$$

44.
$$3^{x+1} = 7$$
$$\ln 3^{x+1} = \ln 7$$
$$(x+1)\ln 3 = \ln 7$$
$$x = -1 + \frac{\ln 7}{\ln 3}$$
$$x \approx 0.7712$$

45.
$$e^{3x-4} = 20$$
$$\ln e^{3x-4} = \ln 20$$
$$(3x-4)\ln e = \ln 20$$
$$3x - 4 = \ln 20$$
$$x = \frac{\ln 20 + 4}{3} \approx 2.3319$$

46.
$$1.03^x = 20$$
$$\ln 1.03^x = \ln 20$$
$$x \ln 1.03 = \ln 20$$
$$x = \frac{\ln 20}{\ln 1.03}$$
$$x \approx 101.3482$$

47.
$$A = P(1+r)^t$$
$$2P = P(1+0.08)^t$$
$$2 = 1.08^t$$
$$\ln 2 = \ln 1.08^t$$
$$\ln 2 = t \ln 1.08$$
$$t = \frac{\ln 2}{\ln 1.08}$$

$t \approx 9$ years to double money in account

48.
$$A = P(1+r)^t$$
$$A = 5000(1+0.06)^4$$

$A = \$6312.38$ in account after 4 years

49.
$$A = P(1+r)^t$$
$$20,000 = 12,000(1+0.07)^t$$
$$\frac{5}{3} = 1.07^t$$
$$\ln 1.07^t = \ln \frac{5}{3}$$
$$t \ln 1.07 = \ln \frac{5}{3}$$
$$t = \frac{\ln \frac{5}{3}}{\ln 1.07}$$
$$t = 7.550041795...$$
$$t \approx 8$$

It would take approximately 8 years for $12,000 to grow to $20,000.

50. $A = P(1+r)^t$

$$A_{\text{Robert}} + 500 = A_{\text{Brother}}$$

$$3500(1+0.05)^t + 500 = 3500(1+0.06)^t$$

$$7(1.06)^t - 7(1.05)^t = 1$$

$$(1.06)^t - (1.05)^t = \frac{1}{7}$$

Solve with a graphing calculator. Graph $y_1 = 1.06^x - 1.05^x$ and $y_2 = \frac{1}{7}$ and use the *intersect* feature.

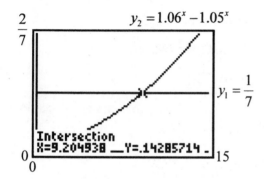

It will take approximately 9 years for Robert's amount to be $500 less than his brother's amount.

51. $\quad A = A_0 e^{rt}$

$$10 = 6e^{0.02t}$$

$$\ln 1.\overline{6} = \ln e^{0.02t}$$

$$\ln 1.\overline{6} = 0.02t \ln e$$

$$t = \frac{\ln 1.\overline{6}}{0.02} = 25.54128119...$$

It will take approximately 26 years.

52. $\quad A = A_0 e^{rt}$

$$16 = 7e^{0.02t}$$

$$\ln \frac{16}{7} = \ln e^{0.02t} \quad \text{(continued)}$$

52. (continued)

$$\ln \frac{16}{7} = 0.02t \ln e$$

$$t = \frac{\ln \dfrac{16}{7}}{0.02} = 41.33392866...$$

It will take approximately 41 years.

53. $\quad A = A_0 e^{rt}$

$$2600 = 2000e^{0.03t}$$

$$\ln 1.3 = \ln e^{0.03t}$$

$$\ln 1.3 = 0.03t$$

$$t = \frac{\ln 1.3}{0.03} = 8.745475482...$$

It will take approximately 9 years.

54. $\quad A = A_0 e^{rt}$

$$95,000 = 40,000e^{0.08t}$$

$$2.375 = e^{0.08t}$$

$$\ln 2.375 = \ln e^{0.08t}$$

$$0.08t = \ln 2.375$$

$$t = \frac{\ln 2.375}{0.08} = 10.81246797...$$

It will take approximately 11 years.

55. $\quad M = \log\left(\frac{I}{I_0}\right)$

$$8.4 = \log\left(\frac{I_A}{I_0}\right) = \log I_A - \log I_0$$

$$-\left(6.7 = \log\left(\frac{I_T}{I_0}\right) = \log I_T - \log I_0\right)$$

$$\overline{\rule{6cm}{0.4pt}}$$

$$1.7 = \log I_A - \log I_T$$

(continue)

55. (continued)

$$\log \frac{I_A}{I_T} = 1.7$$

$$10^{\log \frac{I_A}{I_T}} = 10^{1.7}$$

$$\frac{I_A}{I_T} = 50.11872336...$$

The Alaska earthquake was about 50.1 times more intense than the Turkey earthquake.

56. $W = p_0 V_0 \ln\left(\dfrac{V_1}{V_0}\right)$

(a) $W = 40(15)\ln\left(\dfrac{24}{15}\right) \approx 282$

(b) $100 = p_0 (8) \ln\left(\dfrac{40}{8}\right)$

$$p_0 = \frac{100}{(8)\ln\left(\dfrac{40}{8}\right)} \approx 7.77$$

Chapter 11 Test

1. $f(x) = 3^{4-x}$

x	$y = f(x) = 3^{4-x}$
3	3
4	1

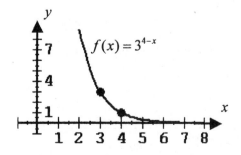

2. $4^{x+3} = 64 = 4^3$

$$x + 3 = 3$$

$$x = 0$$

3. $\log_w 125 = 3 \Leftrightarrow w^3 = 125 = 5^3,\ w = 5$

4. $\log_8 x = -2 \Leftrightarrow x = 8^{-2} = \dfrac{1}{64}$

5. $\log_8 x + \log_8 w - \dfrac{1}{4}\log_8 3$

$$= \log_8 (xw) - \log_8 3^{1/4}$$

$$= \log_8 (xw) - \log_8 \sqrt[4]{3}$$

$$= \log_8 \frac{xw}{\sqrt[4]{3}}$$

6. $\ln 5.99 = 1.7901$

7. $\log 23.6 = 1.3729$

8. $\log_3 1.62 = \dfrac{\log 1.62}{\log 3} = 0.4391$

9. $\log x = 3.7284 \Leftrightarrow x = 10^{3.7284}$

$$x = 5350.569382$$

10. $\ln x = 0.14 \Leftrightarrow x = e^{0.14} = 1.150273799$

11. $\log_8 (x+3) - \log_8 2x = \log_8 4$

$$\log_8 \left(\frac{x+3}{2x}\right) = \log_8 4$$

$$\frac{x+3}{2x} = 4$$

$$x + 3 = 8x$$

$$7x = 3$$

$$x = \frac{3}{7}$$

(continued)

11. (continued)

$$\text{check}: \log_8\left(\frac{3}{7}+3\right)-\log_8\left(2\cdot\frac{3}{7}\right)\overset{?}{=}\log_8 4$$

$$\log_8\left(\frac{24}{7}\right)-\log_8\left(\frac{6}{7}\right)\overset{?}{=}\log_8 4$$

$$\log_8\left(\frac{\frac{24}{7}}{\frac{6}{7}}\right)\overset{?}{=}\log_8 4$$

$$\log_8 4 = \log_8 4$$

12. $\log_8 2x + \log_8 6 = 2$

$$\log_8((2x)(6)) = 2 \Leftrightarrow 12x = 8^2$$

$$12x = 64$$

$$x = \frac{16}{3}$$

$$\log_8 2\cdot\frac{16}{3}+\log_8 6\overset{?}{=}2$$

$$\log_8\left(2\cdot\frac{16}{3}\cdot 6\right)\overset{?}{=}2$$

$$\log_8 64\overset{?}{=}2$$

$$\log_8 8^2\overset{?}{=}2$$

$$2\log_8 8\overset{?}{=}2$$

$$2 = 2$$

13.
$$29 = 16e^{3x+1}$$

$$e^{3x+1} = 0.25$$

$$\ln e^{3x+1} = \ln 0.25$$

$$(3x+1)\ln e = \ln 0.25$$

$$3x+1 = \ln 0.25$$

$$3x = -1+\ln 0.25$$

$$x = \frac{-1+\ln 0.25}{3}$$

14.
$$5^{3x+6} = 17$$

$$\ln 5^{3x+6} = \ln 17$$

$$(3x+6)\ln 5 = \ln 17$$

$$3x+6 = \frac{\ln 17}{\ln 5}$$

$$x = \frac{-6+\dfrac{\ln 17}{\ln 5}}{3} \approx -1.4132$$

15. $A = P(1+r)^t$

$$A = 2000(1+0.08)^5 = 2938.656154\ldots$$

Henry will have $2938.66

16. $A = P(1+r)^t$

$$2P = P(1+0.05)^t$$

$$2 = (1.05)^t$$

$$\ln 2 = \ln 1.05^t$$

$$\ln 2 = t\ln 1.05$$

$$t = \frac{\ln 2}{\ln 1.05} = 14.20669908\ldots$$

Barb can double her money in approximately 14 years.

Cumulative Test for Chapters 1-11

1. $2(-3)+12\div(-2)+3\sqrt{36}$
$$= -6+(-6)+3(6) = 6$$

2. $H = 3bx - 2ay$
$$3bx = H + 2ay$$
$$x = \frac{H+2ay}{3b}$$

3. $y = -\frac{2}{3}x+4$

(continued)

3. (continued)

x	y
0	4
6	0

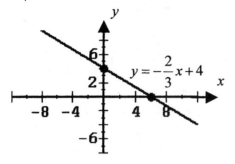

$$y = -\frac{2}{3}x + 4$$

4. $5ax + 5ay - 7wx - 7wy$
$$= 5a(x + y) - 7w(x + y)$$
$$= (x + y)(5a - 7w)$$

5.
$$3x - y + z = 6$$
$$2x - y + 2z = 7$$
$$x + y + z = 2$$

Switch first and third equation
$$x + y + z = 2$$
$$3x - y + z = 6$$
$$2x - y + 2z = 7$$

Add -3 time first equation to second equation and -2 times first equation to third equation
$$x + y + z = 2$$
$$-4y - 2z = 0$$
$$-3y \quad\quad = 3 \Rightarrow y = -1, \text{ from second}$$
equation, $-4(-1) - 2z = 0$
$$2z = 4 \Rightarrow z = 2, \text{ from first equation,}$$
$$x + (-1) + 2 = 2 \Rightarrow x = 1$$

$(1, -1, 2)$ is the solution.

6. $(5\sqrt{2} + \sqrt{3})(\sqrt{5} - 2\sqrt{6})$
$$= 5\sqrt{10} - 10\sqrt{12} + \sqrt{15} - 2\sqrt{18}$$
$$= 5\sqrt{10} + \sqrt{15} - 10\sqrt{4 \cdot 3} - 2\sqrt{9 \cdot 2}$$
$$= 5\sqrt{10} + \sqrt{15} - 20\sqrt{3} - 6\sqrt{2}$$

7. $\quad x^4 - 5x^2 - 6 = 0$
$$(x^2 - 6)(x^2 + 1) = 0$$
$$x^2 - 6 = 0, \ x^2 + 1 = 0$$
$$x^2 = 6, \ x^2 = -1$$
$$x = \pm\sqrt{6}, \ x = \pm i$$

8. $2x - y = 4 \Rightarrow y = 2x - 4$
$$4x - y^2 = 0 \Rightarrow 4x - (2x - 4)^2 = 0$$
$$4x - 4x^2 + 16x - 4 = 0$$
$$x^2 - 5x + 4 = 0$$
$$(x - 4)(x - 1) = 0$$
$$x = 4, \ x = 1$$
$$y = 2(4) - 4 = 4,$$
$$y = 2(1) - 4 = -2$$
$x = 4, \ y = 4; \ x = 1, \ y = -2$ is the solution.

9. $\quad\quad 2x - 3 = \sqrt{7x - 3}$
$$4x^2 - 12x + 9 = 7x - 3$$
$$4x^2 - 19x + 12 = 0$$
$$(x - 4)(4x - 3) = 0$$
$$x = 4, \ x = \frac{3}{4} \text{ which does not check}$$
$x = 4$ is the solution.

10. $\dfrac{5}{\sqrt[3]{2xy^2}} \cdot \dfrac{\sqrt[3]{4x^2y}}{\sqrt[3]{4x^2y}}$
$$= \frac{5\sqrt[3]{4x^2y}}{2xy}$$

11. $f(x) = 2^{3-2x}$

x	$y = f(x) = 2^{3-2x}$
1	2
2	1/2
3	1/8
4	1/32

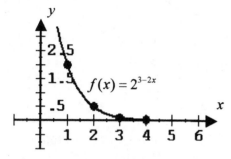

12. $\log_x\left(\dfrac{1}{64}\right) = 3 \Leftrightarrow x^3 = \dfrac{1}{64} = \left(\dfrac{1}{4}\right)^3$

$$x = \frac{1}{4}$$

13. $5^{2x-1} = 25 = 5^2$

$2x - 1 = 2$

$2x = 3$

$$x = \frac{3}{2}$$

14. $\log 7.67 = 0.8847953639$

15. $\log_x = 1.8209 \Leftrightarrow x = 10^{1.8209}$

$x \approx 66.20640403$

16. $\log_3 7 = \dfrac{\ln 7}{\ln 3}$

$= 1.771243749$

17. $\ln x = 1.9638 \Leftrightarrow x = e^{1.9638}$

$x \approx 7.1263558$

18. $\log_9 x = 1 - \log_9(x - 8)$

$\log_9 x + \log_9(x - 8) = 1$

$\log_9(x(x-8)) = 1 \Leftrightarrow x(x-8) = 9^1$

$x^2 - 8x - 9 = 0$

$(x - 9)(x + 1) = 0$

$x = 9,\ x = -1$ which must be rejected

    since it gives $\log_9$(negative)

$x = 9$ is the solution.

19. $\log_5 x = \log_5 2 + \log_5(x^2 - 3)$

$\log_5 x - \log_5(x^2 - 3) = \log_5 2$

$\log_5\left(\dfrac{x}{x^2 - 3}\right) = \log_5 2$

$\dfrac{x}{x^2 - 3} = 2$

$2x^2 - 6 = x$

$2x^2 - x - 6 = 0$

$(x - 2)(2x + 3) = 0$

$x = 2,\ x = -\dfrac{3}{2}$ which must be rejected

    since it gives $\log_5$(negative)

$x = 2$ is the solution.

20. $3^{x+2} = 5$

$\ln 3^{x+2} = \ln 5$

$(x + 2)\ln 3 = \ln 5$

$x + 2 = \dfrac{\ln 5}{\ln 3}$

$x = -2 + \dfrac{\ln 5}{\ln 3} \approx -0.535$

21. $33 = 66e^{2x} \Rightarrow e^{2x} = 0.5 \Rightarrow \ln e^{2x} = \ln 0.5$

$\ln e^{2x} = \ln 0.5$

$2x = \ln 0.5,\ x = \dfrac{\ln 0.5}{2}$

22. $A = P(1+r)^t$

$A = 3000(1+0.09)^4 = 4234.74483...$

They will have $4234.74.

Practice Final Examination

1. $(4-3)^2 + \sqrt{9} \div (-3) + 4$

$= 1^2 + 3 \div (-3) + 4$

$= 1 + (-1) + 4$

$= 4$

2. $\left(\dfrac{2x^3 y^{-2}}{3x^4 y^{-3}}\right)^{-2} = \left(\dfrac{2y}{3x}\right)^{-2} = \left(\dfrac{3x}{2y}\right)^2 = \dfrac{9x^2}{4y^2}$

3. $5a - 2ab - 3a^2 - 6a - 8ab + 2a^2$

$= -a^2 - 10ab - a$

4. $3[2x - 5(x+y)] = 3[2x - 5x - 5y]$

$\qquad\qquad\qquad = 3[-3x - 5y]$

$\qquad\qquad\qquad = -9x - 15y$

5. $F = \dfrac{9}{5}C + 32 = \dfrac{9}{5}(-35) + 32 = -31$

6. $\dfrac{1}{3}y - 4 = \dfrac{1}{2}y + 1$

$\dfrac{1}{6}y = -5$

$y = -30$

7. $A = \dfrac{1}{2}a(b+c)$

$2A = ab + ac$

$ab = 2A - ac$

$b = \dfrac{2A - ac}{a}$

8. $\left|\dfrac{2}{3}x - 4\right| = 2$

$\dfrac{2}{3}x - 4 = 2$ or $\dfrac{2}{3}x - 4 = -2$

$2x - 12 = 6 \qquad 2x - 12 = -6$

$2x = 18 \qquad\quad 2x = 6$

$x = 9 \qquad\qquad x = 3$

9. $2x - 3 < x - 2(3x - 2)$

$2x - 3 < x - 6x + 4$

$7x < 7$

$x < 1$

10. $\qquad\qquad P = 2L + 2W = 1760$

$L + W = 880$

$2W - 200 + W = 880$

$3W = 1080$

$W = 360$

$L = 2W - 200 = 520$

The width is 360 meters and the length is 520 meters.

11. Let x = amount invested at 14%

$0.14x + 0.12(4000 - x) = 508$

$0.14x + 480 - 0.12x = 508$

$0.02x = 28$

$x = 1400$

$4000 - x = 2600$

$1400 was invested at 14% and $2500 at 12%.

12. $x + 5 \le -4$ or $2 - 7x \le 16$

$x \le -9 \qquad\quad -7x \le 14$

$x \ge -2$

13. $|2x-5|<10$

$-10<2x-5<10$

$-5<2x<15$

$-\dfrac{5}{2}<x<\dfrac{15}{2}$

14. $7x-2y=-14$

x	y
0	7
–2	0

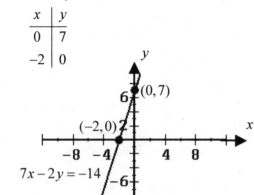

15. $3x-4y\le 6$

Test point: $(0,0)$

$3(0)-4(0)\le 6$

$0\le 6,\ \text{True}$

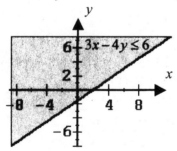

16. $m=\dfrac{y_2-y_1}{x_2-x_1}=\dfrac{-3-5}{-2-1}=\dfrac{8}{3}$

17. $3x+2y=8\Rightarrow y=-\dfrac{3}{2}x+4,\ m=-\dfrac{3}{2}$

$m_{\parallel}=-\dfrac{3}{2},\ y-4=-\dfrac{3}{2}(x-(-1))$

$3x+2y=5$

18. $f(x)=3x^2-4x-3$

$f(3)=3(3^2)-4(3)-3$

$f(3)=12$

19. $f(x)=3x^2-4x-3$

$f(-2)=3((-2)^2)-4(-2)-3$

$f(-2)=17$

20. $f(x)=|2x-4|$

x	$y=f(x)=|2x-4|$
1	2
2	0
3	2

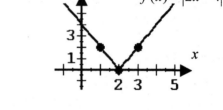

21. $\dfrac{1}{2}x+\dfrac{2}{3}y=1\xrightarrow{\times 6}3x+4y=6$

$\dfrac{1}{3}x+y=-1\rightarrow y=-1-\dfrac{1}{3}x$

$3x+4\left(-1-\dfrac{1}{3}x\right)=6$

$3x-4-\dfrac{4}{3}x=6$

$9x-12-4x=18$

$5x=30$

$x=6$

$y=-1-\dfrac{1}{3}x=-1-\dfrac{1}{3}(6)=-3$

$x=6,\ y=-3$ is the solution.

22. $4x - 3y = 12 \xrightarrow{\times 4} 16x - 12y = 48$

$\quad 3x - 4y = 2 \xrightarrow{\times -3} \underline{-9x + 12y = -6}$

$\qquad\qquad\qquad\qquad\quad 7x \qquad\ = 42$

$\qquad\qquad\qquad\qquad\qquad\quad x = 6$

$4(6) - 3y = 12$

$\quad\ -3y = -12$

$\qquad\quad y = 4$

$x = 6,\ y = 4$ is the solution.

23. Solve the system.

$2x + 3y - z = 16$

$\ x - y + 3z = -9$

$5x + 2y - z = 15$

Multiply first equation by 3 and add to the second equation

$6x + 9y - 3z = 48$

$\underline{\ x - y + 3z = -9}$

$7x + 8y \qquad = 39$

Multiply third equation by 3 and add to the second equation

$\ x - y + 3z = -9$

$\underline{15x + 6y - 3z = 45}$

$16x + 5y \qquad = 36$

Now solve the system

$7x + 8y = 39 \xrightarrow{\times 5} 35x + 40y = 195$

$16x + 5y = 36 \xrightarrow{\times -8} \underline{-128x - 40y = -288}$

$\qquad\qquad\qquad\qquad -93x \qquad\ = -93$

$x = 1$

$16x + 5y = 36 \Rightarrow 16(1) + 5y = 36$

$5y = 20$

$y = 4$

$x - y + 3z = -9 \Rightarrow 1 - 4 + 3z = -9$

$z = -2$

$x = 1,\ y = 4,\ z = -2$ is the solution.

24. Solve the system.

$\quad y + z = 2$

$x \qquad + z = 5$

$x + y \qquad = 5$

Subtract the second equation from the first.

$\quad y \ + z = 2$

$\underline{-x \qquad - z = -5}$

$-x + y \qquad = -3$, add third equation

$\underline{\ x \ + y \qquad = 5}$

$\quad 2y \qquad = 2$

$y = 1$, from $x + y = 4 \Rightarrow x + 1 = 5,\ x = 4$

From $x + z = 5 \Rightarrow 4 + z = 5,\ z = 1$

$x = 4,\ y = 1,\ z = 1$ is the solution.

25. Solve the system.

$2x - y + 5z = -2$

$\ x + 3y - z = 6$

$4x + y + 3z = -2$

Add the first and third equations.

$2x - y + 5z = -2$

$\underline{4x + y + 3z = -2}$

$6x \qquad + 8z = -4$

Add 3 times the first equation to the second equation.

$6x - 3y + 15z = -6$

$\underline{\ x + 3y - z = 6}$

$7x \qquad + 14z = 0$

Now solve the system

$6x + 8z = -4 \xrightarrow{\times 7} 42x + 56z = -28$

$7x + 14z = 0 \xrightarrow{\times -6} \underline{-42x - 84z = 0}$

$\qquad\qquad\qquad\qquad\qquad -28z = -28$

$z = 1$ is the solution.

26. $3y \geq 8x - 12$ $2x + 3y \leq -6$

Test point: $(0,0)$ Test point: $(0,0)$

$3(0) \geq 8(0) - 12$ $2(0) + 3(0) \leq -6$

$0 \geq -12$, true $0 \leq -6$, false

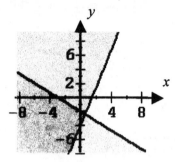

27. $(3x - 2)(2x^2 - 4x + 3)$

$= 6x^3 - 12x^2 + 9x - 4x^2 + 8x - 6$

$= 6x^3 - 16x^2 + 17x - 6$

28. $(25x^3 + 9x + 2) \div (5x + 1)$

$$\require{enclose}\begin{array}{r}5x^2 - x + 2 \\ 5x+1\enclose{longdiv}{25x^3 + 0x^2 + 9x + 2} \\ \underline{25x^3 + 5x^2} \\ -5x^2 + 9x \\ \underline{-5x^2 - x} \\ 10x + 2 \\ \underline{10x + 2} \\ 0\end{array}$$

$(25x^3 + 9x + 2) \div (5x + 1) = 5x^2 - x + 2$

29. $8x^3 - 27 = (2x)^3 - 3^3$

$= (2x - 3)((2x)^2 + (2x)(3) + (3)^2)$

$= (2x - 3)(4x^2 + 6x + 9)$

30. $x^3 + 2x^2 - 4x - 8$

$= x^2(x + 2) - 4(x + 2)$

$= (x + 2)(x^2 - 4) = (x + 2)(x + 2)(x - 2)$

31. $2x^3 + 15x^2 - 8x = x(2x^2 + 15x - 8)$

$= x(2x - 1)(x + 8)$

32. $x^2 + 15x + 54 = 0$

$(x + 9)(x + 6) = 0$

$x + 9 = 0, \quad x + 6 = 0$

$x = -9, \quad\quad x = -6$

33. $\dfrac{9x^3 - x}{3x^2 - 8x - 3} = \dfrac{x(9x^2 - 1)}{(3x + 1)(x - 3)}$

$= \dfrac{x(3x + 1)(3x - 1)}{(3x + 1)(x - 3)}$

$= \dfrac{x(3x - 1)}{(x - 3)}$

34. $\dfrac{x^2 - 9}{2x^2 + 7x + 3} \div \dfrac{x^2 - 3x}{2x^2 + 11x + 5}$

$= \dfrac{(x + 3)(x - 3)}{(2x + 1)(x + 3)} \cdot \dfrac{(2x + 1)(x + 5)}{x(x - 3)}$

$= \dfrac{(x + 5)}{x}$

35. $\dfrac{3x}{x + 5} - \dfrac{2}{x^2 + 7x + 10}$

$= \dfrac{3x(x + 2)}{(x + 5)(x + 2)} - \dfrac{2}{(x + 5)(x + 2)}$

$= \dfrac{3x(x + 2) - 2}{(x + 5)(x + 2)}$

$= \dfrac{3x^2 + 6x - 2}{(x + 5)(x + 2)}$

36. $\dfrac{\dfrac{3}{2x + 1} + 2}{1 - \dfrac{2}{4x^2 - 1}}$

(continued)

36. (continued)

$$= \frac{3(2x-1)+2(2x-1)(2x+1)}{(2x-1)(2x+1)-2}$$

$$= \frac{6x-3+2(4x^2-1)}{4x^2-1-2}$$

$$= \frac{8x^2+6x-5}{4x^2-3}$$

37.
$$\frac{x-1}{x^2-4} = \frac{2}{x+2} + \frac{4}{x-2}$$

$$\frac{x-1}{(x+2)(x-2)} = \frac{2}{x+2} + \frac{4}{x-2}$$

$$x-1 = 2(x-2)+4(x+2)$$

$$x-1 = 2x-4+4x+8$$

$$5x = -5$$

$$x = -1$$

38. $\dfrac{5x^{-4}y^{-2}}{15x^{-1/2}y^3} = \dfrac{1}{3x^{7/2}y^5}$

39. $\sqrt[3]{40x^4y^7} = \sqrt[3]{8x^3y^6 \cdot 5x^1y} = 2xy^2\sqrt[3]{5xy}$

40. $5\sqrt{2} - 3\sqrt{50} + 4\sqrt{98}$

$\quad = 5\sqrt{2} - 3\sqrt{25\cdot 2} + 4\sqrt{49\cdot 2}$

$\quad = 5\sqrt{2} - 15\sqrt{2} + 28\sqrt{2}$

$\quad = 18\sqrt{2}$

41. $\dfrac{2\sqrt{3}+1}{3\sqrt{3}-\sqrt{2}} \cdot \dfrac{3\sqrt{3}+\sqrt{2}}{3\sqrt{3}+\sqrt{2}}$

$\quad = \dfrac{18+2\sqrt{6}+3\sqrt{3}+\sqrt{2}}{27-2}$

$\quad = \dfrac{18+2\sqrt{6}+3\sqrt{3}+\sqrt{2}}{25}$

42. $i^3 + \sqrt{-25} + \sqrt{-16} = -i + 5i + 4i = 8i$

43.
$$\sqrt{x+7} = x+7$$

$$x+7 = x^2+10x+25$$

$$x^2+9x+18 = 0$$

$$(x+6)(x+3) = 0$$

$$x+6 = 0,\ x+3 = 0$$

$$x = -6,\ x = -3$$

$$\text{check: } \sqrt{-6+7} \stackrel{?}{=} -6+5$$

$$\sqrt{1} \stackrel{?}{=} -1$$

$$1 \ne -1$$

$$\sqrt{-3+7} \stackrel{?}{=} -3+5$$

$$\sqrt{4} \stackrel{?}{=} -3+5$$

$$2 = 2$$

$$x = -3 \text{ is the solution.}$$

44. $y = kx^2$

$$15 = k(2)^2 \Rightarrow k = \frac{15}{4} \Rightarrow y = \frac{15}{4}x^2$$

$$y = \frac{15}{4}(3)^2 = 33.75$$

45. $5x(x+1) = 1+6x$

$\quad 5x^2+5x = 1+6x$

$\quad 5x^2-x-1 = 0,\ \text{use quadratic formula}$

$$x = \frac{-(-1)\pm\sqrt{(-1)^2-4(5)(-1)}}{2(5)}$$

$$x = \frac{1\pm\sqrt{21}}{10}$$

46. $5x^2-9x = -12x \Rightarrow 5x^2+3x = 0$

$\quad x(5x+3) = 0$

$$x = 0,\ x = -\frac{3}{5}$$

47. $x^{2/3} + 5x^{1/3} - 14 = 0$, let $x^{1/3} = w$, $x^{2/3} = w^2$

$w^2 + 5w - 14 = 0$

$(w-2)(w+7) = 0$

$w - 2 = 0, \ w + 7 = 0$

$w = 2, \quad w = -7$

$x^{1/3} = 2, \quad x^{1/3} = -7$

$x = 8, \quad\quad x = -343$

48. $3x^2 - 11x - 4 \geq 0$

$3x^2 - 11x - 4 = 0$

$(3x+1)(x-4) = 0$

$3x + 1 = 0$ or $\quad x - 4 = 0$

$x = -\dfrac{1}{3}$ or $\quad x = 4$

Region I: Test $x = -1$

$3(-1)^2 - 11(-1) - 4 = -4 < 0$

Region II: Test $x = 0$

$3(0)^2 - 11(0) - 4 = -4 < 0$

Region III: Test $x = 5$

$3(5)^2 - 11(5) - 4 = 16 > 0$

$x \leq -\dfrac{1}{3}$ or $x \geq 4$.

49. $f(x) = -x^2 - 4x + 5$

parabola, opening downward

$f(0) = -0^2 - 4(0) + 5 = 5 \Rightarrow$ y-int: $(0,5)$

$-x^2 - 4x + 5 = 0 \Rightarrow x^2 + 4x - 5 = 0$

$(x+5)(x-1) = 0 \Rightarrow x = -5, \ x = 1$

x-int: $(-5,0), \ (1,0)$

$-\dfrac{b}{2a} = -\dfrac{-4}{2(-1)} = -2$

$f(-2) = -(-2)^2 - 4(-2) + 5 = 9$

$V(-2,9)$

(continued)

49. (continued)

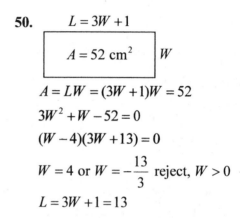

50.

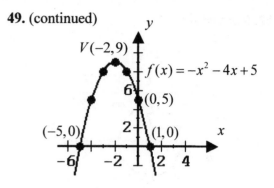

$A = LW = (3W+1)W = 52$

$3W^2 + W - 52 = 0$

$(W-4)(3W+13) = 0$

$W = 4$ or $W = -\dfrac{13}{3}$ reject, $W > 0$

$L = 3W + 1 = 13$

The width is 4 cm and the length is 13 cm.

51. $x^2 + y^2 + 6x - 4y = -9$

$x^2 + 6x + 9 + y^2 - 4y + 4 = -9 + 9 + 4 = 4$

$(x+3)^2 + (y-2)^2 = 2^2$

$C(-3,2), \ r = 2$

52. $\dfrac{x^2}{16} + \dfrac{y^2}{25} = 1$

$\dfrac{x^2}{4^2} + \dfrac{y^2}{5^2} = 1$

ellipse: $C(0,0)$

$a = 4, \ b = 5$

x-int: $(\pm 4, 0)$

y-int: $(0, \pm 5)$

(continued)

52. (continued)

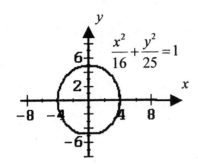

53. $\dfrac{x^2}{4} - \dfrac{y^2}{9} = 1 \Rightarrow \dfrac{x^2}{2^2} - \dfrac{y^2}{3^2} = 1$, hyperbola

$C(0,0)$, $a = 2$, $b = 3$

x-int: $(\pm 2, 0)$

$y_{\text{asymptote}} = \pm \dfrac{3}{2} x$

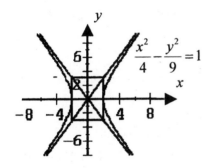

54. $x = (y - 3)^2 + 5$

parabola opening right, $V(5,3)$

$x = (0 - 3)^2 + 5 = 14$, x-int: $(14, 0)$

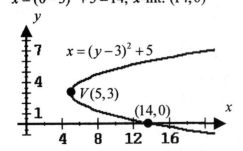

55.

$$x^2 + y^2 = 16$$
$$x^2 - y = 4 \Rightarrow y = x^2 - 4$$
$$x^2 + (x^2 - 4)^2 = 16$$
$$x^2 + x^4 - 8x^2 + 16 = 16$$
$$x^4 - 7x^2 = 0$$
$$x^2(x^2 - 7) = 0$$
$$x^2 = 0, \ x^2 = 7$$
$$x = 0, \ x = \pm\sqrt{7}$$
$$y = x^2 - 4 = 0^2 - 4 = -4$$
$$y = x^2 - 4 = (\pm\sqrt{7})^2 - 4 = 3$$
$(0, -4)$, $(\pm\sqrt{7}, 3)$ is the solution.

56. $f(x) = 3x^2 - 2x + 5$

(a) $f(-1) = 3(-1)^2 - 2(-1) + 5 = 10$

(b) $f(a) = 3a^2 - 2a + 5$

(c) $f(a + 2) = 3(a + 2)^2 - 2(a + 2) + 5$
$$= 3a^2 + 12a + 12 - 2a - 4 + 5$$
$$= 3a^2 + 10a + 13$$

57. $f(x) = 5x^2 - 3$, $g(x) = -4x - 2$

$f[g(x)] = f(-4x - 2)$
$$= 5(-4x - 2)^2 - 3$$
$$= 5(16x^2 + 16x + 4) - 3$$
$$= 80x^2 + 80x + 20 - 3$$
$$= 80x^2 + 80x + 17$$

58. $f(x) = \dfrac{1}{2} x - 7$, $f(x) \to y$

$y = \dfrac{1}{2} x - 7$, $x \leftrightarrow y$

$x = \dfrac{1}{2} y - 7 \Rightarrow y = 2x + 14$, $y \to f^{-1}(x)$

$f^{-1}(x) = 2x + 14$

59. $f(x) = |x|$

$f(x+2) \to f(x)$ shifted left 2 units

$f(x) - 3 \to f(x)$ shifted down 3 units

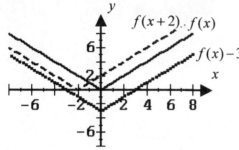

60. $f(x) = 2^{1-x}$

x	$y = f(x) = 2^{1-x}$
-1	4
0	2
1	1

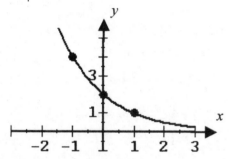

61. $\log_5 x = -4 \Leftrightarrow x = 5^{-4} = \dfrac{1}{5^4} = \dfrac{1}{625}$

62. $\log_4(3x+1) = 3 \Leftrightarrow 3x+1 = 4^3 = 64$

$$3x + 1 = 64$$
$$3x = 63$$
$$x = 21$$

63. $\log_{10} 0.01 = y \Leftrightarrow 10^y = 0.01 = 10^{-2}$

$$y = -2$$

64.
$$\log_2 6 + \log_2 x = 4 + \log_2(x-5)$$
$$\log_2(x-5) - \log_2 x = \log_2 6 - 4$$
$$\log_2 \frac{x-5}{x} = \log_2 6 - 4$$
$$\frac{x-5}{x} = 2^{\log_2 6 - 4}$$
$$\frac{x-5}{x} = 2^{\log_2 6} \cdot 2^{-4}$$
$$\frac{x-5}{x} = \frac{6}{16} = \frac{3}{8}$$
$$8x - 40 = 3x$$
$$5x = 40$$
$$x = 8$$